Plant Propagation Techniques and Impacts on Sustainable Agriculture

Wasan Salih Hussain
https://orcid.org/0000-0002-1537-9979
Mosul University, Iraq

Amjad Abdul-Hadi Mohammed
https://orcid.org/0000-0002-5821-7523
Mosul University, Iraq

Rana Tariq Yahya
Mosul University, Iraq

IGI Global
Scientific Publishing
Publishing Tomorrow's Research Today

Vice President of Editorial — Melissa Wagner
Director of Acquisitions — Mikaela Felty
Director of Book Development — Jocelynn Hessler
Production Manager — Mike Brehm
Cover Design — Jose Rosado

Published in the United States of America by
IGI Global Scientific Publishing
701 East Chocolate Avenue
Hershey, PA, 17033, USA
Tel: 717-533-8845 | Fax: 717-533-7115
Website: https://www.igi-global.com E-mail: cust@igi-global.com

Library of Congress Cataloging-in-Publication Data

Names: Hussain, Wasan Salih, 1979- editor | Mohammed, Amjad Abdul-Hadi, 1980- editor | Yahay, Rana Tariq, 1980- editor
Title: Plant propagation techniques and impacts on sustainable agriculture / edited by Wasan Salih Hussain, Amjad Abdul-Hadi Mohammed, Rana Tariq Yahay.
Description: Hershey, PA : IGI Global Scientific Publishing, 2026. | Includes bibliographical references and index. | Summary: "This book aims to explain how plant propagation practices have contributed to increasing agricultural efficiency and achieving self-sufficiency"-- Provided by publisher.
Identifiers: LCCN 2025042327 (print) | LCCN 2025042328 (ebook) | ISBN 9798337356266 hardcover | ISBN 9798337356273 paperback | ISBN 9798337356280 ebook
Subjects: LCSH: Plant propagation | Sustainable agriculture
Classification: LCC SB119 .P57 2026 (print) | LCC SB119 (ebook)
LC record available at https://lccn.loc.gov/2025042327
LC ebook record available at https://lccn.loc.gov/2025042328

British Cataloguing in Publication Data
A Cataloguing in Publication record for this book is available from the British Library.

Table of Contents

Detailed Table of Contents

seeds, or those that do not maintain their traits when planted from seeds. Breeding techniques focus on developing plant varieties with a higher capacity to absorb nutrients from the soil, reducing the need for chemical fertilizers and producing plants resistant to pests and diseases.,

Plant propagation is the process of reproducing plants, either sexually or asexually, to produce new individuals. This practice is crucial in horticulture, agriculture, and conservation. Sexual propagation typically involves seeds, while asexual methods include cuttings, grafting, and layering. Each method has its advantages; for example, seeds can produce genetic diversity, whereas asexual methods allow for the cloning of desirable traits. The significance of plant propagation extends beyond mere cultivation. It plays a vital role in conserving endangered species, enhancing food security through the cultivation of crops, and improving landscapes and gardens. Additionally, propagation enables the preservation of genetic resources, which is essential in adapting to climate change and pest challenges. Understanding the techniques and significance of plant propagation empowers gardeners, farmers, and conservationists alike, ensuring a sustainable future for plant diversity.

Propagation, the process by which plants reproduce and spread, is profoundly affected by various environmental factors. Key elements such as light, temperature, water availability, soil conditions, and air movement play a vital role in determining the success and efficiency of both natural and artificial propagation methods. Light intensity and duration influence photosynthesis and germination, while temperature regulates enzymatic activity and growth rates. Adequate water and humidity support seed germination and root development, and soil composition affects nutrient availability and microbial interactions. Additionally, wind and biotic factors, including herbivores, pollinators, and beneficial soil organisms, significantly impact propagation outcomes. Understanding and managing these environmental variables is essential for improving propagation success in agriculture, horticulture, and conservation practices. This chapter explores the complex interplay of these factors, offering insights into optimizing propagation strategies for different plant species and ecological conditions.

The scientific foundations of plant propagation integrate classical knowledge with cutting-edge biotechnology. Sexual propagation ensures diversity and adaptability, while vegetative methods maintain clonal fidelity. Tissue culture and totipotency provide tools for mass production and conservation, guided by hormonal and genetic regulation. Anatomical, histological, and physiological processes form the basis of successful regeneration, while environmental factors determine practical outcomes. As global challenges intensify, the refinement of propagation science will remain central to agricultural innovation, biodiversity conservation, and sustainable development.

Sexual propagation through seeds and seedlings remains a cornerstone of agricultural and horticultural production. It ensures genetic diversity, enhances adaptability to changing environmental conditions, and contributes to sustainable crop improvement. The success of this approach depends on the integration of proper seed quality management, advanced cultivation techniques, and effective disease control strategies. By improving seed systems, optimizing germination conditions, and maintaining seedling health, agricultural productivity and biodiversity conservation can be secured. Therefore, sexual propagation continues to play a decisive role in food security and the future of sustainable agriculture.

Reproduction in living organisms is an important means for the continuation of life and the transmission of genetic traits between generations and their preservation from disappearance, especially those good genetic traits. The importance of biological diversity is due to the strength of different ecosystems. Biodiversity is represented in three forms: the multiplicity of living species, the diversity of genetic structures, and the multiplicity of ecosystems. Genetic diversity is one of the components of

Preface

INTRODUCTION

The global demand for sustainable agricultural practices has never been more urgent. With increasing pressures from climate change, population growth, and diminishing natural resources, the need to develop efficient, environmentally sound, and productive plant propagation techniques has become a cornerstone of modern agricultural science. This book, *Plant Propagation Techniques and Impacts on Sustainable Agriculture*, emerges from this context, seeking to bring together diverse perspectives and current research findings that explore both traditional and advanced propagation methods as engines of agricultural resilience and sustainability.

Plant propagation—the science and art of producing new plants—serves as the foundation of all crop improvement, biodiversity conservation, and food security strategies. Through both sexual and asexual methods, propagation ensures the survival, adaptability, and enhancement of plant species across generations. As scientific innovation accelerates, propagation techniques have evolved to include advanced biotechnological tools such as micropropagation, tissue culture, and plant growth-promoting bacteria (PGPB) applications. Each of these innovations contributes not only to higher yields and resource efficiency but also to the larger global vision of achieving sustainable agricultural systems.

This edited volume represents a collaborative endeavor among researchers and practitioners from Iraq, India, Pakistan, Libya, Egypt, Morocco, and Turkey—reflecting the shared international commitment to agricultural progress. Together, the chapters provide an integrated exploration of propagation methods, environmental factors, genetic conservation, and technological innovations. The book is designed for researchers, educators, policymakers, and agricultural professionals seeking both foundational knowledge and applied insights into how propagation underpins sustainable agriculture and ecological balance.

CHAPTER OVERVIEW

Chapter 1: Introduction to Plant Propagation and Its Importance

This opening chapter provides the historical and scientific background of plant propagation, examining both traditional and modern techniques—from seeds and cuttings to hydroponics and aeroponics. The authors emphasize how propagation underpins biodiversity, genetic conservation, and the growth of global agricultural industries.

Chapter 2: Impact of Propagation Techniques on Agricultural Production

Focusing on the role of propagation as a strategic tool in crop improvement, this chapter explores how asexual and breeding techniques enhance resilience, reduce dependency on chemical inputs, and promote sustainable productivity.

Chapter 3: An Overview of Plant Propagation and Its Significance

This chapter elaborates on the dual importance of sexual and asexual propagation in maintaining genetic diversity and conserving plant resources, linking propagation practices directly to global food security and biodiversity preservation.

Chapter 4: Environmental Factors Influencing Propagation

The authors examine how light, temperature, water, soil, and other ecological variables influence propagation success, providing insights into optimizing environmental conditions for different propagation methods.

Chapter 5: Scientific Foundations of Plant Propagation

This chapter delves into the anatomical, physiological, and genetic principles that govern propagation, including tissue totipotency, hormonal regulation, and the integration of biotechnology in improving propagation outcomes.

Chapter 6: Sexual Propagation—Seeds and Seedling Production

A detailed analysis of seed-based propagation, this chapter discusses seed quality, germination optimization, and disease management, emphasizing how these factors sustain agricultural productivity and genetic diversity.

Chapter 7: The Impact of Plant Reproduction on Biodiversity

Here, the authors discuss the genetic and ecological significance of plant reproduction, differentiating between sexual and asexual reproduction in their contributions to genetic variation and ecosystem stability.

Chapter 8: Modern and Traditional Propagation Techniques for Fruit Crops

This chapter compares conventional propagation methods such as grafting and layering with modern biotechnological approaches, highlighting their respective roles in fruit crop productivity, quality, and disease resistance.

Chapter 9: Modern Techniques in Plant Propagation

An in-depth exploration of cutting-edge propagation tools—such as micropropagation, biotechnological interventions, and controlled environmental conditions—demonstrating how modern techniques enhance agricultural efficiency and sustainability.

Chapter 10: Plant Propagation and Sustainable Agriculture

This chapter ties propagation practices to the broader goals of sustainable development, showcasing their role in promoting climate resilience, improving resource use, and contributing to the United Nations' Sustainable Development Goals (SDGs).

Chapter 11: Application of Plant Growth-Promoting Bacteria (PGPB) in Propagation Stages

Focusing on microbial biotechnology, this chapter illustrates how PGPB enhance germination, root formation, and stress tolerance, thereby reducing dependence on chemical fertilizers and supporting ecological agriculture.

Chapter 12: Tissue Culture and Micropropagation

The concluding chapter explores tissue culture and micropropagation as the pinnacle of modern propagation techniques, explaining their applications in large-scale plant production, conservation, and biotechnology research.

CONCLUSION

This book underscores the vital interconnection between plant propagation and sustainable agriculture. From classical seed-based methods to advanced biotechnological applications, the chapters collectively highlight how the refinement of propagation practices supports global food systems, enhances environmental stewardship, and preserves genetic resources for future generations. By examining propagation as both a science and a sustainability strategy, this volume bridges the gap between traditional agricultural wisdom and modern innovation.

The contributions from diverse international scholars reinforce the notion that plant propagation is not confined to a single region or method but represents a universal pursuit toward agricultural resilience. Through shared research and experience, this collection advances understanding of how propagation techniques can address pressing global challenges such as soil degradation, water scarcity, and climate change.

Ultimately, *Plant Propagation Techniques and Impacts on Sustainable Agriculture* aspires to serve as both a scholarly reference and a practical guide—empowering researchers, educators, and practitioners to apply propagation science in ways that strengthen ecosystems, secure food production, and sustain the balance between human progress and nature's integrity.

Wasan Hussain
Mosul University, Iraq

Amjad Mohammed
Mosul University, Iraq

Rana Yahay
Mosul University, Iraq

Chapter 1
Introduction to Plant Propagation and Its Importance

Ban Edan
https://orcid.org/0009-0005-1183-8532
Universty of Al-Hamdaniya, Iraq

Muhammad Umer
https://orcid.org/0009-0005-4898-5077
PMAS-Arid Agriculture University, Rawalpindi, Pakistan

Taha Edan
Duhuok Agriculture Extention, Iraq

ABSTRACT

Plant propagation is a basic practice that is applied in the field of agriculture and horticulture. This practice is linked with biodiversity, food security, plant production, and conservation. Therefore, this chapter revolves around the history as well as the knowledge of today's plant propagation techniques. These techniques encircle around the sexual and asexual methods that include the seeds, grafting, use of tubers, bulbs, rhizomes, corms, as well as tissue culture. This book chapter is not just confined to this, but it also provides the traces of the historic details and the evolution process from the ancestors to today's world. Key improvements, such as micro propagation and seed priming, also came under consideration, and futuristic techniques which are under development, like hydroponics and aeroponics. Therefore, in conclusion, we can say that this chapter emphasizes the indispensable role of propagation in sustainable agriculture, crop improvement, genetic conservation, and the expanding global nursery and seed industries.

DOI: 10.4018/979-8-3373-5626-6.ch001

1. INTRODUCTION

According to the conventional classifications in archaeological and anthropological theory, there are fundamentally only two primary subsistence strategies for deriving a livelihood from the natural environment: foraging and production. (Ingold, 2024).

Plant propagation is the process of multiplying or reproducing plants. The shift toward crop commercialization has driven the development of diverse propagation techniques, each possessing distinct advantages and limitations. Plants exhibit variable physiological and morphological responses to different propagation methods. These techniques aim to achieve crop uniformity, induce precocious fruiting, enhance yield, improve resistance to pests and diseases, and introduce specific desirable traits into successive generations. Consequently, plant propagation represents a domain that intertwines both scientific rigor and artistic skill, offering substantial economic potential through commercial nurseries. Moreover, propagation plays a critical role in the conservation of plant genetic resources and the preservation of endangered species. (Thokchom and Singh, 2024; Rathour and Kumar, 2023; Awotedu et al., 2021).

1.1 History

Although it is difficult to accurately document the origins of plant propagation, it is fair to assume that it coevolved with agriculture around 10,000 years ago. The inadvertent planting of seeds collected during foraging and harvesting activities, which slowly evolved into deliberate agricultural methods, may have been the cause of the earliest forms of propagation. Early horticulture in Egypt, Babylon, China, and other areas is described as having a reasonably advanced knowledge of the cultivation of both ornamental and food plants, along with propagation methods that made it possible to multiply them effectively. Babylon and Assyria were famous for their landscaped parks and terraced gardens. The intentional and widespread cultivation of ornamental plants required expert understanding of propagation techniques. (Beyl & Trigiano, 2016). Roman literature, including works by Pliny the Elder and Columella, extensively addressed agricultural practices and the significance of cultivars. Columella provided comprehensive guidelines on the cultivation of vineyards and olive trees. The Roman society held fruit trees in high esteem, regarding them as divine gifts, and possessed a thorough understanding of various propagation methods, including grafting. (Cofer, 2015).

This knowledge persisted throughout the medieval period, with Arab scholars such as Ibn al-Awwam contributing detailed writings in the field of horticulture, particularly focusing on grafting and plant propagation techniques (Shopov, 2021). During the Renaissance, experts such as Gallo and De Ceres expanded propagation techniques, with Gallo advocating for citrus cultivation in greenhouses and providing detailed grafting methodologies. Over the centuries, these practices became increasingly refined, albeit largely grounded in empirical knowledge. The eighteenth and nineteenth centuries witnessed significant scientific advancements, including enhanced understanding of plant physiology and genetics, followed by the discovery of photosynthesis. These breakthroughs transformed plant propagation into a scientifically controlled discipline characterized by greater precision and reliability. Overall, plant propagation evolved from an artisanal craft and practical experimentation into a scientific specialization founded on rigorous principles, shaped across diverse civilizations and culminating in modern botanical science. (Sansavini, 2014).

2. MODES OF PROPAGATION

The following are the different modes of propagation that are used worldwide to propagate vegetable crops. (Figure 1)

2.1 Asexual Propagation

In vegetable crops, propagation refers to asexual or vegetative reproduction, which uses the plant's vegetative components, such as stems, leaves, roots, tubers, or altered versions of those parts to produce new plants.

2.1.1 Cuttings

A cutting is a **segment of a** plant taken **vegetatively** from the **mother** plant, which undergoes regeneration to develop into a complete, independent new plant. (Awotedu et al., 2021).

Vegetables such as pointed gourd, sweet potato, and cassava are commercially propagated through stem cuttings. (Choudhary et al., 2017). This approach is helpful to propagators since it allows them to grow more plants in a simple and cost-effective manner. Historically, leaf, leafbud, eye, cane, and tip cuttings are the common categories used to classify foliage plant cuttings (Chen and Stamp, 2006).

2.1.2 Stem Tubers

A stem tuber is a specialized structure formed by the swelling of rhizomes or creeping (aboveground) stems. The upper side of the tuber develops buds that give rise to shoots and foliage, while the lower side generates roots. Stem tubers typically develop laterally on the parent plant, usually close to the soil line. Potato is predominantly propagated through the planting of stem tubers. A successful potato crop depends on the genetic integrity of the cultivars and the high quality of the healthy seed tubers. However, the most costly component of potato farming is seed tubers. A perfect seed tuber weighs between 30 and 40 grams, is free of diseases, and has well-developed sprouted, and weigh between 30 and 40 grams. The use of whole tubers is generally preferred for planting. In mountainous regions, however, tubers are often cut into sections and planted in late winter, as the moderate temperatures reduce the risk of rot. The two main goals of cutting large tubers are to lower seed costs and ensure consistent sprouting. Tubers should be sliced longitudinally through the crown eye, with each piece preferably weighing 30 to 40 grams. Seed tubers are typically sliced with a knife, followed by treatment with a fungicide before planting. The knife must be sterilized with a potassium permanganate solution before cutting. The high cost of planting materials, the lack of excellent seed tubers, the logistical challenges associated with transporting bulky potato seeds, and the infiltration of viruses into seed tubers represent some of the major constraints in using tubers as planting material. The production of small tubers in aeroponic systems and net houses is preceded by the in vitro multiplication of virus-free microplants and minitubers, constitutes a core component of high-tech seed production. These advanced methodologies help to overcome the limitations inherent in traditional seed tuber production systems. (Murlidhar et al. 2017; Bag et al.2015).

Figure 1. Different modes of Plant propagation.

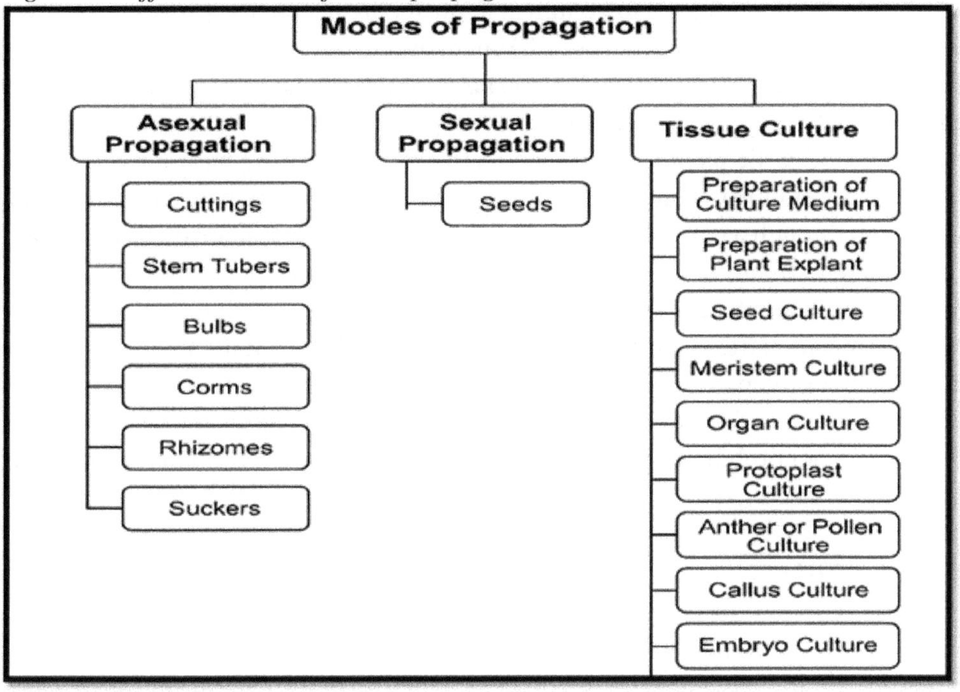

2.1.3 Bulbs

The term 'bulb', which describes a bulbous plant's adapted, shortened, and often subterranean shoots, is frequently used to refer to them. Generally speaking, bulbous plants are those species that have underground storage organs, such as rhizomes, tubers, bulbs, and corms. But to a botanist, bulbs are subterranean organs distinguished by scales. (De Klerk, 2012). A bulb is a rounded and enlarged portion of the stem, essentially a brief stem that holds leaves and serves as a storage organ for nutrients during dormant periods. A bud and several layers of fleshy, scale-like leaves make up the bulb. The main source of stored food for the growing new plant comes from these leaves. Onion and garlic are propagated through bulbs. Onion propagation can be achieved using small bulbs, known as “sets,” typically measuring 1.5 to 2.0 cm in diameter. Garlic, being a sterile crop, is propagated exclusively through vegetative means. Most garlic cultivars do not produce viable seeds; therefore, propagation is conducted using cloves obtained by dissecting the bulb. Cloves with a diameter of 8 to 10 mm yield higher productivity and superior quality. It is advisable to select larger cloves from the outer layers of the bulb. To establish one hectare, nearly 500 kg of cloves ranging from 8 to 10 mm in diameter are necessary. Cloves are planted with 15 to 20 cm between rows, 8 to 10 cm between each plant, and 5 to 8 cm deep. To ensure optimal growth, cloves should be planted with the basal plate facing down and the pointed end facing up. Garlic is susceptible to a complex of viral infections, but virus-free garlic plants have been effectively created using meristem tip culture (apical shoot culture). (PenaIglesias and Ayuso, 1982), meristem tip culture (Li et al, 1995) and through a combination of thermotherapy and meristematic tissue culture (meristematic tissue culture) (Senula et al, 2000). A number of variables, including explant type, photoperiod, light quality, incubation temperature, plant growth regulators, and more, might influence bulb multiplication. Bulblet development in many bulbous plants is often favored by a higher incubation temperature (typically 25 °C or 30 °C). For plant growth regulators, appropriate quantities of auxins, cytokinins, or combinations of the two have shown promising ability in initiating bulblets (Ren et al.,2023)

2.1.4 Corms

A corm is a brief, upright, thick underground plant stem that serves as a storage organ for some plants to endure winter or other difficult circumstances. It has a firm tissue composition and makes up the base of the flowering stem. Around the larger corm, smaller cormels arise on top of the older corm, creating a new large corm. After the foliage has withered, the corm is carefully dug up and left to dry in indirect light for two to three weeks. Subsequently, the small cormels are removed, and the new corm is gently separated from the old corm. (Hanson, 2007).

2.1.5 Rhizomes

A rhizome is a thick, horizontal plant stem that grows continuously either underground or at the soil surface, capable of producing roots and new shoots. Ginger (Zingiber officinale) and turmeric (Curcuma longa) are both members of the Zingiberaceae family and are old plants that were cultivated centuries ago. Typically, turmeric and ginger are propagated by dividing their fleshy rhizome roots and sowing them. (Prasath et al. 2017).

2.1.6 Suckers

Root suckers or root sprouts are vegetative outgrowths that develop from buds found on roots or below stems. Additionally, these shoots can originate from buds close to the base of the parent plant, producing new separate plants. The generation of such suckers allows many trees and shrubs to reproduce spontaneously. Jerusalem artichoke (Cynara scolymus L.) is usually propagated vegetatively using underground shoots known as "offoli," or from suckers and dried stems or branches from the last growth season. Shoots should be selected from high-yielding plants and harvested when they reach approximately 20 cm in length. They are carefully severed from the parent plant at the root level using a sharp knife, then gently loosened from the soil and root mass. The detached shoots are planted in a freshly prepared bed at a depth of 15 to 20 cm, with spacing of 1.2 to 1.8 meters between individual plants. (Chauhan et al., 2021).

2.1.7 Grafting

The cultivation of vegetables using grafted seedlings has reduced notable crop losses from soil-borne illnesses, which are aggravated by continuous monoculture. (Kumar et al. 2015).

In several fruit-bearing vegetables, such as tomatoes, eggplants, peppers, watermelons, cucumbers, and melons, the utilization of grafted seedlings has become widespread. Grafting is regarded as an environmentally friendly technique for increasing the yields of susceptible cultivars and controlling soilborne diseases. (Lee and Oda 2003). This technique is environmentally sustainable for Vegetable cultivation, as the use of resistant rootstocks reduces reliance on agricultural chemical pesticides. (Rivard et al. 2008). Grafting improves yield and increases resistance to biotic and abiotic stresses. To promote resistance to extreme heat and cold, grafts were frequently employed. Fruit-bearing vegetables use a variety of grafting methods. Tomatoes and eggplants are commonly grafted using the splice and tube grafting methods. The tongue approach grafting method is primarily employed for cucurbits, particularly cucumbers. The wedge grafting technique, which has recently become popular for melons and watermelons, is regarded as the simplest. (Venema 2008).

2.2 Sexual Propagation

Seed propagation, also known as sexual reproduction, involves the formation of seeds as a result of the successful fertilization of gametes from both parent organisms. This technique is widely used and is one of the oldest and simplest ways to propagate crop plants. High-quality seeds are a necessary requirement for successful vegetable production. The majority of vegetables are propagated using true seeds, with the exception of the following: cassava, ginger, garlic, turmeric, yam, taro, asparagus, pointed gourd, sweet potatoes, and potatoes. Currently, true potato seeds (TPS) are also employed commercially in the production of potatoes. (Awotedu et al., 2021).

2.2.1 Seeds

Seed propagation is a crucial method in field farming, forestry, vegetable production, and flowering plants that are cultivated for ornamental use. Direct seeding into the soil is a common propagation method for vegetable crops such as okra, peas, beans, radishes, carrots, turnips, maize, coriander, and spinach. Plants that areimmune to transplant shock or seeds that are very expensive are typically first sown in nursery beds before being transferred to the main fields. A site for growing vegetables should be level and have adequate soil drainage. The seedbed should have a loose yet smooth physical texture to ensure good seed-to-soil contact and a steady supply of moisture for the seeds. In addition, it shouldn't get too dry while allowing for adequate soil aeration. The topsoil should have a texture that prevents crusting and be crumb-free. Soil resistance from crusting, which can occur as a result of bad seedbed preparation or adverse weather conditions during planting, can be quite significant. (Chauhan et al., 2021). (Figure 2)

Figure 2. Illustration of different methods of Plant Propagation

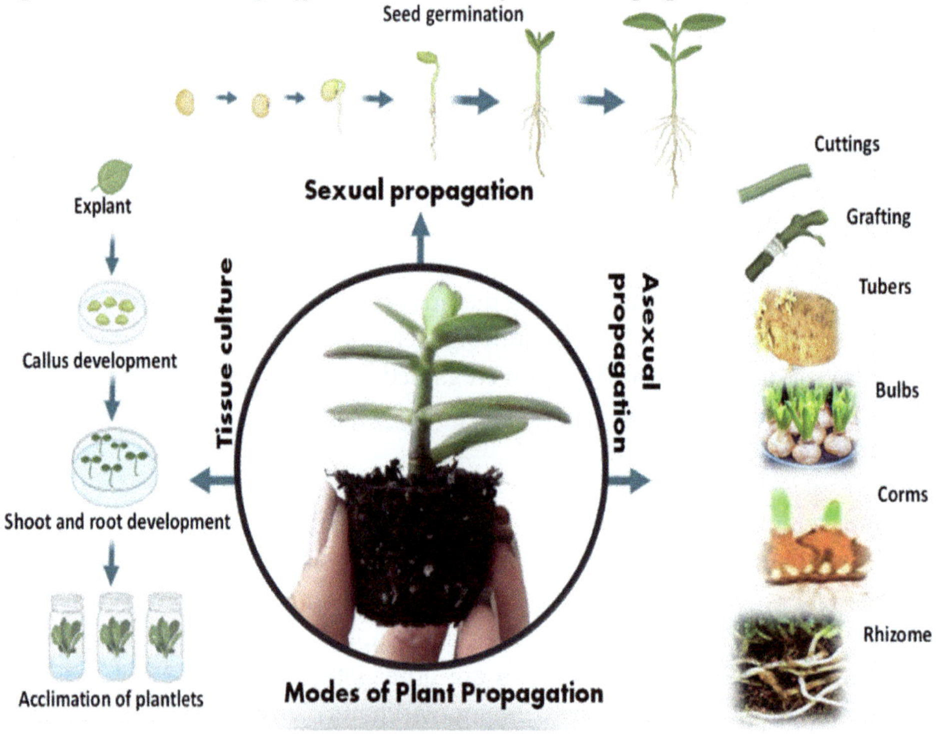

Table 1. Different Plants and their mode of propagation. (Thamburaj & Singh, 2000; Ağaoğlu, 2010; Megersa, 2017)

S.No	Plant (Crop/fruit/vegetables)	Mode of Propagation
1	Radish, carrot, peas, beans, cucurbits, turnip, coriander, okra, sweet corn, spinach	Direct sown crops (seed)
2	Cabbage, cauliflower, brinjal, tomato, capsicum, chili, knol khol, onion.	Transplanted crops (seed)
3	Onion, garlic, ginger and turmeric	Plantingbulbs and rhizomes
4	Potato, yam, taro	Planting tubers and corms
5	Pointed gourd, sweet potato and cassava	Planting vines and cuttings
6	Asparagus, rubarb and globe artichoke	Divisions of crown/suckers
7	quince, pomegranate, tea, olive, berry, vine, fig and berry	Cuttings
8	Apple	Grafting
9	pineapple	Sucker

2.3 Tissue Culture

Plant tissue culture is an advanced method used to cultivate plant cells and tissues in a nutritionally designed medium under sterile (aseptic) circumstances. (George et al., 2008).

2.3.1 Tissue Culture Media

The medium can be either solid or liquid. The objectives of plant tissue culture include the mass propagation of high-quality planting materials from mother plants in a short period, while reducing space and cost per plant. It also aims to obtain disease-free plants, rapid propagation of plants that are difficult to grow by conventional methods, genetic improvement of commercial plants, somatic hybridization, and the production of numerous plants without the need for seeds or pollinators. (Espinosa-Leal, Puente-Garza, and García-Lara 2018; George and Manuel 2013; Ak 2018).

2.3.2 Types of Tissue Culture

2.3.2.1 ORGANISED STRUCTURES

The phrase "organ culture" refers broadly to those different kinds of culture in which there is an organized pattern of growth It can be maintained around-the-clock and contains the aseptic isolation of such distinct entire plants buildings that resemble leaf primordia, immature flowers, and as well as how they develop in vitro.

This is done in order to the most crucial forms of organs for plant reproduction. The components of culture include:

1. Meristem cultures: in which the plants are cultivated quite little each composed of the apical excised shoot apex the meristematic dome, which may or may not have one or two leaves primordial. The shoot apex is usually developed to produce one shot only.
2. Shoot tip, or shoot cultures: began with excised shoot tips, also known as buds, that were larger than the shoot apices used to create meristem cultures with numerous leaf primordia. These shoot apices are typically grown in a manner that allows them to produce several shoots.
3. Node cultures: Individual lateral bud node cultures, each supported by a little piece of stem tissue; stem fragments can be cultivated with one or more nodes. Each bud is cultivated in order to produce a single shoot.
4. Isolated root cultures. The development of roots, a root system that branches but is not connected to the plant's shoots be acquired.
5. embryo cultures, whether fertilized or unfertilized Seed embryos are dissected out of zygotic tissue. grown in vitro until the formation of fruits or seeds. They have developed into seedlings. Embryo culture is significantly different from somatic embryogenesis.

2.3.2.2 Unorganized Structure

The phrase “tissue culture” is frequently employed as a catchall term. The word used to refer to all types of plant cultures grown in vitro despite its strict meaning, it should only apply to cultures of cell clusters that lack organization. The reality is that the The following sorts of cultures are the most prevalent:

1. Callus (or tissue) cultures. The development and maintaining somewhat disorganized cell groups resulting from the disorganized and uncoordinated the development of tiny plant structures, fragments of plant tissue, or cells that have been cultivated before.
2. Suspension Cultures or cell Cultures. Plant populations cells and tiny clumps of cells, dispersed in a stirred,which is a liquid medium that is aerated.
3. Protoplast cultures. These involve the cultivation of plant cells that have been isolated after the removal of their cell walls.
4. Anther cultures: This method uses intact anthers containing immature pollen microspores. The main objective is to produce haploid plants, either through the direct formation of somatic embryos from pollen or, in some cases, via callus formation followed by organogenesis. There are also **pollen cultures**, which

rely on pollen grains separated from the anthers and used directly. (George et al.,2008)

2.3.3 Plant Tissue Culture Techniques and Procedures

1. Preparation of Culture Medium: The process involves formulating concentrated solutions of various macro- and micronutrients. These solutions are appropriately diluted and mixed, followed by precise adjustment of the pH level. A gelling agent is incorporated to solidify the medium when preparing a solid matrix. Subsequently, the medium undergoes sterilization via autoclaving (steam sterilization). Finally, the medium is dispensed into culture vessels such as Petri dishes or culture tubes, where it solidifies.
2. Preparation of plant explant: The procedure begins with selecting an appropriate plant tissue segment for micropropagation. The chosen sample is then surface-sterilized and aseptically dissected into small explants. These explants are inoculated onto the prepared culture medium. To sustain the proliferation of cultured cells, callus tissues are periodically subcultured onto fresh media during the multiplication phase. (Pant & Husen, 2022; El-Hassani & Ekinci 2024; Elasi, Ak, & Ekinci, 2024).
3. In seed culture, seeds are germinated in a sterile environment containing a suitable nutrient medium. In contrast, meristem culture focuses on cultivating the tissues responsible for plant growth, specifically those located in apical and lateral buds. Organ culture involves the cultivation of various plant parts such as root and shoot tips, leaves, flowers, and immature fruits. Protoplast culture is a technique based on the cultivation of plant cells that have had their cell walls removed, typically obtained through enzymatic digestion or mechanical methods. Anther or pollen culture involves the use of anthers, which contain pollen grains—the key agents in the pollination process. Callus culture refers to the irregular proliferation of plant cells in a growth regulator-enriched medium, such as auxins and cytokinins, leading creation of a disorganized mass of undifferentiated cells known as callus. Cell culture entails the isolation of individual cells from a living organism and their growth in a liquid medium under controlled environmental conditions. Embryo culture relies on cultivating plant embryos in an artificial environment. These methodologies are fundamental for plant propagation, genetic improvement, and the production of disease-free plants. (Pasternak & Steinmacher 2024; Shishkova, Rost, & Dubrovsky, 2008).

3. ADVANTAGES AND DISADVANTAGES OF ASEXUAL AND SEXUAL METHODS OF PROPAGATION

3.1 Advantages of Using the Asexual Method

- Crop plants that cannot readily produce seeds can be propagated effectively using this method.
- Plants propagated through asexual reproduction are genetically identical to the parent plant, meaning they retain the exact genetic traits of the original source.
- By the technique of top-grafting (using grafting and budding), old crop plants with low economic productivity can be transformed into superior plants with higher yields.
- The advantages offered by the original roots (rootstocks) and the upper shoots (scions) can be utilized through asexual propagation methods.
- Plants propagated vegetatively exhibit uniformity at maturity and produce high-quality yields.
- Plants propagated asexually are typically smaller in size, which facilitates agricultural operations such as pesticide spraying, harvesting, and other related activities.

3.2 Disadvantages of Using the Asexual Method

- This limits the development of new cultivars, unlike sexual reproduction.
- Cutting and grafting procedures require specialized skills; therefore, this method of propagation is considered costly.
- • Plants that are propagated asexually generally live shorter lives than those that are propagated sexually.
- Plants that are propagated vegetatively are more vulnerable to biotic and abiotic stresses. (Chauhan et al., 2021)

3.3 Advantages of Using the Sexual Method

- Vigorous plants produced through seed propagation have a longer lifespan and are resistant to biotic stresses (like insects, pests, and diseases) as well as abiotic stresses (like extreme environmental conditions).
- Seed propagation is regarded as a straightforward, practical, and simple-to-use technique. However, some plants, like papaya, marigold, hot and sweet peppers, and tomatoes, cannot reproduce sexually.
- This technique is the main way to produce genetic variation in plants.

- This method allows the development of only a restricted range of vegetable cultivars and hybrids
- The initial plant material required for budding and grafting must be derived from seed-grown plants.
- This method allows for easy seed storage in a small space for extended periods.

3.4 Disadvantages of using Sexual methods.

- Sexually propagated plants show differences and are not genetically identical to one another because of genetic segregation.
- Plants produced from seed propagation undergo a long juvenile phase, resulting in delayed flowering and fruit production.
- Seed-grown plants tend to have vigorous growth, which makes agricultural operations such as harvesting, weeding, and spraying more labor-intensive.
- Sexual propagation cannot take advantage of the benefits offered by scions and rootstocks, as demonstrated by asexual propagation.
- This approach cannot be used to propagate crop kinds that do not generate seeds. (Chauhan et al., 2021).

4. PLANT PHYSIOLOGY CONCEPTS AND PLANT PROPAGATION

A thorough understanding of basic physiological principles is necessary to explain the success or failure of different plant propagation methods since plant physiology concentrates on how plants operate. Regardless of whether propagation takes place in the field, greenhouse, or laboratory, a variety of genetic and physiological variables affect how plants respond. The plant's metabolic and physiological processes are greatly influenced by the environmental factors around it. Light intensity, photoperiod, light quality (intensity, day length, and quality, such as light sources and wavelength variations), temperature, water balance, seasonal variations, and the photoperiodic reaction are all examples of these variables. The appropriate propagation methods must be carefully chosen by plant propagators, and this decision is heavily influenced by the kind of plant, whether it be annual, biennial, or perennial. (Beyl, & Sharma, 2015).

4.1 Fundamental Physiological Processes Influencing Propagation Efficiency

The following are the basic plant processes that impact plant propagation

4.1.1 Light and Carbohydrate Metabolism (Photosynthesis and respiration)

Photosynthesis serves as the plant's "light-harvesting mechanism" and represents the fundamental step in food production. During this process, plants absorb carbon dioxide (CO_2) through stomata, which is then reduced using water molecules that are split by light energy absorbed by chlorophyll complexes located within specialized organelles known as chloroplasts. This photosynthetic system, particularly in higher plants, utilizes electrons released from the photolysis of water to reduce CO_2, ultimately producing simple sugars such as glucose and releasing oxygen into the atmosphere.

In contrast, cellular respiration is almost the reverse of photosynthesis. In this process, plants utilize glucose and oxygen to release stored energy, resulting in the production of water vapor and carbon dioxide. It is noteworthy that photosynthesis is a light-dependent process taking place exclusively in chloroplasts, cellular respiration can occur in both light and dark conditions, taking place within the cytoplasm and mitochondria, regardless of the cell's location within the plant. Temperature directly influences both photosynthesis and respiration in plants. Typically, the efficiency of photosynthesis increases with rising temperatures until it reaches an optimum range usually between 25°C and 30°C—beyond which it declines rapidly. On the other hand, With each 10°C rise in temperature, the rate of cellular respiration usually doubles, within a range of 5°C to 35°C. However, if temperatures continue to rise, respiration may outpace the photosynthetic rate, leading to an energy deficit. In extreme cases, excessively high temperatures can damage or denature cellular enzymes, resulting in a breakdown of metabolic function. (Posch et al., 2019). Temperature is a key factor in the success of plant propagation. For instance, cuttings taken from stock plants may have reduced rooting potential if high nighttime temperatures cause an increase in respiration that depletes carbohydrate reserves. In a different situation, utilizing bottom heat is a standard procedure for propagating hardwood cuttings in the winter or rooting cuttings of evergreen plants in late autumn. Bottom heating gently warms the root zone, encouraging adventitious root development and callus formation. At the same time, the air temperature surrounding the upper part of the cutting is kept relatively low, which helps suppress premature shoot growth until the newly formed root system is capable of supporting the aerial parts. (Preece, 2003).

4.1.2 Influence of Light Intensity and Photoperiodism on Flowering Initiation

Light has a major impact on flowering and seed production, which is important for both propagators and plant breeders. This phenomenon is known as "photoperiodism." However, the term can be somewhat misleading because the critical factor is not the duration of light exposure per se, but rather the duration of the continuous dark interval. Plants that react to day length are divided into two categories: short-day plants and long-day plants. Short-day plants, such as chrysanthemums and poinsettias, flower only when the continuous night period exceeds a certain critical threshold. In contrast, long-day plants like clover start to flower when the continuous night duration is less than a certain threshold length. Plants are also categorized as "obligate" or "facultative" according to their photoperiodic reaction. Obligate short-day plants, like African marigolds, and obligate long-day plants, like fennel and flax, both need the critical nightlength condition to be met to bloom. Although facultative long-day plants, such as lilies, will eventually flower after reaching a particular size, exposure to the proper photoperiod speeds up or improves their flowering response. Similarly, facultative short-day plants, such as dahlias, will eventually flower after reaching a particular size, but exposure to the appropriate photoperiod accelerates or enhances their flowering response. The light-sensitive pigment "phytochrome," which regulates photoperiods, is found in two main forms: the active form (Pfr) and the inactive form (Pr). The Pr form is most sensitive to red light at about 660 nanometers, whereas the Pfr form is more responsive to far-red light at approximately 730 nanometers. The inactive Pr form of plants changes into the active Pfr form when exposed to intense red light, like bright sunlight. This form activates light-related reactions within the plant. On the other hand, exposure to far-red light, such as that seen in the evening or in the shade, causes the active Pfr to slowly return to the inactive Pr form. This reversion to the inactive state happens gradually in the dark. Phytochrome is associated with the regulation of various gene groups and is essential for initiating or repressing processes like gene transcription, metabolism, photosynthesis, chloroplast development, cellular signaling, transport, growth and development, hormone regulation, and stress and defense responses. (Tepperman et al., 2006). (Figure 3)

Figure 3. Influential Physiological Parameters that Affect Plant Propagation.

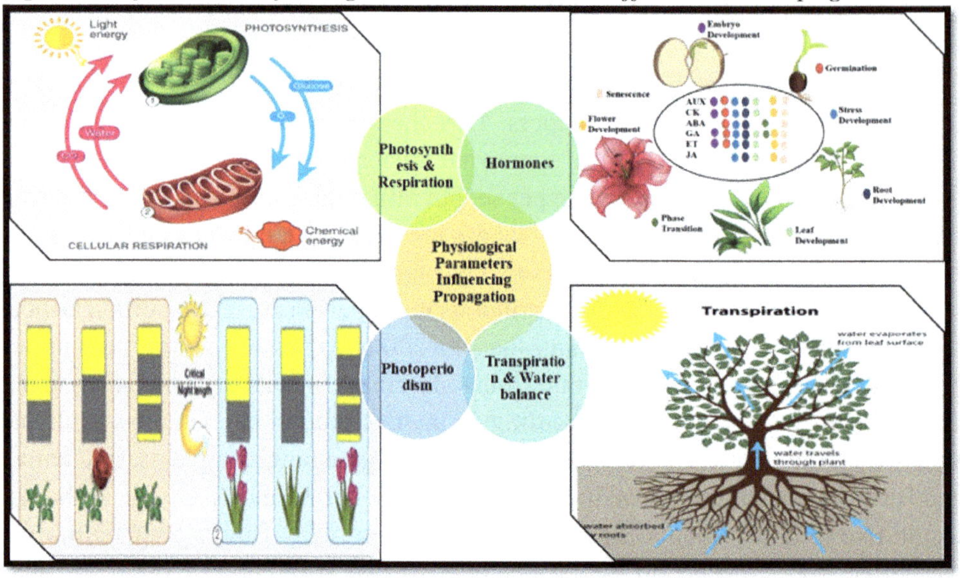

4.1.3 Other Major Processes Impacting Plant Propagation

The success of plant propagation depends critically on several other physiological factors. Partial light reduction or shading is frequently used to encourage root development, especially during the rooting stage, among these techniques. Processes such as transpiration and water balance are also fundamental physiological pillars in propagation. Controlling water loss is critically important during the root initiation period, as it prevents desiccation of the planted tissues and ensures high survival rates. To accomplish this, environmental variables like temperature, humidity, and light intensity need to be tightly controlled to preserve the plant's internal water balance. To guarantee the successful propagation and healthy growth of new plants, this environmental management is crucial. Additionally, plant hormones are essential for controlling essential developmental processes throughout propagation. By manipulating the levels of auxins, cytokinins, and other phytohormones, growers can enhance root initiation in cuttings, stimulate bud development, and improve the survival rates of propagated plant materials. Achieving the appropriate hormonal balance is a critical factor in the success of propagation. Another important factor influencing propagation success is the physiological age of the plant whether it is in the juvenile or mature phase. Juvenility refers to the early stage of growth characterized by vigorous vegetative development, a high capacity for root formation, and the absence of flowering. During this phase, plant tissues exhibit greater responsiveness to propagation techniques such as cutting and tissue culture, making it the optimal period for harvesting propagation materials.

During the transition from the juvenile to the adult stage, plants initiate the formation of reproductive organ such as flowers, and their capacity for vegetative propagation declines due to physiological and morphological changes in their tissues. Mature tissues typically exhibit reduced rooting potential, which often necessitates the use of rejuvenation techniques such as pruning, grafting, or micropropagation (tissue culture) to restore their propagation capability. Distinguishing between the juvenile and mature phases is therefore crucial when selecting appropriate plant materials for propagation, as this understanding significantly enhances propagation success rates. In addition, several other physiological factors influence propagation outcomes, including the timing of flowering, the chronological age of the parent plant, nitrogen status in the tissues, and the season in which propagation is performed. All of these factors collectively contribute to determining the effectiveness and success of plant propagation. (Beyl & Sharma, 2015).

5. IMPORTANCE OF PLANT PROPAGATION

The propagation of plants is extremely crucial and is generally recognized across the globe. The significance of plant propagation techniques has grown to new heights due to advancements in their methods. Improvements in plant propagation, such as

5.1 Seed priming

It is a pre-sowing procedure that enables the seed to be hydrated in a controlled manner up to the moment just prior to radicle initiation. Primed seeds can germinate more quickly and consistently than unprimed seeds, especially in unfavorable circumstances. Osmopriming with salt or polyethylene glycol solutions, solid matrix priming with materials like peat or vermiculite, and drum priming, which adds moisture while tumbling seeds, are all methods. Priming is now used commercially on a lot of flower and vegetable seeds. (Pill, 2020)

5.2 Mist Propagation

The use of intermittent mist to retain moisture around cuttings is changing the propagation industry. Regularly spraying short bursts of mist at intervals can help maintain leaf moisture without drowning the rooting medium. This lowers the risk of illness and allows cuttings to root more rapidly. Computerized mist controllers can enhance misting intervals depending on light, humidity, and temperature factors. (Preece, 2003)

5.3 Vegetative Propagation of Difficult Species

Research is ongoing to determine propagation techniques for difficult species. Rooting can be enhanced by using revitalized tissue for cuttings, such as by harvesting stump shoots or hedging stock plants. Formerly challenging plants like coast redwood can now be propagated commercially thanks to the stringent selection of superior clones, enhanced environmental conditions, and thoughtfully timed cutting collection (Nahar, 2025).

5.4 Breeding for Propagation Traits

Genomic advancements are making it possible to identify genes associated with traits important for propagation, such as adventitious rooting, graft compatibility, seed and bud dormancy, precocious flowering, and others. This allows marker-assisted breeding to improve the propagation efficiency of several plants. Some studies even seek to give valuable genotypes that were previously unable to be cloned through a particular propagation technique, such as somatic embryogenesis, the ability to do so. (Awotedu et al., 2021)

5.5 Micropropagation Advances

Numerous advances are increasing the effectiveness of micropropagation and its use among a greater variety of organisms. The propagation process can be scaled up with less handling thanks to the use of bioreactors, which are large containers for liquid culture. Multiplication rates are boosted by momentary immersion systems that intermittently immerse plant tissue in a liquid medium. The success rates of novel culture media compositions, which are occasionally tailored to a specific crop, have increased. LEDs can minimize energy consumption and optimize light spectra for each developmental stage when used for lighting in tissue culture laboratories. (Cardoso et al., 2018)

In addition to all these significant advancements in plant propagation, we can assert that anything that uses plants in any manner depends on the ability to propagate them. Seed germination is the most critical step in the global food chain, and this dependence has led to the development of a large industry. As a result, the cultivation of crops, including vegetables, on a large scale is possible. The worldwide domestic seed market was valued at about $45 billion annually in 2012. Although many seeds can be sown mechanically, some need to be planted by hand to avoid harm. The nursery industries utilize a variety of plant propagation techniques worldwide, including stool bedding apple rootstocks in England, T-budding dogwoods in Tennessee, and Junebudding peaches in Georgia. Nurserymen have a thorough understanding of how to propagate various plants, including ornamental trees, fruit, shrubs, and grasses (Beyl & Trigiano, 2016). The floral and nursery industries have been the two fastest expanding areas of horticulture in recent years. The vegetable production industry, another branch of horticulture that depends on effective seed germination (Jaskani & Khan, 2021). In the fifth century, the Chinese showed that vegetable plants could be propagated via grafting. Today, Korea and Japan are creating robots and grafting machines to propagate large quantities of grafted peppers, eggplants, and tomatoes. (Kubota et al., 2008; Awazade & Verma, 2024) The forest industry also demonstrates a strong interest in plant propagation methods that improve tree performance, such as the use of superior clone cuttings and the creation of seed orchards with enhanced characteristics. (Manohar et al., 2024). Numerous methods have highlighted nonzygotic embryogenesis, which involves creating an embryo that is identical to the original using the plant's somatic tissue. (Corredoira et al., 2019). Worldwide, molecular genetics methods are being investigated for their ability to increase auxin production in aspen, which could speed up the growth of transgenic trees. Additionally, cloned strategies are being looked at for species like pine and eucalyptus. (Dubouzet et al., 2013). Traditional and contemporary plant propagation methods are extremely crucial for the conservation of rare and endangered plants. These methods are occasionally employed to reintroduce endangered plants to both their historical and contemporary environments. In vitro seed germination or tissue culturing are examples of plant propagation methods that can quickly produce a large quantity of specific species that thrive in the vitro setting. Plant propagators must be exceptionally proficient in propagation methods to germinate the seeds and grow the plants to share with other botanical gardens to boost the quantity of rare and endangered specimens (Madzivire, 2021).

6. CONCLUSION

Plant propagation is a key process that helps grow more plants, improve their quality, and protect different plant species. It's important for keeping ecosystems healthy, supporting biodiversity, and making sure we have enough food to meet the needs of a growing population. Without proper propagation methods, many valuable plant species could be lost, affecting both nature and agriculture. There are two main ways to propagate plants: sexual and asexual. Both methods are widely used today in farming and horticulture to increase crop production, grow disease-resistant plants, and save rare or endangered species. These techniques also help preserve the genetic diversity of plants, which is vital for long-term agricultural sustainability. Plant propagation connects science with real-world uses in areas like gardening, forestry, and environmental protection. It plays a crucial role in restoring degraded lands, maintaining green spaces in urban areas, and supporting plant-based industries. As the world faces big challenges like climate change, land degradation, and food shortages, finding better and more efficient ways to grow and multiply plants is becoming more important than ever.

Future Perspective

The future of plant propagation is moving forward by mixing old and new methods. While traditional techniques are still used, new technologies are changing the way we grow plants. In modern farming, methods like aeroponics, hydroponics, and tissue culture are being used to grow plants without soil and in small spaces. These techniques are useful for saving space and resources. Scientists are also using genetic engineering and marker-assisted breeding to help plants grow better. These methods help improve important traits like stress tolerance, faster rooting, and stronger growth. On top of that, artificial intelligence (AI) is now helping farmers grow plants more accurately and efficiently. As we move forward, combining these tools and ideas will help take plant propagation to the next level.

REFERENCES

Ağaoğlu, Y. S., Çelik, H., Çelik, M., Fidan, Y., Gülşen, Y., Günay, A., & Yanmaz, R. (2010). *Genel bahçe bitkileri.* Ankara Üniversitesi Ziraat Fakültesi.

Ak, B. E. (2018). The Importance of in vitro micropropagation of fruit crops. In *1st International GAP agriculture and livestock congress proceedings book* (pp. 716-723).

Awazade, A. S., & Verma, D. (2024). Advancing Vegetable Grafting: A comprehensive review of techniques, challenges, and the future of automated solutions. *Journal of Scientific Research and Reports, 30*(11), 517–530. DOI: 10.9734/jsrr/2024/v30i112580

Awotedu, B. F., Omolola, T. O., Akala, A. O., Awotedu, O. L., & Olaoti-Laaro, S. O. (2021). Vegetative propagation: A unique technique of improving plants growth. *World News of Natural Sciences, 35*, 83–101.

Bag, T. K., Srivastava, A. K., Yadav, S. K., Gurjar, M. S., Diengdoh, L. C., Rai, R., & Singh, S. (2015). Potato (Solanum tuberosum) aeroponics for quality seed production in north eastern Himalayan region of India. *Indian Journal of Agricultural Sciences, 85*(10), 1360–1364. DOI: 10.56093/ijas.v85i10.52303

. Beyl, C. A., & Sharma, G. C. (2015). Plant physiology concepts important for propagation success. *Plant propagation concepts and laboratory exercises*, 29-46.

Beyl, C. A., & Trigiano, R. N. (Eds.). (2016). *Plant propagation concepts and laboratory exercises.* CRC press. DOI: 10.1201/b17340

. Cardoso, J. C., Sheng Gerald, L. T., & Teixeira da Silva, J. A. (2018). Micropropagation in the twenty-first century. *Plant cell culture protocols*, 17-46.

Chauhan, A., Sharma, D., Kumar, R., Shiwani, K., & Sharma, N. (2021). *Methods of propagation in vegetable crops. Recent trends in propagation of forest and horticultural crops.* Taran Publication.

Chen, J., & Stamps, R. H. (2006). *Cutting propagation of foliage plants. Cutting propagation: A guide to propagating and producing floriculture crops.* Ball Publishing.

. Choudhary, R. C., Meena, V. K., Choudhary, M., & Tak, J. K. (2017). Propagation Methods of Horticultural Crops by Specialized Organs. *of the Book: Advancement and Innovations in Agriculture*, 115.

Cofer, C. M. (2015). *The Ara Pacis Augustae and the Ancient Understanding of Grafting* (Doctoral dissertation, Bryn Mawr College)

Corredoira, E., Merkle, S. A., Martínez, M. T., Toribio, M., Canhoto, J. M., Correia, S. I., Ballester, A., & Vieitez, A. (2019). Non-zygotic embryogenesis in hardwood species. *Critical Reviews in Plant Sciences*, *38*(1), 29–97. DOI: 10.1080/07352689.2018.1551122

De Klerk, G. J. (2012). Micropropagation of bulbous crops: Technology and present state. *Floriculture & Ornamental Biotechnology*, *6*(1), 1–8.

Dubouzet, J. G., Strabala, T. J., & Wagner, A. (2013). Potential transgenic routes to increase tree biomass. *Plant Science*, *212*, 72–101. DOI: 10.1016/j.plantsci.2013.08.006 PMID: 24094056

ELASI. M. F., Ak, B. E., & Ekinci, H. (2024). Some researches on micropropagation of Orchid (Phalaenopsis spp.) in vitro conditions. In *5 th International Conference on Engineering and Applied Natural Sciences August* (pp. 25-26).

Espinosa-Leal, C. A., Puente-Garza, C. A., & García-Lara, S. (2018). In vitro plant tissue culture: Means for production of biological active compounds. *Planta*, *248*(1), 1–18. DOI: 10.1007/s00425-018-2910-1 PMID: 29736623

George, E. F., Hall, M. A., & De Klerk, G. J. (2008). *Plant propagation by tissue culture* (3rd ed.). The Back Ground Springer.

George, E. F., Hall, M. A., & Klerk, G. J. D. (2008). The components of plant tissue culture media I: macro-and micro-nutrients. In *Plant Propagation by Tissue Culture: Volume 1. The Background* (pp. 65-113). Dordrecht: Springer Netherlands

George, P., & Manuel, J. (2013). Low cost tissue culture technology for the regeneration of some economically important plants for developing countries. *International Journal of Agriculture Environment and Biotechnology*, *6*(Special Issue), 703–711.

Hanson, B. (Ed.). (2007). *Buried Treasures: Tasty Tubers of the World: how to Grow and Enjoy Root Vegetables, Tubers, Rhizomes, and Corms* (Vol. 188). Brooklyn Botanic Garden.

Ingold, T. (2024). Growing plants and raising animals: an anthropological perspective on domestication. In *The origins and spread of agriculture and pastoralism in Eurasia* (pp. 12–24). Routledge. DOI: 10.4324/9781003579861-3

Jaskani, M. J., & Khan, I. A. (2021). Horticulture: an overview. *University of Agriculture Faisalabad*, 3-22.

Kubota, C., McClure, M. A., Kokalis-Burelle, N., Bausher, M. G., & Rosskopf, E. N. (2008). Vegetable grafting: History, use, and current technology status in North America. *HortScience*, *43*(6), 1664–1669. DOI: 10.21273/HORTSCI.43.6.1664

. Kumar, P., Rana, S., Sharma, P., & Negi, V. (2015). Vegetable grafting: a boon to vegetable growers to combat biotic and abiotic stresses. *Himachal Journal of Agricultural Research*, 1-5.

. Lee JungMyung, L. J., & Oda, M. (2003). Grafting of herbaceous vegetable and ornamental crops.

. Li ChangHua, L. C., Li XiaoChuan, L. X., Zhao MeiHua, Z. M., & Zhao JunLiang, Z. J. (1995). Studies on the virus-free technology of meristem-tip and tissue culture in Allium sativum.

Madzivire, P. K. (2021). *In Vitro Propagation of Haemanthus pumilio and H. albiflos (Amaryllidaceae) and the population genetics of H. pumilio* (Doctoral dissertation, Stellenbosch: Stellenbosch University).

Manohar, K. A., Shukla, G., Shahina, N. N., Sivasankarreddy, K., Ravuther, S. S., & Chakravarty, S. (2024). Conventional Versus Non-Conventional Methods of Propagation of Forest Tree Species: Applications and Limitations. In *Biotechnological Approaches for Sustaining Forest Trees and Their Products* (pp. 483–517). Springer Nature Singapore. DOI: 10.1007/978-981-97-4363-6_18

Megersa, H. G. (2017). Propagation methods of selected horticultural crops by specialized organs [review]. *J Hortic*, *4*(198), 2376–0354. DOI: 10.4172/2376-0354.1000198

Murlidhar, J., Sadawarti, Pandey, K. K. Singh, S. P. Singh, Y. P. (2017). A review on comparison of tissue culture v/s conventional system of seed potato production. *Indian Journal of Hill Farming.*, *30*(2), 160–167.

Nahar, K. (2025). Propagation Technologies: Seedling, Stem Cutting and Grafting. In *Jatropha curcas L: A Potential 2G Energy Crop to Produce Biofuel in Bangladesh: Agronomy, Biotechnology, Biodiesel and Byproducts* (pp. 33–55). Springer Nature Switzerland. DOI: 10.1007/978-3-031-77642-7_3

Pant, M., & Husen, A. (2022). Micropropagation in mature trees by manipulation of phase change, stress, and culture environment. In *Environmental, Physiological and Chemical Controls of Adventitious Rooting in Cuttings* (pp. 421-437). Academic Press. DOI: 10.1016/B978-0-323-90636-4.00002-7

Pasternak, T. P., & Steinmacher, D. (2024). Plant growth regulation in cell and tissue culture in vitro. *Plants*, *13*(2), 327. DOI: 10.3390/plants13020327 PMID: 38276784

Peña-Iglesias, A., & Ayuso, P. (1982, September). Characterisation of Spanish garlic viruses and their elimination by in vitro shoot apex culture. In *IV Conference on recent Advances in Vegetable Virus Research 127* (pp. 183-193). .

Pill, W. G. (2020). Low water potential and presowing germination treatments to improve seed quality. In *Seed quality* (pp. 319–359). CRC Press. DOI: 10.4324/9781003075226-10

Posch, B. C., Kariyawasam, B. C., Bramley, H., Coast, O., Richards, R. A., Reynolds, M. P., Trethowan, R., & Atkin, O. K. (2019). Exploring high temperature responses of photosynthesis and respiration to improve heat tolerance in wheat. *Journal of Experimental Botany*, *70*(19), 5051–5069. DOI: 10.1093/jxb/erz257 PMID: 31145793

Prasath, D. K., Kandiannan, R., Chitra, V. A., Muhammed Nissar, J., Suresh, K., & Babu, N. (2017). Quality seed production in ginger and turmeric: Present status and future prospects. *Shodh Chintan*, *9*, 163–170.

Preece, J. E. (2003). A century of progress with vegetative plant propagation. *HortScience*, *38*(5), 1015–1025. DOI: 10.21273/HORTSCI.38.5.1015

Rivard, C. L., & Louws, F. J. (2008). Grafting to manage soilborne diseases in heirloom tomato production. *HortScience*, *43*(7), 2104–2111. DOI: 10.21273/HORTSCI.43.7.2104

Sansavini, S. (2014). *A short history of scholarship in horticulture and pomology. Horticulture: Plants for People and Places* (Vol. 3). Social Horticulture.

Senula, A., Keller, E. R. J., & Leseman, D. E. (2000). Elimination of viruses through meristem culture and thermotherapy for the establishment of an in vitro collection of garlic (*Allium sativum*). *Acta Horticulturae*, (530), 121–128. DOI: 10.17660/ActaHortic.2000.530.12

Shishkova, S., Rost, T. L., & Dubrovsky, J. G. (2008). Determinate root growth and meristem maintenance in angiosperms. *Annals of Botany*, *101*(3), 319–340. DOI: 10.1093/aob/mcm251 PMID: 17954472

. Shopov, A. (2021). Grafting in Sixteenth-Century Mamluk and Ottoman Agriculture and Literature. *History and Society During the Mamluk Period (1250–1517)*, 381-406.

Tepperman, J. M., Hwang, Y. S., & Quail, P. H. (2006). phyA dominates in transduction of red-light signals to rapidly responding genes at the initiation of Arabidopsis seedling de-etiolation. *The Plant Journal*, *48*(5), 728–742. DOI: 10.1111/j.1365-313X.2006.02914.x PMID: 17076805

Thamburaj, S., & Singh, N. (2000). *Textbook of Vegetables, Tubercrops and Spices*. Indian Council of Agricultural Research.

Thokchom, R., & Singh, S. S. (2024). Plant propagation techniques. In *Horticulture research* (1st ed., pp. 125–137). ND Global Publication House.

Venema, J. H., Dijk, B. E., Bax, J. M., van Hasselt, P. R., & Elzenga, J. T. M. (2008). Grafting tomato (Solanum lycopersicum) onto the rootstock of a high-altitude accession of Solanum habrochaites improves suboptimal-temperature tolerance. *Environmental and Experimental Botany*, *63*(1-3), 359–367. DOI: 10.1016/j.envexpbot.2007.12.015

Chapter 2
Impact of Propagation Techniques on Agricultural Production

Wasan Salih Hussain
https://orcid.org/0000-0002-1537-9979
Mosul University, Iraq

Abdullah Khder Mohammad
Mosul University, Iraq

Alaa Khalid Ibrahim
Northern Technical University, Iraq

ABSTRACT

Breeding techniques are not just means of increasing plant numbers; they are strategic tools for "programming" plants to be more robust, resilient, and resource-efficient. By directing these techniques toward producing varieties adapted to environmental challenges and reducing reliance on chemical inputs, we can achieve sustainable agriculture that ensures food security while preserving the environment and ensuring food security for current and future generations. Asexual propagation methods (such as cuttings, grafting, and tissue culture) allow the precise genetic characteristics of parent plants to be preserved, ensuring that new generations of plants will possess the same excellent traits. This is vital for hybrid varieties, those that do not produce seeds, or those that do not maintain their traits when planted from seeds. Breeding techniques focus on developing plant varieties with a higher capacity to absorb nutrients from the soil, reducing the need for chemical fertilizers and producing plants resistant to pests and diseases.,

DOI: 10.4018/979-8-3373-5626-6.ch002

Introduction

In an era where the world faces increasing challenges in food security and climate change, the importance of adopting sustainable agricultural technologies emerges as a strategic solution to ensure a sustainable and productive agricultural future. In this context, we highlight sustainable agriculture, which is the optimal solution to address the environmental and economic challenges related to the agricultural sector. By raising awareness, providing government support, and providing appropriate technology, more sustainable agriculture can be achieved that meets present needs without compromising the future of the planet. This is a means to develop the agricultural sector and improve crop productivity. There are many unconventional methods of agriculture, such as aeroponics, which involves growing plants suspended in the air without a growth medium, and obtaining food and moisture through spray distributed by sprinklers or nozzles (Hussein and Abbas,2020).

There are also other methods of agriculture, such as hydroponics and Sandponic farming (Hussein and Mohammad, 2024).

Sustainable breeding techniques ensure crop availability by producing high-quality seeds and planting materials, transferring the distinctive genetic traits of the parent plant through cuttings, and diversifying genetic resources to address environmental challenges and climate change, thereby enhancing food security and reducing dependence on external resources.

Reproduction is defined as one of the applied branches of botany, which is concerned with increasing the number of economic plants to maintain them to meet human needs and requirements for food, clothing, and shelter. Although most field crops reproduce sexually, primarily through seeds, there are some crops that reproduce asexually. Sexual reproduction is a common method of reproduction among higher plants, and in particular, most field crops reproduce through seeds. Although some crops reproduce by seeds only a few times, sexual reproduction is the most common method. Some plants can reproduce sexually without undergoing meiosis, gamete formation, or fertilization. Common methods of asexual reproduction include corms, bulbs, rhizomes, stolons, tubers, and other vegetative parts. For example, potatoes are propagated through tubers, and sugarcane is propagated through cuttings.

Horticultural Plants Propagation

Propagation is defined as increasing the number of individuals and plant species by sexual (seeds) or asexual (vegetative) means in order to preserve and spread them.

Horticultural plants are propagated in two ways:

First - Sexual Propagation:

This refers to the use of seeds containing sexual embryos to propagate plants. These seeds are the result of pollination and fertilization in plants.

Some plant seeds can be planted for propagation after being harvested directly from the fruit, as in the case of various vegetable crop seeds, provided that the environmental conditions are favorable for germination.

Another category of plants has seeds that cannot germinate after being harvested from the fruit, even if grown under suitable environmental conditions, as in the case of some types of fruit, such as stone fruit. Apple seeds require special treatment before planting, such as exposure to low temperatures. Another category requires seed treatment, i.e., scarification, to facilitate germination.

Advantages of Sexual Propagation (Advantages and Disadvantages of Reproductive Methods).

Humans Unlike plants that plant cannot move around their environment, which means they need to use other strategies to transport pollen to ovules to form grains. For cross-pollinated plants, male plants do not have to expend energy on grain production and can therefore direct more energy toward producing and dispersing high-quality pollen. Similarly, since female plants do not have to produce pollen, they can direct more of their energy toward producing high-quality ovules, meaning these plants can pass on more abilities to their offspring to improve their chances of survival. Furthermore, because dioeciously plants require two different plants to reproduce, their offspring will have more genetic diversity from their two parents. This is crucial when the environment changes, as offspring with this greater genetic diversity will be better able to adapt to the new environment thanks to their genetic diversity. On the other hand, self-pollinating plants, which obtain their genes from only one parent, will suffer from low genetic diversity and may therefore be susceptible to many problems related to their adaptation to environmental changes. Therefore, this (abundance of genetic diversity) is considered an advantage for plants in which cross-pollination occurs! I previously learned that only 5% of plant species reproduce this way, which means that these plants are not numerous.

1. The ability to propagate large numbers of plants from a single plant or produce plants on a large scale (mass propagation), as in forest trees.
2. Plants produced from seeds, i.e., sexually propagated, are stronger than their asexually propagated counterparts (vegetative), especially in fruit trees, which

are characterized by their greater resistance to environmental conditions and climatic factors due to the spread and size of the root system.

3. Sexual propagation is a means of obtaining disease-free plants, especially viral diseases, as is the case with fruit trees, particularly citrus fruits.
4. It is a means of obtaining varieties and types of seeds in fruit and vegetable trees.
5. It is a means of obtaining new varieties and types of plants through hybridization between different varieties or species.

Disadvantages of sexual propagation:

1- Trees produced from seed differ in characteristics from the parent plant from which the seeds were taken, as in palm trees, citrus fruits, and other fruit species.
2- Trees propagated by seed are often delayed in reaching fruiting. For example, palm trees propagated by seeds yield fruit after 10-20 years, or sometimes more, while those propagated by offshoots require 6-12 years, or sometimes less.
3- Trees propagated by sex exhibit undesirable traits, such as the appearance of thorns, as these impede various maintenance processes.

Second: Asexual Propagation (Vegetative Propagation):

This refers to the use of any part of the plant other than the sexual embryo of the seed. The part used for propagation may be the stem, leaf, bud, a portion of leaf tissue, the growing tip, etc.

Vegetative Propagation Methods:

1- Cuttings: These are taken from branches or stems one year or older and divided into stem cuttings, leaf cuttings, or rootstock cuttings (apples and olives).
2- Budding: A bud is taken from the plant, called the scion, and placed on another plant, called the rootstock (stone fruit, citrus fruits, roses, gardenias).
3- Grafting: This refers to the fusion of a plant stem with a branch or stem of another plant of the same variety.
4- Propagation by specialized parts such as corms (gladiolus), bulbs and tubers (potatoes), or tuberous roots (sweet potatoes and dahlias).
5- Propagation by offshoots (palm trees and bananas).
6- Propagation by plant tissue culture: This method uses very small plants (bud, part of the growing tip, pollen grain, anther, or part of the ovary) grown under completely sterile conditions in media containing all growth requirements.

7- Propagation by layering: This method is used for plants that have difficulty forming roots on their cuttings.

Advantages of asexual propagation:

1. Ensuring the production of plants with similar characteristics to the parent plant (palm propagated by offshoots).
2. Vegetative propagated plants begin fruiting earlier than sexually propagated plants. Grafted oranges produce fruit 4-6 years after grafting, while olives produce fruit 6-8 years after grafting.
3. Propagating plants that do not produce seeds under normal conditions, such as bananas, currants, garlic, and navel oranges.
4. Sexual propagation is a means of overcoming certain environmental conditions by combining two plants through the grafting process. One is resistant to environmental conditions, called the stock, and the other is resistant, called the scion.
5. The size of the resulting plants can be controlled to facilitate harvesting and processing.

Disadvantages of asexual propagation:

1- Asexually propagated plants are less resistant to environmental conditions than their sexually propagated counterparts.
2- Asexual propagation is a means of transmitting various diseases, especially viral diseases.
3- It requires more experience and skill on the part of the person performing the propagation process, as well as more specialized supplies than sexual propagation.

The Importance of Vegetative Propagation in Plants

- Vegetative propagation is used in plants for many purposes.
- To propagate seedless plant species.
- To propagate hybrid plants without alteration, as not all of their seeds produce plants identical to the parents.
- To propagate plants with low seed germination rates.
- To increase the speed of plant propagation and accelerate fruiting.
- To adapt and acclimatize plants to new and different environments. For example, pear tree roots grow well in sandy soil, but they can be successfully

transplanted to such soil by grafting onto peach trees, which thrive in such soil.

- Vegetative propagation of various types prevents infection by certain parasites that attack the roots of some plants but not others. For example, European grape roots are usually susceptible to a type of parasite that does not affect American grapes.

The role of propagation in supporting nutrient cycling systems and maintaining soil fertility.

Propagation supports nutrient cycling systems and maintains soil fertility by increasing organic matter through cover crops and green manure, fixing nitrogen with legumes, improving soil structure and structure through the deep roots of trees and native crops, and enhancing the biodiversity of microorganisms that decompose organic matter. Practices such as crop rotation and reduced tillage also contribute to reducing nutrient depletion and maintaining soil structure.

1. Soil Cover

Keeping the soil covered is a fundamental principle of conservation agriculture. Crop residues are left on the soil surface, but cover crops may be needed if the gap between harvesting one crop and establishing the next is too long. Cover crops improve the stability of a conservation agriculture system, not only by improving soil properties, but also by promoting increased biodiversity in the agro-ecosystem. (FAO, 2022) While commercial crops have market value, cover crops are primarily grown for their impact on soil fertility or as livestock feed. In areas where small amounts of biomass are produced, such as semi-arid regions or areas of eroded and degraded soils, cover crops are beneficial because they: protect the soil during fallow periods; mobilize and recycle nutrients; improve soil structure and break up compacted layers and hard beds; and allow for rotation in monocultures.

Different plants, with diverse rooting systems, explore different depths in the soil. They may also have the ability to absorb different amounts of nutrients and produce distinct root exudates (organic acids), resulting in benefits for both the soil and living organisms.

Vegetation is important in conservation agriculture to protect the soil from the effects of raindrops, to keep the soil shaded, and to maintain the highest possible moisture content. We have seen its importance for nutrient recycling; however, it also has a physical and possibly allelopathic effect on weeds, reducing their incidence and leading to a reduction in the use of agrochemicals and, consequently, production costs.

Straw residues act as a cushion that reduces pressure on the soil under wheels and hooves and therefore plays an important role in reducing soil compaction.

2. Seed Selection, Treatment, and Propagation in Organic Systems

The crop performance can be impacted by seed quality directly, which affects germination, seedling growth, resistance to disease, and yield. In organic agriculture, where not used of pesticides and synthetic fertilizers are, the essential starting with seeds to have high quality to ensure plant health and productivity. Studies we shown that the quality of poor seeds can reduce germination by 30% to 50%, which can result in lower plant populations and increased susceptibility to pests and diseases.

The selection of Organic seed depends on purity and adaptability to ecological conditions. High quality organic seeds exhibit superior viability and resistance to stresses factors, that may be reducing dependency on external inputs.

The selection of organic seeds follows principles of ecological that depend on biodiversity, stability of genetic, and ecological adaptability. In contrast to conventional systems that focus on high yield hybrids or seeds that are genetically modified, farming of organic relies on naturally resilient varieties that grow without chemical interventions. are principles of selection of organic seeds guiding, include: genetic diversity, purity of seed, and local conditions adaptation.

For that, Organic seed selection must be free from treatments of chemicals, organisms that genetically modified, and any synthetic that are in line with the ecological principles of sustainability. Seed choice ensures improved to resilience crop, better efficiency of nutrients, and higher environmental adaptation to conditions. Seeds selection in organic farming follows principles, including purity of genetics, capacity of germination, and local climatic suitability for conditions. Studies have shown that farms using organically adapted seed varieties foster long-term sustainability and biodiversity.

Biotechnology Applications in Plant Propagation for Sustainability

Agricultural biotechnology is a set of technologies that utilize genetic modification and engineering in crops, livestock, forestry, fisheries, aquaculture, and agricultural industries. These technologies aim to improve plants or animals or develop microorganisms that are more easily and cost-effectively resistant to diseases, insects, and pesticides. These technologies also help adapt to climate change and conserve natural resources, such as biodiversity, land, and water, which are essential for food and agricultural production (Khatik 2023).

Environmental biotechnology plays a crucial role in addressing global environmental challenges by using biological processes to achieve sustainable solutions. This is achieved through three approaches:

1. Molecular Biology Techniques:

Plant molecular genetics is a branch of genetics that studies the molecular structure and function of genetic material (DNA) in plants, including how genetic information is stored and transmitted, and how genes express plant traits. Research in this field focuses on genetic engineering to produce crops resistant to diseases and pests, improve their characteristics, and develop techniques for diagnosing plant diseases. Using molecular biology techniques, selection and the search for new and improved genetic combinations are among the most important breeding methods. Therefore, this task becomes difficult for plant breeders if the sources of genetic variation are limited. However, if these sources are available, their task is also somewhat difficult, which is to conduct screening and selection, especially if the number of plants and genetic combinations is very large. The genetic factors (genes) responsible for traits are located in the genetic material (DNA). Therefore, by understanding the molecular structure of the gene and studying the systems that control various genetic traits, we can identify the factors that control the appearance or absence of certain traits despite the presence of the gene or genes for these traits. This is known as gene expression. This facilitates the identification of varieties that are tolerant and resistant to environmental conditions in a short time, accelerating breeding programs to produce new varieties of major grain crops that are tolerant to environmental stresses such as drought, soil salinity, and high temperatures, and that are high-yielding and high-quality. Extensive studies of viruses and bacteria that infect plants have provided a significant opportunity to advance molecular studies of these organisms.

Molecular and Plant Pathology

Numerous viral genes have been isolated and transferred either to plant hosts to attempt to confer a degree of resistance to the virus, or injected into the bacterial genetic material to enable the bacteria to produce other proteins suitable for study. The first attempts in this direction were in 1956, demonstrating that viral RNA in TMV is responsible for infecting plants and also for producing complete virus particles. In the 1970s, extensive studies were conducted on the crown gall bacteria

Agrobacterium tumefaciens in dicotyledons. It was discovered that tumors are caused by a small segment of DNA called tumor DNA (TiDNA), located on the plasmid carrying the disease genes. This segment was injected into the tumor-inducing plasmid (Ti-plasmid), also known as the chromosomes of plant cells. It contains two genes responsible for overgrowth. Scientists were subsequently able to replace these two genes with others, whether from bacteria, viruses, or even animals. This enabled them to multiply foreign genes in plant cells. This discovery confirmed the possibility of introducing a foreign gene into a plant and then, through tissue culture, producing an entire plant containing these genes. This marked the beginning of genetic engineering and its applications in agriculture. This led to the development of plants with new genetic traits, including resistance to diseases and insects, and improved quality and quantity of production (Al wakel,2010). As for how DNA is introduced into host cells, this is done in various ways, including using viruses as a vector, or bombarding the plant with DNA. This involves using a special bombardment device filled with DNA and forcibly inserting it into the host cells. Some DNA molecules may successfully bind to the plant's chromosomes and replicate with them, thus producing a plant with new genetic traits.

It has been possible to produce plants resistant to the action of cell wall-degrading bacteria and fungi by introducing the genes of these bacteria and fungi responsible for producing cell wall-degrading enzymes into plant cells. These genes replicated in the host cells, resulting in plants resistant to the degradation effect of these organisms.

Molecular biology applications have had a significant impact on the design of modern methods for diagnosing plant diseases in very short periods of time, even when these organisms are present in very limited numbers or mixed with other, closely related pathogens. These methods include:

- 1Monoclonal antibodies
- 2Fatty acid profiles of pathogens
- 3 Analysis of fragments of nucleic acids produced by specific enzymes
- 4Determination of nucleotide sequences of the pathogen

In the 1980s, it was possible to label a portion of DNA in microorganisms with radioactive isotopes or colorimetric compounds, becoming a means of identifying pathogens through the design of special kits.

In general, the use of molecular biology in diagnosing plant diseases is still in its infancy, despite important previous discoveries, and its high cost currently stands as an obstacle to implementation at the field level.

2. Tissue Culture:

Plant tissue culture is a scientific technique for growing plant cells, tissues, or organs in a laboratory under sterile conditions using a specially formulated medium that provides the nutrients and hormones necessary for their growth. The technique aims to rapidly propagate plants, produce genetically identical copies, obtain disease- and virus-free plants, produce plants with desirable traits, create entire plants from genetically modified cells, and use them to preserve rare species. Plant tissue culture relies on the ability of plant tissue to form a complete new plant when provided with a suitable growth medium and environment. This ability of plant cells or tissues is called pluripotency. In the propagation of plants that have problems with seed germination (non-traditional plants), or that have short seed viability. by seed these orchids can be propagated, it takes approximately (2-3) years to produce mature orchids. Therefore, plant tissue culture has become an ideal method for protecting this species from extinction and achieving commercial viability.

Microscopic propagation can be performed in three different ways:

- Using an apical or axillary bud.
- Using an adventitious bud.
- By forming a somatic embryo using somatic plant cells.

The success of each technique depends on the genetic background, culture media, and incubation conditions.

Using tissue culture technology, which involves culturing any part of a plant (leaf, embryo, cell, root, stem, or protoplast, which is a cell without a cell wall) and growing it in an artificial medium containing all the necessary auxins, hormones, vitamins, and nutrients. Tissue culture is used as a rapid means of propagation (Altpeter et al., 2016).

Traditional vegetative propagation is not fast enough to meet the growing demand for plants, especially ornamental and vegetable plants, date palms, and bananas. As a result, the prices of these plants are constantly increasing in most countries around the world, prompting many to use tissue culture technology to provide these numbers of plants at an affordable price within a limited space, in addition to obtaining disease-free strains (Efferth, 2019).

Genetic Engineering:

Genetic engineering has been used for hundreds of years to improve plant varieties and achieve desirable traits through breeding and selection programs. These programs can take 10-12 years to implement, and sometimes extensive crossbreeding is performed between genetically distant and diverse individuals, such as different species and genera. Such crossbreeding is sometimes difficult or impossible due to the numerous genetic barriers within the cells of these strains. This requires significant effort and financial costs.

It also requires that the crops to be improved reproduce sexually so that hybridization can be done between them. Therefore, genetic engineering came with the hope of improving genetic traits in a short period of time and at lower costs by transferring the desired genes, trying to find new genetic combinations (genotypes) that carry in their cells the best genetic factors that control their production and resistance to the adverse environmental conditions surrounding them (Shetty et al., 2018).

It is now possible to transfer genes not only from one plant species or genus to another, but also to transfer genes from microorganisms to higher plants or animals. Also, through genetic engineering, we can resist viral diseases that are difficult to control in any other way.

Biotechnology has several goals in various fields, including the following: Providing consumers with nutrient-rich, long-lasting foods, or foods that contain lower levels of certain naturally occurring toxic substances found in some food plants. Developers are using biotechnology to try to reduce saturated fats in cooking oils, reduce allergens in foods, and increase disease-fighting nutrients in foods. Using genetically modified crops to produce new medicines, which could lead to improved plant-based pharmaceutical industries that could reduce production costs by using a sustainable resource. Developing genetically modified plants for a purpose known as phytoremediation, where plants detoxify contaminants in the soil or absorb and accumulate contaminants from the soil so that the plants can be harvested and disposed of safely, is intended to improve soil quality at a contaminated site.

Producing more resilient crops that thrive even in the harshest environments and require less fuel, labor, fertilizer, and water, helping to reduce stress on land and soil. Protecting crops from deadly viral diseases, as biotechnology has genetically engineered plants to resist viral diseases that are otherwise difficult to control. Increasing crop quality, leading to increased production and financial returns. Providing improved quality traits such as increased beta-carotene levels in rice to help reduce vitamin A deficiency and improving oil compositions in canola, soybeans, and corn.

Plant propagation and its role in the sustainable harvest of medicinal and aromatic crops

Plant propagation plays a pivotal role in the sustainable harvest of medicinal and aromatic plants by providing effective methods for increasing production, preserving desirable species traits, and ensuring resource sustainability through sexual and vegetative propagation. This supports the pharmaceutical and aromatic industries, reduces pressure on wild plants, and improves the quality of the produced crops.

Plants play a significant role in human life, making life possible since the beginning of time. They also provide shelter, food, clothing, and medicine. Therefore, it was not surprising that man turned to the plant kingdom to treat his ailments. Undoubtedly, he began experimenting with plants and found in them what was a cure for his ailments, so he kept them and prescribed them to his family and loved ones. As for what was harmful, he avoided it. In this way, man was the doctor, the pharmacist, and the patient all at the same time.

Since ancient times, humans used medicinal plants in their natural state, but with the increase in population and their gathering in cities, these plants were traded in concentrated form, and extracts, pills, and tablets containing active plant substances appeared. With the development of civilization and scientific progress, a new era of science was born at the beginning of the ninth century, which is "Modern Pharmacognosy." People all over the world knew the therapeutic benefits of plant extracts and herbs and preferred to use them since ancient times. People's attention did not shift from the real, proven benefits of this ancient natural treatment until the last century, when medical sciences began to search for drug treatment among industrial chemical preparations, supporting their discoveries with large-scale advertising campaigns. Due to the quick, temporary results of chemical treatments, this trend succeeded in deceiving people for some time - even after a period that may be short or long. In many cases, disease conditions appeared in another part of the body that were worse than the original disease condition, and thus the beginnings of the science of treatment using herbs appeared recently (Qandil and Ibrahim 2007).

Using agricultural residues to improve propagation processes and production systems.

Propagation processes and production systems can be improved by using agricultural residues by converting it into organic fertilizer to improve soil quality and nutrition, or converting it into animal feed to provide a source of protein and nutrients for animals. It can also be used to produce energy through biogas, or as raw materials in other industries such as paper and wood panels. These practices contribute to sustainable agriculture, reduce environmental pollution, and decrease dependence on natural resources.

Agricultural residues is a vital product of food production processes in various plants, remaining in fields after the main product or crop is harvested. Some tend to call it agricultural residues, but it has significant economic value and sometimes constitutes significant wealth that is often lost due to lack of exploitation. There is a long list of these materials, the availability and availability of which varies from country to country. Perhaps the most important include wheat straw, barley straw, rice field residues, rice husks, corn field residues, corn cobs, sunflower stalks, vegetable crop residues, and others. These residues, if annual plants, some bushes, and pruning products in orchards, forests, and trees are added together, will take on a more general title, which is what is called "lignocellulosic materials," which is an expression for materials that contain cellulose and lignin in their composition, other than tree wood.

Agricultural residues recycling systems and methods vary depending on the type of residues, the available methods and technologies, and the purpose of recycling. One of the most important uses is converting residues into organic fertilizers or animal feed, and using it to produce clean energy. This contributes to achieving clean agriculture, protecting the environment from pollution, improving agricultural products, and providing rural employment opportunities, thereby improving economic and environmental conditions and raising health and social standards.

In Iraq, there are no accurate statistics on the quantities of residues in agricultural fields because we still do not consider it a significant issue. Iraq, with its environmental diversity and vast agricultural land, produces millions of tons of these materials annually. Although some of them are used in limited areas, the vast majority is not only left unused, thus wasting resources, but is also burned on farms, thus becoming a factor in environmental pollution, leaving a negative impact on the soil (Kirkby and Fattore, 2006), the air, and living organisms.

It can be used for growing on agricultural residues. This method is used in lands suffering from soil problems such as severe salinity or soil diseases, whether in greenhouses or open fields. The residues is stacked on the soil, similar to rice husks, and vegetables are grown on them. Rice husks, cotton stalks, or sawdust can also be used to grow and produce mushrooms. Agricultural residues is used to control plant diseases, whether directly by mixing it with the soil or after converting it into organic fertilizers or compost. Some beneficial microorganisms are loaded onto these residues and added to the soil to combat pathogens present in the soil, as occurs in inhibiting the growth of the fungus *Phytophthosa Pythium* by compost, and using green fertilizers to inhibit the fungus *Rhizoctonia solani*, which causes root rot. Converting agricultural residues into organic fertilizers improves soil properties and increases growth rates. Because it makes the soil environment more alkaline, it becomes unsuitable for the growth of nematodes. In addition, the microorganisms that grow on it play a role in combating these harmful organisms (Youngquist et al., 1993).

Therefore, it is necessary to focus on raising rural awareness of how to deal with agricultural residues by developing their knowledge in this vital field. Among the most prominent areas of benefiting from agricultural residues are the production of unconventional feed, the production of organic fertilizers, and energy production. The benefit from agricultural residues can be maximized by using it as raw materials for many important industries, which increases its economic value and opens up job opportunities for young graduates. As a result of the rising prices of animal and poultry feed due to the increasing demand for it, interest has increased in trying to exploit agricultural residues in many different ways as an attempt to fill the deficit in animal feed and its high prices (Zaka et al., 2018). Agricultural residues is considered a source of income for the farmer in addition to the value of the main crop output (Sadh et al., 2018).

The Role of Plant Breeding in Enhancing Food Security

Many challenge Food security was faces for (nine – ten) billion people at 2050, this problem may be solve for that there are more options, including: new technologies use to increase production of crops and investment in research to reduce residues(Rteba and Hussain,2024), It increases agricultural production costs and reduces the farmer's economic profit margin, which negatively affects the standard of living of farmers and threatens the survival of small farms. Food security is the ability of people, at all times, to access sufficient, safe, and nutritious food to lead healthy and active lives. This concept is based on four key dimensions: physical availability of food, economic and physical access to it, proper utilization of food, and its stability over time. Food security is affected by factors such as conflict, poverty, climate change, and lack of infrastructure, and also includes combating food waste and ensuring proper nutrition for sustainable development(Murd,2010).

More than 50 million people in the Arab region or one in ten suffer from malnutrition. This number is expected to increase unless appropriate national policy measures are taken to achieve food security for all. Tracking food security is a complex task, but it is essential to help communities in the Arab region become more resilient to climate change, curb the depletion of natural resources, shift towards preserving terrestrial ecosystems, and adopt more sustainable consumption and production patterns. Food security is directly linked to plant reproduction, as the availability of sufficient and healthy food depends on the ability of plants to reproduce and produce abundant and healthy crops. Plant reproduction is influenced by multiple factors, such as soil health and nutrient availability, while pests and diseases can reduce crop production and create food insecurity. Therefore, improving plant health and reproduction through efficient agricultural practices and the use of appropriate genetic resources is essential to achieving global food security. Seed production or plant sustainability means increasing food supply, ensuring its availability for the population.

Plant breeding plays a pivotal role in enhancing food security by increasing crop productivity and ensuring a sustainable and sufficient food supply to meet the growing needs of the population. It also improves crop quality, reduces crop loss, and supports the local economy by providing job opportunities and increasing farmers' income. Strategies adopted in this area include adopting advanced agricultural technologies, innovation, focusing on plant genetic resources, sustainable pest management, and enhancing soil and seed health to ensure sustainable food systems.

Given that climate change poses a major threat to the agricultural and environmental sectors, including water shortages, rising temperatures, and increased soil salinity, which in turn threatens food security for society, biotechnology plays a significant role as one of the modern solutions to address these challenges. This is achieved by developing crops that are resistant to environmental conditions and improving production using advanced technologies such as genetic engineering and tissue culture. Sustainable development is one of the most important topics formulated from a human or peoples' rights perspective. The relationship between development and the environment emphasizes that the realization of the human or peoples' right to development cannot be fully realized unless it is coupled with the provision of a healthy environment. The human right to a healthy environment requires rational and targeted programs to achieve it. International efforts to reconcile environmental protection, sustainable development, and biodiversity conservation have contributed to this. Environmental conservation issues have become a top priority for the international community, especially after technological advances and the emergence of new biotechnology techniques to improve food and medicine, which have impacted human life, health, nutrition, and the environment.

The Industrial Revolution led to the emergence of many modern innovations, which led to scientific and technological progress. This resulted in significant results in many fields, enabling humans to utilize the Earth and its ecosystems, utilize its renewable and non-renewable resources, and develop new energy sources. Preserving biodiversity means preserving biological resources, investing in them sustainably, and reclaiming and improving these resources within a healthy environment. Therefore, cooperation among countries is essential to ensure the conservation of biodiversity, the sustainable use of biological resources, and the sharing of benefits resulting from the use of genetic resources.

The Role of Breeding in the Conservation of Genetic Resources

Crop cultivation has a long history in human civilization, with records indicating that the cultivation of cereal plants began around 8000 BC. This early plant breeding was primarily developed to provide a steady supply of food.

As human understanding of genetics advances, improved crop varieties can be achieved more rapidly. Artificial selection can be more targeted, and crop varieties can be optimized for favorable traits more quickly, producing better, more vigorous, or more palatable plants.

However, traditional plant breeding methods are slow and do not always produce the desired crop varieties. Later, biotechnology tools have made it easier to engineer desirable traits into plants that are difficult to breed using traditional methods. For example, improving nutritional deficiencies in plants is difficult through artificial selection, and vitamin A and iron deficiency pose a particular challenge. Rice, for example, lacks the genes for beta-carotene, a precursor to vitamin A. However, it does contain genes for geranyl pyrophosphate, which can be sequentially converted to beta-carotene using four enzymes. Rice was engineered using genes for two enzymes derived from daffodils and the remaining two from the bacterium *Erwinia uredovora*. The resulting crop is known as golden rice. Given that rice is the staple food for more than half the world's population, such bioengineered crops could potentially play a role in preventing blindness among children caused by vitamin A deficiency, or improving the health of rice-dependent countries.

Many commonly grown crop plants now have some degree of genetic modification introduced using biotechnology. For example, corn, papaya, and many potato varieties have been modified to resist herbicides, diseases, or pests. Genetic modification can be applied to reduce the production of allergens, as is the case with soybeans.Breeding plays a pivotal role in the conservation of genetic resources by increasing genetic diversity, enabling species to adapt to changing environmental conditions and cope with diseases and threats, and reducing the risk of inbreeding that is detrimental to survival. This can be achieved through in situ conservation (in its natural habitat) and ex situ conservation (in dedicated facilities).

Genetic resources are genetic material of actual or potential value, meaning they are any material of plant, animal, microbial, or other origin that contains functional units of heredity. Genetic resources therefore refer to that part of genetic diversity that has or could have a practical use, such as in plant breeding.

Genetic resources are one of the three levels of biological diversity identified in Article 2 of the Convention on Biological Diversity (CBD) in Rio de Janeiro, 1992.

The Importance of Reproduction in Biodiversity

Biodiversity, or biodiversity, encompasses all life on Earth, from genes to entire ecosystems. It is essential for climate stability, the provision of food, water, and medicine, and the prevention of climate crises. Humans depend heavily on biodiversity for their basic needs, with more than a billion people deriving their livelihoods from forests alone. However, biodiversity is under global threat due to human activities and climate change, leading to species loss and ecosystem degradation.

Reproduction is essential for biodiversity because it generates genetic diversity that allows organisms to adapt to changing environments, preventing species extinction and maintaining the continuity of organisms and ecosystem services such as the provision of food, water, and clean air, and supporting the evolution of species and their ability to survive in the face of pests, diseases, and climate change. Plants differ in their methods of reproduction. They either reproduce by seeds or by vegetative parts, as mentioned previously. Vegetative reproduction in plants results from the lack of seed formation, or because seeds only form under special conditions, or because of their short lifespan, as well as the wide range of genetic variations within the plant. All commercial varieties of sugarcane, bananas, and potatoes are generally sterile due to their failure to flower due to genetic factors and unsuitable climatic conditions. The nature of agricultural processes also significantly influences the failure of some plants, such as garlic and ornamental plants, to flower. All these obstacles call for the use of other methods for propagating these plants, including vegetative parts.

Sexual reproduction, by its nature, provides plant breeders with genetic variations, while vegetative propagation involves no genetic changes at all unless a genetic mutation occurs in the plant population. Grafting, in its various forms, usually occurs between varieties of the same plant species. However, grafting can sometimes be performed between varieties of different plant species. Hybrids obtained through this method of grafting are called chimeras. Chimeras are composed of tissues from two parent plants, which are naturally different in their genetic makeup. As a result of the grafting process, their different tissues overlap or accumulate. When these tissues multiply, the accumulated tissues either separate or overlap. Chimeric hybrids are propagated only vegetative. An example of chimeric hybrids is what is observed in grapevines, where clusters with black fruits and others with yellow fruits are formed as a result of vegetative hybridization between varieties with different genetic makeup. When wind or animals transfer pollen from one plant to another, the advantage of cross-pollination is that it helps maintain the genetic diversity of species. Genetic diversity becomes beneficial when conditions change and organisms need to adapt to this change. Some plants have characteristics that prevent self-pollination, such as pollen and ovules that develop at different times.

Pollinators are animals that transfer pollen between plants. Many pollinators are insects, such as bees, butterflies, moths, and beetles. Some birds, including hummingbirds, also play a role. Similarly, some mammals, such as bats and rodents, transfer pollen between plants. The colors and scents of flowers often attract pollinators. Pollen sticks to the pollinator's body as it feeds on the flower's leaves. The seeds produced by fertilization contain genetic material from both parents. As a result, the offspring are not genetically identical to either parent plant. This genetic diversity enables them to survive in the event of a change in the environment. Offspring inherit one set of chromosomes from each parent. Therefore, each offspring has a combination of chromosomes (and alleles) from both parents. Sexual reproduction also results in genetic variation among siblings . During reproduction, genetic recombination occurs through processes such as crossing over during meiosis, resulting in the formation of new gene combinations. Genetic diversity is increased through sexual reproduction, as offspring inherit a unique set of genes from both parents.

References

Ali, H. J.(1988). Fundamentals of Breeding and Genetics of Field Crops. Ministry of Higher Education and Scientific Research, University of Baghdad. Directorate of Dar Al-Kutub for Printing and Publishing, University of Mosul. PP:76.

Altpeter, F., Springer, N. M., Bartley, L. E., Blechl, A. E., Brutnell, T. P., Citovsky, V., Conrad, L., Gelvin, S. B., Jackson, D., Kausch, A. P., Lemaux, P. G., Medford, J. I., Orozo-Cardenas, M., Tricoli, D., VanEck, J., Voytas, D. F., Walbot, V., Wang, K., Zhang, Z. J., & Stewart, C. N.Jr. (2016). Advancing crop transformation in the era of genome editing. *The Plant Cell, 28*(7), 1510–1520. DOI: 10.1105/tpc.16.00196 PMID: 27335450

Efferth, T. (2019). Biotechnology applications of plant callus cultures. *Engineering (Beijing), 5*(1), 50–59. DOI: 10.1016/j.eng.2018.11.006

Khatik, N. (2023). Role of Plant Biotechnology in Agricultural Sustainability and Improvement of Crops. *Journal of Agricultural Research, 11*(6), 929–936.

Kirkby C A and Fattore, A. (2006). Effect of Rice Stubble Burning on Soil Health . A report for the Rural Industries Research and Development Corporation. RIRDC Publication

Qandil, A. M. A., & Ibrahim, A. K. (2007). *Production of medicinal, aromatic and ornamental plants*. Open Education Center, Faculty of Agriculture, Ain Shams University.

Rteba, M. M., & Hussain, W. S. (2024). Sustainable Agriculture: Way to Achieve Food Security. In Achieving Food Security Through Sustainable Agriculture (pp. 182-202). IGI Global.

Sadh, P. K., Duhan, S., & Duhan, J. S. (2018). Agroindustrial wastes and their utilization using solid state fermentation. *Bioresources and Bioprocessing*, *5*(1), 1–15. DOI: 10.1186/s40643-017-0187-z

Shetty, M. J., Chandan, K., Krishna, H. C., & Aparna, G. S. (2018). Genetically modified crops: An overview. *Journal of Pharmacognosy and Phytochemistry*, *7*(1), 2405–2410.

Youngquist, J., Krzysik, A., English, B., & Spelter, H. P, Chow (1993): Forest Agricultural fibers for use in building components. Proc7268 Madison, WI: product Society: 123 – 134.

Zaka, M., Morsi, Y. El-Din Mohamed, B. and Amin, I. F. (2018) The Economic Impacts of Recycling Agricultural Waste on Agricultural Production in Egypt, Journal of the Association of Arab Universities for Agricultural Sciences, AinShams University, 26(2B) (Special Issue): 1289-1302.

Chapter 3
An Overview of Plant Propagation and Its Significance

Oindrilla Ghosh
https://orcid.org/0000-0001-9609-0781
Christ University, India

Binod Kumar
https://orcid.org/0009-0002-9721-1367
Radha Govind University, India

ABSTRACT

Plant propagation is the process of reproducing plants, either sexually or asexually, to produce new individuals. This practice is crucial in horticulture, agriculture, and conservation. Sexual propagation typically involves seeds, while asexual methods include cuttings, grafting, and layering. Each method has its advantages; for example, seeds can produce genetic diversity, whereas asexual methods allow for the cloning of desirable traits. The significance of plant propagation extends beyond mere cultivation. It plays a vital role in conserving endangered species, enhancing food security through the cultivation of crops, and improving landscapes and gardens. Additionally, propagation enables the preservation of genetic resources, which is essential in adapting to climate change and pest challenges. Understanding the techniques and significance of plant propagation empowers gardeners, farmers, and conservationists alike, ensuring a sustainable future for plant diversity.

DOI: 10.4018/979-8-3373-5626-6.ch003

INTRODUCTION

Plant propagation is a fundamental aspect of horticulture and agriculture that refers to the various techniques used to create new plants from existing ones (Prip, 2024). It encompasses both sexual reproduction, which involves the production of seeds, and asexual methods, where new plants arise from vegetative parts of a parent plant (Acemoglu et al., 2016). This practice is not just about producing more plants; it is a vital process that supports biodiversity, enhances food security, and fosters the cultivation of ornamental species (Cullet, 2004).

Understanding plant propagation is crucial for multiple reasons (Calicioglu et al., 2019). Firstly, it ensures the continuity of plant species, especially in the face of environmental challenges, including climate change, disease, and habitat loss (Ruzzante et al., 2021). Secondly, propagation techniques allow for the efficient cultivation of food crops, enabling farmers and gardeners to produce a diverse array of plants with desirable traits, such as higher yields, pest resistance, and better adaptability to varying climates (Gupta, 2004). Lastly, the decorative and therapeutic qualities of plants mean that propagation methods play a significant role in landscape architecture, interior design, and even mental health practices (Smidt & Jokonya, 2022).

Moreover, plant propagation serves a ecological purpose by maintaining and restoring ecosystems (Prasad et al., 2012). Many plant species are crucial for supporting wildlife and maintaining soil health (Kolb, 2018). By facilitating the propagation of native and endangered species, conservationists can restore degraded habitats and enhance biodiversity (Flandrin et al., 1999). This not only benefits wildlife but also reinforces the resilience of ecosystems against climate variations (Boccaletti, 2022).

In contemporary practices, technology and research have greatly influenced plant propagation techniques (Anderson, 2023). Advances in biotechnology, such as tissue culture and genetic modification, have enabled rapid multiplication of plants with specific desirable traits, facilitating the swift response to agricultural demands (Monteiro et al., 2016). Social factors, including the growing trend towards sustainability and organic farming, have spurred interest in traditional propagation methods, such as seed saving and grafting, which are often perceived as more ecologically sound (Robinson & Carson, 2015).

Despite the advancements in modern techniques, traditional methods of plant propagation remain pivotal, particularly for home gardeners and small-scale farmers who may lack access to advanced technology (Sharma et al., 2020). Techniques such as layering, cuttings, and division have stood the test of time due to their simplicity and effectiveness (Rapinski et al., 2023).

HISTORIC EVOLUTION

The history of plant propagation is a fascinating journey that reflects humanity's deepening relationship with nature and the evolution of agricultural practices over millennia (Cullet, 1999). From primitive foraging to advanced scientific techniques, the methods of propagating plants have significantly transformed, driven by the necessity for food, aesthetic appreciation, and conservation (Anjanappa & Gruissem, 2021).

Initially, early humans practiced a rudimentary form of plant propagation through seed collection and storage (Prasanna et al., 2020). Archaeological evidence suggests that as early as 10,000 years ago, in regions like the Fertile Crescent, humans began to domesticate wild plants (Howitt & Miskelly, 2017). This early agricultural era marked the beginning of intentional propagation, where seeds were sown at the onset of seasonal rains to ensure continued food supply (Hallman et al., 2002). Simple methods such as scattering seeds or planting tubers enabled communities to stabilize their food sources (Pimentel et al., 1989).

As agriculture evolved, so too did propagation techniques (Byrne, 2014). Ancient civilizations recognized the benefits of cultivating specific plant varieties with desirable traits, leading to controlled breeding practices (Kruft, 2001). The Egyptians, Greeks, and Romans all practiced horticulture, employing rudimentary techniques for grafting and layering to propagate fruit trees and ornamental plants (Azadi et al., 2016). These early techniques showcased a growing understanding of plant biology and genetics, albeit without the scientific framework we have today (Sagoff, 2001).

The Middle Ages saw further developments in plant propagation as monastic gardens became centers for agricultural innovation (Stein, 2005). Monks played a crucial role in preserving various plant species, particularly herbs used for medicinal purposes, through techniques like cuttings and division (Smith, 2009). The Renaissance period marked another significant leap, as botanical knowledge expanded and books on gardening and plant propagation began to surface, introducing a formalized approach to horticulture (Granieri, 2015).

The Age of Exploration in the 15th and 16th centuries facilitated the exchange of plant species across continents, prompting a need for effective propagation methods to acclimatize non-native species to new environments (Pultrone, 2012). The incorporation of plants such as potatoes, tomatoes, and maize into European agriculture was largely dependent on successful propagation techniques that allowed for adaptation to different climates and soil conditions (Egwu & Egwu, 2018).

In the 19th and 20th centuries, the advent of scientific exploration and botanical research paved the way for more sophisticated methodologies (Borowiak, 2004). Innovations such as tissue culture emerged, revolutionizing the way plants were propagated (Sell, 2009). This technique allows for the mass production of plants under controlled sterile conditions, enabling the propagation of disease-free plants and significantly enhancing crop yields (Prasad et al., 2012).

Today, the evolution of plant propagation continues, driven by modern science, technology, and social needs (Correa, 2006). The intersection of genetic engineering, hybridization, and sustainable practices reflects an ongoing commitment to meeting global demands (Verkey, 2007). Moreover, as awareness of biodiversity and environmental sustainability grows, there is a renewed interest in traditional propagation techniques alongside modern advancements, reinforcing the importance of this ancient practice in contemporary horticulture (Touch et al., 2024).

LITERATURE REVIEW

Plant propagation is a critical process that involves the production of new plants from various sources, obeying two primary methods: sexual and asexual propagation (He, 2023). The literature on plant propagation dates back centuries, illustrating traditional methods that were employed by early agronomists and horticulturists (McAfee, 2003). Recent advancements have deepened our understanding of plant reproductive biology and the genetic mechanisms underlying propagation (Grech et al., 2020).

Several studies underscore the ecological significance of plant propagation, particularly in conservation biology (Grain, 2015). Research shows that successful propagation techniques can aid in the restoration of endangered species and habitats (Alliance, 2022). For example, studies published in journals like "Conservation Biology" highlight the importance of seed banks and propagation for biodiversity preservation (Mahoney, 2018). Moreover, the role of propagation in food security has gained attention, with literature focusing on sustainable agricultural practices to ensure crop diversity and resilience in the face of climate change (Kaiser et al., 2020).

A noteworthy aspect in recent literature is the integration of technology into plant propagation methods (Gulati, 2019). Innovations such as molecular markers, genetic engineering, and biotechnological techniques are being explored to enhance propagation efficiency and develop plants with desirable traits (Imran et al., 2021). Various studies in journals like "Plant Cell, Tissue and Organ Culture" document the advances in tissue culture, which enables mass propagation of disease-free and genetically superior plants (Rodda, 2020).

THEORETICAL FRAMEWORKS

In the context of plant propagation, theoretical frameworks serve as essential guidelines for understanding the various techniques, methods, and scientific principles involved in the propagation of plants (Padmanabhan et al., 2017). These frameworks guide research, innovation, and practical applications in horticulture, agriculture, and plant biology (Olusegun & Olubiyi, 2017). Broadly, the theoretical frameworks related to plant propagation can be divided into several categories such as genetic, physiological, ecological, and technological frameworks (Brush, 2007).

Genetic Framework

At the forefront of plant propagation is the genetic framework, which examines the role of genetics in the growth and reproduction of plants (Delmer et al., 2003). Understanding genetic variation allows researchers and practitioners to select the best parent plants for breeding programs, thereby enhancing traits such as disease resistance, yield, and environmental adaptability (McEowen, 2003). This framework also encompasses the use of molecular markers and techniques like genetic engineering to produce plants with desired characteristics, making it a pivotal area in modern plant propagation research (Herrera-Estrella & Alvarez-Morales, 2001).

Physiological Framework

The physiological framework focuses on the biological processes involved in plant propagation (Lesser, 1987). This includes understanding the plant's life cycles, growth patterns, and responses to environmental conditions (Shiva, 1996). Critical to this framework is the study of photoperiodism, dormancy, water transport, and nutrient uptake all of which influence how plants can be effectively propagated (Winter, 2010). Knowledge of physiological processes such as rooting and adventitious shoot formation is vital for methods like cutting propagation or tissue culture (Lor, 2004). This framework emphasizes the need for scientific principles grounded in plant biology to optimize propagation techniques (Mgbeoji, 2001).

Ecological Framework

The ecological framework addresses plant propagation within the context of ecosystems and biodiversity (Reid, 2009). It examines how plants interact with their environment, including relationships with pollinators, soil characteristics, and climate factors (Burger, 2014). Understanding these ecological aspects is crucial for sustainable propagation practices, such as selecting native species for restoration projects, which helps maintain genetic diversity and ecosystem resilience (Dal Monico, 2019). This framework argues for a holistic approach that considers not just the plants themselves, but also their roles in larger systems, thereby supporting conservation efforts and promoting the health of natural habitats (Nomani & Rahman, 2016).

Technological Framework

Lastly, the technological framework encompasses the advancements and methods used in plant propagation (Edler & Fagerberg, 2017). This includes traditional techniques such as seed sowing and cuttings, advanced methods like tissue culture, and biotechnological applications such as genetic modification (Andersen, 1996). Technology has revolutionized plant propagation by increasing efficiency, reducing costs, and enabling the mass production of plants (Halewood & Lapeña, 2016). Innovations like automated propagation systems and controlled environments (e.g., greenhouses) enhance plant growth and survival rates, showcasing the intersection of technology and biology (Alkhoraif, 2024).

ADVANTAGES OF PLANT PROPAGATION

Increased Yield and Food Security

One of the primary advantages of plant propagation is its ability to significantly enhance agricultural productivity (Campi & Nuvolari, 2021). Through techniques such as seed selection, cloning, and tissue culture, farmers and horticulturists can produce high-yielding varieties that are more resilient to diseases and adverse environmental conditions (Lawson & Bikundo, 2019). This increase in yield is crucial for ensuring food security in a world facing rising populations and climate challenges (De Schutter, 2009). By propagating plants that are tailored to specific growing conditions, we can not only optimize output but also reduce the reliance on chemical inputs, thereby promoting sustainable agricultural practices (Lawson, 2015).

Conservation of Genetic Resources

Another significant advantage of plant propagation lies in its role in conserving genetic diversity (Chandrasekhar & Mehrotra, 2016). The ability to propagate endangered or rare plant species can aid in their preservation and restoration efforts (Baron et al., 2019). By utilizing techniques such as seed banking, tissue culture, and cultivation in controlled environments, conservationists can prevent extinction and restore threatened ecosystems (Ezell & Cory, 2019). This aspect of plant propagation is vital for maintaining biodiversity, which is crucial for the resilience of ecosystems and the services they provide, including pollination, nutrient cycling, and carbon sequestration (Ramankutty et al., 2018).

Customization and Adaptation

Plant propagation allows for customization and adaptation to specific environments or purposes (Lavall et al., 2022). For instance, landscapers and urban planners can propagate native plants that are better suited for local conditions, thereby promoting environmental stability and reducing the need for irrigation and fertilization (Mermer, 2018). Additionally, commercial growers can create proprietary hybrids that meet market demands for specific plant traits, such as color, size, or fragrance (Gollin, 2008). This flexibility not only meets consumer needs but also supports the economic viability of nurseries and agricultural enterprises (Prip, 2024).

CHALLENGES

Plant propagation, the process of creating new plants from existing plants, is crucial for horticulture, agriculture, and conservation (Adhikari et al., 2021). While its significance is indisputable, several challenges can hinder effective propagation techniques, affecting biodiversity, food security, and economic sustainability (Ali-Aali & Teece, 2013).

One of the primary challenges in plant propagation is genetic variation (Okediji, 2008). In traditional propagation methods, especially asexual techniques like cuttings or grafting, there is a tendency to reduce genetic diversity (Chon, 2011). This can make crops more susceptible to diseases and pests, leading to larger-scale failures in agricultural production (Netanel, 2008). A lack of genetic diversity can also weaken the resilience of plant populations in the face of environmental changes (Subeesh & Mehta, 2021). For example, monocultures, where a single crop variety is planted extensively, can lead to devastating consequences when a disease or pest targets that particular variety (Ingram & Maye, 2020).

Another significant challenge is the availability of quality seeds and plant materials (Tantalaki et al., 2019). In many regions, especially in developing countries, farmers struggle to access hybrid or improved varieties that can withstand climatic changes and yield better harvests (Shandilya & Khanduja, 2020). The quality of propagation materials can vary greatly, and subpar seeds can lead to lower germination rates, poor plant health, and reduced productivity (Sawhney, 2003). In addition, the reliance on commercial seed companies can create economic challenges for farmers, who may have to pay premium prices for seeds instead of using traditional methods to grow their own (Thrupp & Giling, 1996).

Environmental factors also pose a significant challenge in plant propagation (Altieri et al., 2012). Temperature, humidity, soil type, and water availability all influence the success of propagating plants (Prip, 2024). Changes in climate can lead to unpredictability in these essential factors, creating challenges for both conventional and advanced propagation methods (Cullet, 2004). For instance, seedlings that generally thrive in certain climates might fail if the weather patterns shift unexpectedly, leading to crop failures (Acemoglu et al., 2016). In some cases, growers may need to invest in sophisticated technologies, such as climate-controlled greenhouses, to mitigate these challenges, adding an economic burden (Calicioglu et al., 2019).

Pests and diseases can also disrupt propagation efforts (Gupta, 2004). Pathogens can be transmitted through various means seeds, soil, or contaminated tools harming not only individual plants but entire crops (Prasad et al., 2012). For instance, plant viruses can severely affect seedling development and fruit production (Flandrin et al., 1999). Furthermore, the spread of invasive species can lead to competition for resources, further complicating the propagation of native species (Ruzzante et al., 2021).

In addition to biological and environmental factors, sociopolitical influences also play a role in the challenges of plant propagation (Monteiro et al., 2016). Policies surrounding seed ownership and intellectual property rights can impede farmers from using traditional propagation methods or saving seeds from their harvests, which complicates agricultural practices (Robinson & Carson, 2015). Furthermore, societal attitudes toward specific crops can affect market demand, influencing university and government agricultural research funding (Clapp, 2022).

ROLE OF AI IN PLANT PROPAGATION

Artificial Intelligence (AI) is revolutionizing countless industries, and plant propagation is no exception (Smidt & Jokonya, 2022). As agricultural practices increasingly integrate technology, AI provides innovative solutions that can enhance efficiency, improve outcomes, and mitigate existing challenges in plant propagation (Kolb, 2018).

One of the most significant roles of AI in plant propagation is data analysis (Boccaletti, 2022). Advanced algorithms can process vast amounts of data from various sources, such as soil sensors, climate data, and pest management systems (Anderson, 2023). For example, AI tools can analyze soil nutrient levels and help determine the most compatible plant varieties for specific environments, thus optimizing the growth conditions needed for successful propagation (Sharma et al., 2020). By analyzing environmental factors and predicting optimal propagation conditions, farmers can enhance seed germination rates and growth success (Rapinski et al., 2023).

AI can also significantly improve crop monitoring and management (Cullet, 1999). Technologies such as satellite imagery and drones equipped with AI capabilities allow for real-time monitoring of plant health and environmental conditions (Howitt & Miskelly, 2017). These tools can identify signs of stress or disease in plants early on, enabling farmers to take proactive measures (Hallman et al., 2002). By utilizing AI in this way, farmers can implement targeted treatments rather than broad-spectrum applications, thus promoting sustainable practices and reducing chemical usage (Prasanna et al., 2020).

Moreover, AI can play a vital role in breeding programs, facilitating the selection of plants with desirable traits (Pimentel et al., 1989). Traditional breeding methods can be time-consuming and labor-intensive (Byrne, 2014). AI-driven tools can analyze genetic data to predict how specific traits will be inherited, expediting the process of developing new cultivars (Azadi et al., 2016). Machine learning models can also assist in identifying the best combinations for hybrid plants, allowing for the rapid development of new varieties that are more resilient to diseases and pests, a key factor in addressing food security challenges (Anjanappa & Gruissem, 2021).

AI can also enhance propagation through automation (Kruft, 2001). Robotic systems, powered by AI, can undertake various tasks such as planting seeds, monitoring growth conditions, and even harvesting crops (Sagoff, 2001). By automating labor-intensive tasks, farmers can save time and reduce costs, allowing for a greater focus on strategic planning and resource management (Egwu & Egwu, 2018). Automated systems can be programmed to adjust their actions based on real-time feedback from the environment, further improving efficiency and effectiveness in the propagation process (Nabuuma et al., 2022).

Augmented Reality (AR)

Augmented Reality (AR) is revolutionizing the way individuals and professionals engage with plant propagation (Stein, 2005). By superimposing digital information onto the real world, AR applications provide interactive experiences that enhance the understanding of plant growth and propagation techniques (Smith, 2009). For instance, nursery staff and horticulturalists can use AR tools to visualize different plant species' growth requirements, such as sunlight, water, and nutrient needs (Granieri, 2015). As users point their devices at specific plants, AR can overlay information about ideal temperature, pH levels, and necessary propagation techniques like cuttings, layering, or grafting (Pultrone, 2012).

In educational settings, AR applications make learning about plant propagation engaging (Borowiak, 2004). Students can use AR headsets or smartphones to interact with 3D models of plant anatomy, observing root systems and reproductive structures in detail (Sell, 2009). This immersive experience not only aids retention of knowledge but also allows learners to simulate various propagation methods without the expense and time of real-world trials (Prasad et al., 2012). Additionally, AR can assist in troubleshooting propagation issues, providing instant feedback on plant health (Correa, 2006). It can guide users through diagnosing problems like pests or diseases by analyzing visual cues in real time (Verkey, 2007). This technology democratizes horticultural education, making it accessible to both novice gardeners and seasoned professionals (Touch et al., 2024).

With the rise of AR, businesses also find new marketing avenues, enabling customers to visualize how healthy plants will appear in their homes or gardens before making a purchase (He, 2023). By fostering a deeper understanding and appreciation for plant propagation, AR enhances both individual skills and collective knowledge, emphasizing the significance of plants in our ecosystems and daily lives (Barrett, 2021).

Virtual Reality (VR)

Virtual Reality (VR) technology offers an innovative platform for deepening understanding and skills in plant propagation (Grech et al., 2020). Through immersive experiences, users can engage with complex agricultural concepts in a simulated environment, enabling them to practice techniques without the constraints of time, space, or resources (McAfee, 2003). For instance, VR simulations allow individuals to virtually enter greenhouses or nurseries, where they can experiment with various propagation methods such as seed germination, tissue culture, and hydroponics (Alliance, 2022). This hands-on experience facilitates better comprehension of the intricate processes involved in plant care and growth (Grain, 2015).

Moreover, VR can serve as a powerful educational tool for agricultural students and professionals (Kaiser et al., 2020). Training programs can incorporate realistic VR scenarios that simulate the challenges of plant propagation, such as environmental stressors or pest infestations (Imran et al., 2021). This experiential learning cultivates decision-making skills and problem-solving abilities, essential for effective horticulture and sustainable agriculture (Mahoney, 2018). As learners navigate through these realistic experiences, they can develop an intuitive understanding of the propagation cycle, from seed selection to transplanting techniques (Gulati, 2019).

Furthermore, VR's potential extends beyond individual learning to collaborative research and training sessions (Rodda, 2020). Multiple users can enter a shared VR environment for joint experiments or workshops, fostering teamwork among students, researchers, and experienced growers (Alkhoraif, 2024). By breaking geographical barriers, VR can facilitate partnerships between institutions, allowing diverse knowledge sharing and innovation in plant propagation strategies (Pandey et al., 2024).

Mixed Reality (MR)

Mixed Reality (MR) combines the best of both AR and VR, enabling users to engage with real and digital elements simultaneously in the realm of plant propagation (Alkhoraif, 2024). This innovative technology enhances the way individuals learn, teach, and apply propagation techniques (Padmanabhan et al., 2017). By allowing users to interact with holograms of plants superimposed on their physical environment, MR creates a richer, more engaging experience that deepens understanding of plant growth and care (Olusegun & Olubiyi, 2017).

One of the most significant applications of MR is in educational settings (Lor, 2004). Students and aspiring horticulturists can use MR devices to visualize complex biological processes, such as photosynthesis or root development, in real-time (Mgbeoji, 2001). By manipulating 3D holograms of plants, learners can experiment with different propagation methods, such as cuttings or budding, observing the outcomes of their actions interactively (Reid, 2009). This type of immersive experience encourages active learning and discovery, which can help solidify theoretical knowledge about plant propagation (Winter, 2010).

In professional contexts, MR can also aid in planning and execution of propagation strategies within nurseries or botanical gardens (Burger, 2014). Employees can use mixed reality headsets to access immediate, contextual information about plant species, care techniques, and environmental conditions, all while tending to plants in real-time (Brush, 2007). For instance, MR systems can guide users through identifying problems such as nutrient deficiencies or pest infestations on actual plants, allowing for efficient problem-solving and management (Nomani & Rehman, 2016).

Moreover, MR technology facilitates remote collaboration among horticulturists, researchers, and educators (Dal Monico, 2019). Experts can share their insights and methodologies while projecting their visualizations to practitioners in different locations (Delmer et al., 2003). By fostering communication and knowledge sharing, MR enhances the global exchange of ideas, innovations, and best practices in plant propagation (El- Bilali et al., 2021).

CASE STUDIES

Plant propagation involves the reproduction of plants to produce new plants from a variety of sources including seeds, cuttings, grafts, and divisions (McEowen, 2003). It is fundamental for agriculture, horticulture, and conservation efforts

(Herrera-Estrella & Alvarez-Morales, 2001). Various case studies exemplify the diverse techniques and the significant role of plant propagation in different contexts (Lesser, 1987).

1. Case Study: The Restoration of Endangered Species

In the context of plant conservation, a notable case study is the propagation of the endangered California native plant, *'Abronia umbellata'*. This plant has been threatened by habitat loss due to urban development and climate change (Shiva, 1996). Conservationists have employed both seed propagation and vegetative cuttings to bolster its population (Edler & Fagerberg, 2017).

The seeds were collected during the flowering season and stored under controlled conditions to ensure viability (Halewood & Lapeña, 2016). To enhance germination rates, a stratification process was implemented which mimicked natural seasonal changes (Andersen, 1996). Once sprouted, seedlings were cultivated in controlled environments before being transplanted back into their native habitats (Campi & Nuvolari, 2021). The initiative not only preserved a genetic pool but also increased awareness about local biodiversity and the significance of preserving native flora (Singh & Kolar, 2024).

2. Case Study: Commercial Propagation Techniques

In commercial horticulture, the propagation of *Petunia* plants provides valuable insight into industrial practices (Lawson & Bikundo, 2019). Growers utilize both seed propagation and tissue culture techniques (De Schutter, 2009). Seed propagation is typically employed for large-scale production due to its cost-effectiveness (Lawson, 2015). However, tissue culture has become a preferred method as it allows for the rapid multiplication of genetically identical plants, ensuring uniformity in flower color and size (Chandrasekhar & Mehrotra, 2016).

A leading nursery in the United States adopted tissue culture to meet rising consumer demand for diverse Petunia varieties (Baron et al., 2019). The process began with isolating meristem tissue from desirable mother plants, which underwent in vitro culturing in sterile media (Ezell & Cory, 2019). This method drastically reduced the time from propagation to sale and minimized disease transmission, significantly boosting productivity and profitability (Abiri et al., 2023).

3. Case Study: Urban Agriculture and Community Gardens

Urban agriculture initiatives have also highlighted the importance of plant propagation in community settings (Ramankutty et al., 2018). A project in New York

City focused on creating community gardens as a response to food deserts (Mermer, 2018). Through workshops, residents learned about plant propagation techniques such as seed saving and vegetative propagation to grow food plants in small spaces (Ali-Aali & Teece, 2013).

Participants were encouraged to propagate heirloom seeds and share the resulting plants with neighbors (Gollin, 2008). This not only charged local economies by reducing food prices but also fostered community engagement and sustainability (Okediji, 2008). The practice of saving and propagating seeds preserved genetic diversity while promoting food sovereignty (Gebresenbet et al., 2023).

Implementation of Regulatory Frameworks

The propagation of plants is subject to regulatory frameworks that aim to promote sustainability, protect biodiversity, and ensure safety in the use of genetically modified organisms (GMOs) (Adhikari et al., 2021). These regulations are critical for maintaining environmental integrity while allowing for agricultural advancements (Chon, 2011).

Legislative Measures

Governments worldwide have established laws governing plant propagation, especially concerning the use of GMOs (Ingram & Maye, 2020). For instance, in the European Union, strict regulations control the propagation of genetically modified crops (Netanel, 2008). These regulations require comprehensive risk assessments and labeling, ensuring that consumers are informed about the products they buy (Thrupp & Giling, 1996). Compliance with these regulations thus serves to address environmental concerns while enabling further research into biotechnology (Tantalaki et al., 2019).

Biodiversity and Conservation Laws

Many nations have enacted laws aimed at promoting biodiversity through the regulation of plant propagation practices (Sawhney, 2003). The Convention on Biological Diversity (CBD) is an international treaty that encourages countries to develop strategies concerning the conservation and sustainable use of biological diversity (Altieri et al., 2012). Under this framework, countries implement national guidelines on plant propagation techniques that minimize the risk of introducing invasive species (Subeesh & Mehta, 2021).

For instance, in Australia, the Environment Protection and Biodiversity Conservation Act (EPBCA) necessitates that any plant propagation activities involving native species comply with conservation objectives (Clapp, 2022). This includes adherence to ethical sourcing of seeds through controlled harvests and propagating endangered species under supervision to prevent further habitat degradation (Shandilya & Khanduja, 2020).

Community Guidelines and Education

Practical implementation of regulations often involves community engagement and education programs (Barrett, 2021). Local governments may provide resources and guidelines on best practices in plant propagation, especially in regions where community gardening is popular (El- Bilali et al., 2021). Workshops often emphasize sustainable practices such as organic propagation techniques and the significance of native plant species, fostering a sense of stewardship among local populations (Lavall et al., 2022).

ETHICAL CONSIDERATIONS

Plant propagation is the process of creating new plants from existing ones, encompassing various methodologies including seed sowing, cuttings, grafting, and tissue culture (Nabuuma et al., 2022). While it can significantly impact environmental sustainability, biodiversity, and agricultural practices, several ethical considerations emerge in this field (Singh & Kolar, 2024).

One major ethical consideration is the potential for biodiversity loss (Abiri et al., 2023). Intensive agricultural practices, particularly monoculture systems, can lead to a genetic bottleneck (Acemoglu et al., 2016). Propagating only a handful of popular cultivars puts numerous native plant species at risk of extinction (Calicioglu et al., 2019). As dominant species replace rare varieties, ecosystems can suffer from a decline in resilience and ecological integrity (Ruzzante et al., 2021). Additionally, unethical propagation practices can lead to the overharvesting of certain species from their natural habitats, contributing to the degradation of those ecosystems and threatening fragile plant populations (Gebresenbet et al., 2023).

Moreover, the commercialization of certain plant species raises ethical issues (Smidt & Jokonya, 2022). For instance, during the development of genetically modified organisms (GMOs) or hybrids, corporations control important genetic material, often prioritizing profit over ecological health (Kolb, 2018). This practice not only poses risks to food security but also limits farmers' rights to save seeds and propagate their crops (Prasad et al., 2012). Patents on plant varieties can stifle innovation and enforce dependency on specific companies for agricultural inputs, raising questions about fairness, equity, and sustainability (Boccaletti, 2022).

Environmental justice also emerges in discussions surrounding plant propagation (Monteiro et al., 2016). Communities in developing regions often rely on diverse local species for food, medicine, and culture (Robinson & Carson, 2015). When multinational corporations engage in commercial propagation, they may exploit local knowledge and resources without adequate compensation or recognition (Sharma et al., 2020). This acts not only as an ethical violation but perpetuates economic inequalities (Rapinski et al., 2023). By involving local communities and respecting traditional practices in the propagation of native plants, stakeholders can promote fairness and foster a sense of stewardship toward the environment (Anderson, 2023).

The introduction of non-native plants through propagation presents further ethical issues (Anjanappa & Gruissem, 2021). While these plants can sometimes provide immediate agricultural benefits, they may disrupt local ecosystems by outcompeting native species and altering habitats (Prasanna et al., 2020). Weighing the pros and cons of introducing a new plant species necessitates a thorough understanding of ecological interactions and consequences (Grech et al., 2020). Conservationists and ethicists emphasize the importance of considering long-term impacts on biodiversity, advocating for responsible propagation strategies that prioritize native species and ecosystem restoration (Prip, 2024).

CONCLUSION

Plant propagation plays a crucial role in sustaining biodiversity, promoting food security, and supporting economic development (Touch et al., 2024). Its significance is underscored by the increasing challenges posed by climate change, habitat loss, and the growing global population's demand for food (Alliance, 2022). At its core, responsible plant propagation offers solutions to these challenges while preserving ecological integrity and enhancing human livelihoods (Kaiser et al., 2020).

Moreover, education and awareness are paramount. Stakeholders, including farmers, policymakers, and communities, must be informed of the ecological and ethical implications of their propagation choices (Rodda, 2020). Collaborations that unite diverse perspectives can foster innovation and lead to practices that respect the complexity of ecosystems while addressing global needs.

Thus, the significance of plant propagation extends far beyond mere cultivation; it embodies a commitment to sustainability, ethics, and ecological integrity (Imran et al., 2021). By navigating the intricate balance between human needs and environmental responsibilities, we can harness the power of plant propagation to foster resilient ecosystems and prosperous communities, ensuring a healthier planet for future generations (Gulati, 2019). The road ahead demands sincere collaboration and informed decision-making, engaging all stakeholders in the shared journey toward a sustainable future (Pandey et al., 2024).

Suggestions and Recommendations

Plant propagation is a critical aspect of horticulture, agriculture, and conservation (He, 2023). Understanding the various techniques and their applications can significantly enhance plant productivity and biodiversity.

Diversify Techniques: It's essential to employ a mix of propagation methods, including sexual (seeds) and asexual (cuttings, grafting, layering) propagation. Each method has its advantages and challenges (Alkhoraif, 2024). For example, seeds are vital for genetic diversity, while asexual methods can preserve desirable traits. Training practitioners in multiple techniques can increase resilience against pests and diseases.

Utilize Native Plants: Focusing on local flora in propagation efforts can enhance ecological balance. Native plants often require less water and fertilizer and are more resistant to local pests and diseases (Campi & Nuvolari, 2021). Initiatives to encourage the propagation of indigenous species can foster restoration projects and improve local ecosystems.

Adopt Technological Advancements: Incorporating technology into propagation techniques can optimize efficiency and success rates. For instance, using automated misting systems for cuttings or employing tissue culture techniques can increase propagation scale and control over environmental factors (Lavall et al., 2022). Researching and adopting biotechnological methods can also open doors for creating improved plant varieties that are more resilient to climate change.

Educate and Train: Implementing community outreach and education programs about the benefits and techniques of plant propagation is crucial. Workshops, online courses, and hands-on training can empower individuals and communities, promoting sustainable practices (Prip, 2024). Engaging local schools and organizations can foster interest in horticulture and conservation, creating a broader base of knowledgeable practitioners.

Implement Sustainable Practices: Emphasizing sustainability in propagation is essential to ensure the long-term viability of plant populations (Adhikari et al., 2021). Practices such as organic propagation methods, resource-efficient watering systems, and the use of local, sustainable materials can minimize environmental impacts and promote biodiversity.

Research and Development: Investing in R&D concerning plant propagation techniques can yield significant advancements. There is a need for studies focusing on the genetics of specific species to identify the best propagation strategies, especially in crops that are underutilized or endangered (Ingram & Maye, 2020).

Research should also explore the interactions between plants and their ecosystems to improve propagation success rates.

Document Traditional Knowledge: Many cultures possess traditional knowledge regarding plant propagation. These practices should be documented and studied, as they can provide insights into effective and sustainable methods (Subeesh & Mehta, 2021). Collaborating with indigenous communities can enrich the understanding of plant propagation and lead to the development of hybrid techniques that respect traditional practices while utilizing modern advances.

Future Research Directions

As the field of plant propagation evolves, it is imperative to identify future research avenues that can enhance techniques, sustainability, and biodiversity (Tantalaki et al., 2019). A few promising directions include:

Genomic Studies: Advances in genomics present new opportunities for understanding plant genetics at an unprecedented level (Shandilya & Khanduja, 2020). Research into the genomic basis of traits related to propagation can reveal insights into seed viability, rooting potential in cuttings, and disease resistance. This knowledge can facilitate the development of targeted breeding programs, optimizing propagation methods for various species.

Climate-Resilient Varieties: With climate change impacting agricultural practices globally, there is a pressing need for research focused on developing climate-resilient plant varieties (Clapp, 2022). Studies should explore how specific traits can be propagated and enhanced to withstand extreme weather conditions, pests, and diseases.

Micropropagation Techniques: Research into optimizing micropropagation continues to hold promise. Standardizing protocols for a broader range of species can enhance the efficiency of producing disease-free plants at scale. Investigating the role of plant growth regulators, nutrient media variations, and environmental conditions can refine these techniques further (Nabuuma et al., 2022).

Propagation of Endangered Species: Focused research on the propagation of endangered or threatened plants can contribute to conservation efforts. Exploring seed banks and developing ex-situ conservation techniques, such as cryopreservation, can aid in the preservation of genetic material from at-risk species (Barrett, 2021).

Improved Asexual Methods: New studies could aim to refine traditional asexual propagation methods to increase their effectiveness. Researching optimal conditions for rooting, such as light, humidity, and temperature, can improve consistency and success rates for various plant species, particularly in commercial horticulture (Pandey et al., 2024).

Rooting Hormones and Biostimulants: Investigating various natural rooting hormones and biostimulants underscores an emerging area of research. Understanding the interactions and synergistic effects of these substances can lead to improved cloning success, significant for large-scale plant propagation industries (El- Bilali et al., 2021).

Public Policy and Regulation: Future research should also examine the regulatory framework surrounding plant propagation, including intellectual property rights concerning genetically modified organisms (GMOs) and traditional knowledge associated with indigenous plant propagation. Understanding these policies can guide sustainable practices while encouraging innovation in propagation techniques (Singh & Kolar, 2024).

REFERENCES

Abiri, R., Rizan, N., Balasundram, S. K., Shahbazi, A. B., & Abdul-Hamid, H. (2023). Application of digital technologies for ensuring agricultural productivity. *Heliyon*, *9*(12), e22601. DOI: 10.1016/j.heliyon.2023.e22601 PMID: 38125472

Acemoglu, D., Autor, D., Dorn, D., Hanson, G. H., & Price, B. (2016). Import competition and the great US employment sag of the 2000s. *Journal of Labor Economics*, *34*(S1), S141–S198. DOI: 10.1086/682384

Adhikari, K., Bikundo, E., Chacko, X., Chapman, S., Humphries, F., Johnson, H., Keast, E., Lawson, C., Malbon, J., Robinson, D., Rourke, M., Sanderson, J., & Tranter, K. (2021). What Should Farmers' Rights Look Like? The Possible Substance of a Right. *Agronomy (Basel,* 11(2), 367. DOI:DOI: 10.3390/agronomy11020367

Al-Aali, A. Y., & Teece, D. J. (2013). Towards the (strategic) management of intellectual property: Retrospective and prospective. *California Management Review*, *55*(4), 15–30. DOI: 10.1525/cmr.2013.55.4.15

Alkhoraif, A. (2024). The impact of innovation governance and policies on government funding for emerging science and technology sectors in Saudi Arabia. *Journal of Infrastructure. Policy and Development*, *8*(14), 8515. DOI: 10.24294/jipd8515

Alliance, O. S. (2022). *Putting the power of seed into the hands of growers*. Policy.

Altieri, M. A., Funes-Monzote, F. R., & Petersen, P. (2012). Agroecologically efficient agricultural systems for smallholder farmers: Contributions to food sovereignty. *Agronomy for Sustainable Development*, *32*(1), 1–13. DOI: 10.1007/s13593-011-0065-6

Andersen, R. (1996, June). The history of farmers' rights: A guide to central documents and literature. In *Conference on Plant Genetic Resources in Leipzig* (Vol. 17, p. 50).

Anderson, K. (2023). Agriculture's globalization: Endowments, technologies, tastes and policies. *Journal of Economic Surveys*, *37*(4), 1314–1352. DOI: 10.1111/joes.12529

Anjanappa, R. B., & Gruissem, W. (2021). Current progress and challenges in crop genetic transformation. *Journal of Plant Physiology*, *261*, 153411. DOI: 10.1016/j.jplph.2021.153411 PMID: 33872932

Azadi, H., Samiee, A., Mahmoudi, H., Jouzi, Z., Rafiaani Khachak, P., De Maeyer, P., & Witlox, F. (2016). Genetically modified crops and small-scale farmers: Main opportunities and challenges. *Critical Reviews in Biotechnology, 36*(3), 434–446. PMID: 25566797

Baron, J., Contreras, J. L., Husovec, M., Larouche, P., & Thumm, N. (2019). Making the rules: The governance of standard development organizations and their policies on intellectual property rights. *JRC Science for Policy Report. EUR*, 29655.

Barrett, C. B. (2021). Overcoming global food security challenges through science and solidarity. *American Journal of Agricultural Economics, 103*(2), 422–447. DOI: 10.1111/ajae.12160

Boccaletti, G. (2022). *Water: A biography.* Vintage.

Borowiak, C. (2004). Farmers' rights: Intellectual property regimes and the struggle over seeds. *Politics & Society, 32*(4), 511–543. DOI: 10.1177/0032329204269979

Brush, S. B. (2007). Farmers' rights and protection of traditional agricultural knowledge. *World Development, 35*(9), 1499–1514. DOI: 10.1016/j.worlddev.2006.05.018

Burger, J. (2014). Indigenous peoples, extractive industries and human rights.

Byrne, P. (2014). Genetically Modified (GM) Crops: Techniques and Applications. *Colorado State.*

Calicioglu, O., Flammini, A., Bracco, S., Bellù, L., & Sims, R. (2019). The future challenges of food and agriculture: An integrated analysis of trends and solutions. *Sustainability (Basel), 11*(1), 222. DOI: 10.3390/su11010222

Campi, M., & Nuvolari, A. (2021). Intellectual property rights and agricultural development: Evidence from a worldwide index of IPRs in agriculture (1961-2018). *The Journal of Development Studies, 57*(4), 650–668. DOI: 10.1080/00220388.2020.1817395

Chandrasekhar, S., & Mehrotra, N. (2016). Doubling Farmers' Incomes by 2022: What Would It Take? *Economic and Political Weekly*, 10–13.

Chon, M. (2011). Global intellectual property governance (under construction). *Theoretical Inquiries in Law, 12*(1), 349–380. DOI: 10.2202/1565-3404.1271

Clapp, J. (2022). The rise of big food and agriculture: corporate influence in the food system. In *A research agenda for food systems* (pp. 45–66). Edward Elgar Publishing.

Correa, C. M. (2006). Considerations on the standard material transfer agreement under the fao treaty on plant genetic resources for food and agriculture. *The Journal of World Intellectual Property, 9*(2), 137–165. DOI: 10.1111/j.1422-2213.2006.00272.x

Cullet, P. (1999). Revision of the TRIPS Agreement concerning the protection of plant varieties: Lessons from India concerning the development of a sui generis system. *The Journal of World Intellectual Property*, *2*(4), 617–656. DOI: 10.1111/j.1747-1796.1999.tb00081.x

Cullet, P. (2004). *Food security and intellectual property rights in developing countries.* RIBios (Réseau interdisciplinaire biosécurité): Institut universitaire d'études du développement (IUED).

Dal Monico, S. (2019). Biopiracy, or the misappropriation of traditional knowledge for profit: a human rights perspective.

De Schutter, O. (2009). *Report of the Special Rapporteur on the right to food.* UN.

Delmer, D. P., Nottenburg, C., Graff, G. D., & Bennett, A. B. (2003). Intellectual property resources for international development in agriculture. *Plant Physiology*, *133*(4), 1666–1670. DOI: 10.1104/pp.103.028993 PMID: 14681529

Edler, J., & Fagerberg, J. (2017). Innovation policy: What, why, and how. *Oxford Review of Economic Policy*, *33*(1), 2–23. DOI: 10.1093/oxrep/grx001

Egwu, C. C. N., & Egwu, C. (2018). Genetically Modified Crops Controversy: The Actualization of Intellectual Property Rights Regime.

El Bilali, H., Strassner, C., & Ben Hassen, T. (2021). Sustainable agri-food systems: Environment, economy, society, and policy. *Sustainability (Basel)*, *13*(11), 6260. DOI: 10.3390/su13116260

Ezell, S., & Cory, N. (2019). *The way forward for intellectual property internationally.* Information Technology and Innovation Foundation.

Flandrin, J. L., Montanari, M., & Sonnenfeld, A. (Eds.). (1999). *Food: a culinary history from antiquity to the present.* Columbia University Press.

Gebresenbet, G., Bosona, T., Patterson, D., Persson, H., Fischer, B., Mandaluniz, N., Chirici, G., Zacepins, A., Komasilovs, V., Pitulac, T., & Nasirahmadi, A. (2023). A concept for application of integrated digital technologies to enhance future smart agricultural systems. *Smart Agricultural Technology*, *5*, 100255. DOI: 10.1016/j.atech.2023.100255

Gollin, M. A. (2008). *Driving innovation: intellectual property strategies for a dynamic world.* Cambridge University Press., DOI: 10.1017/CBO9780511805394

GRAIN. L. V. C. (2015). *Seed laws that criminalize farmers: resistance and fight back.* https://grain.org/en/article/5142-seed -laws -that -criminalise -farmers -resistance-and-fightback

Granieri, M. (2015). Genetically modified seeds, intellectual property protection and the role of law in transnational perspective. In *Law, Development and Innovation (pp. 89–113). Springer International Publishing.*

Grech, A., Howse, E., & Boylan, S. (2020). A scoping review of policies promoting and supporting sustainable food systems in the university setting. *Nutrition Journal, 19*(1), 1–13. DOI: 10.1186/s12937-020-00617-w PMID: 32912299

Gulati, R. (2019). BIOPIRACY, A BIOLOGICAL THEFT? [IJOLS]. *International Journal of Legal Studies, 5*(1), 317–350.

Gupta, A. K. (2004). *WIPO-UNEP study on the role of intellectual property rights in the sharing of benefits arising from the use of biological resources and associated traditional knowledge* (Vol. 769). WIPO.

Halewood, M., & Lapeña, I. (2016). Farmers' varieties and farmers' rights: Challenges at the crossroads of agriculture, taxonomy and law. In *Farmers' Crop Varieties and Farmers' Rights* (pp. 15-38). Routledge.

Hallman, W. K., Adelaja, A. O., & Schilling, B. J. (2002, August). Public perceptions of agricultural biotechnology in the United States in 2001. In *XXVI International Horticultural Congress: Biotechnology in Horticultural Crop Improvement: Achievements, Opportunities and 625* (pp. 445-455).

He, Y. (2023). Unraveling the interplay between food security, agriculture, trade policy, and energy consumption: An environmental sustainability insight. *Energy & Environment,* 0958305X231195604.

Herrera-Estrella, L., & Alvarez-Morales, A. (2001). Genetically modified crops: Hope for developing countries? *EMBO Reports, 2*(4), 256–258. DOI: 10.1093/embo-reports/kve075 PMID: 11306538

Howitt, C. A., & Miskelly, D. (2017). Identification of grain variety and quality type. In *Cereal Grains* (pp. 453–492). Woodhead Publishing., DOI: 10.1016/B978-0-08-100719-8.00017-6

Imran, Y., Wijekoon, N., Gonawala, L., Chiang, Y. C., & De Silva, K. R. D. (2021). Biopiracy: Abolish Corporate Hijacking of Indigenous Medicinal Entities. *TheScientificWorldJournal, 2021*(1), 8898842. DOI: 10.1155/2021/8898842 PMID: 33679261

Ingram, J., & Maye, D. (2020). What are the implications of digitalisation for agricultural knowledge? *Frontiers in Sustainable Food Systems, 4*, 66. DOI: 10.3389/fsufs.2020.00066

Kaiser, N., Douches, D., Dhingra, A., Glenn, K. C., Herzig, P. R., Stowe, E. C., & Swarup, S. (2020). The role of conventional plant breeding in ensuring safe levels of naturally occurring toxins in food crops. *Trends in Food Science & Technology, 100*, 51–66. DOI: 10.1016/j.tifs.2020.03.042

Kolb, M. (2018). What is globalization? And how has the global economy shaped the United States. *Peterson Institute for International Economics*.https://www. piie .com/microsites/globalization/what-is-globalization

Kruft, D. (2001). Impacts of genetically-modified crops and seeds on farmers. *The Agricultural Law Resource and Reference Center,* 1-11.

Lavall, T. P., Fontanela, C., & Marocco, A. D. A. L. (2022). Perspectives for the Effectiveness of the National Intellectual Property Strategy: The 2021-2023 Biannal Plan and the Role of the Technological Innovation Centers. *Direito e Desenvolvimento, 13*, 70.

Lawson, C. (2015). Implementing Farmers' Rights: Finding Meaning and Purpose for the International Treaty on Plant Genetic Resources for Food and Agriculture Commitments? *Available at SSRN* 2626668.

Lawson, C., & Bikundo, E. (2019). The Substance and Content of Farmers' Rights–A Framework? *Griffith University Law School Research Paper.*

Lesser, W. (1987). The impacts of seed patents. *North Central Journal of Agricultural Economics, 9*(1), 37–48. DOI: 10.2307/1349340

Lor, P. (2004). Storehouses of knowledge? The role of libraries in preserving and promoting indigenous knowledge. *Indilinga, 3*(1), 45–56.

Mahoney, S. E. (2018). Owning the World's Seed Supply: How Seed Industry Mergers Threaten Global Food Security. *Geo. Envtl. L. Rev., 31*, 563.

McAfee, K. (2003). Corn culture and dangerous DNA: Real and imagined consequences of maize transgene flow in Oaxaca. *Journal of Latin American Geography, 2*(1), 18–42. DOI: 10.1353/lag.2004.0008

McEowen, R. A. (2003). Legal Issues Related to the Use and Ownership of Genetically Modified Organisms. *Washburn Law Journal, 43*, 611.

Mermer, J. E. (2018). The impact of intellectual property in fostering innovation. *Available at SSRN* 3379331. DOI: DOI: 10.2139/ssrn.3379331

Mgbeoji, I. (2001). Patents and traditional knowledge of the uses of plants: Is a communal patent regime part of the solution to the scourge of bio piracy? *Indiana Journal of Global Legal Studies*, 163–186.

Monteiro, C. A., Cannon, G., Levy, R., Moubarac, J. C., Jaime, P., Martins, A. P., & Parra, D. (2016). NOVA. The star shines bright. *World Nutrition*, *7*(1-3), 28–38.

Nabuuma, D., Reimers, C., Hoang, K. T., Stomph, T., Swaans, K., & Raneri, J. E. (2022). Impact of seed system interventions on food and nutrition security in low and middle-income countries: A scoping review. *Global Food Security*, *33*, 100638. DOI: 10.1016/j.gfs.2022.100638

Netanel, N. W. (2008). Introduction: The WIPO Development Agenda and its development policy context.

Nomani, M. Z. M., & Rahman, F. (2016). Bio Piracy of Traditional Knowledge Related Geographical Indications: A Select Study of Some Indian Cases. [MIPR]. *Manupatra Intellectual Property Reports*, *3*(3), 135–152.

Okediji, R. L. (2008). WIPO-WTO relations and the future of global intellectual property norms. *Netherlands Yearbook of International Law*, *39*, 69–125. DOI: 10.1017/S016767680800069X

Olusegun, O. O., & Olubiyi, I. A. (2017). Implications of genetically modified crops and intellectual property rights on agriculture in developing countries. *Journal of African Law*, *61*(2), 253–271. DOI: 10.1017/S0021855317000134

Padmanabhan, A., Reddy, R. S., & Sharma, S. (2017). *MODERN BIOTECHNOLOGY AND INDIA'S GOVERNANCE IMPERATIVES.* Carnegie Endowment for International Peace.

Pandey, S. C., Modi, P., Pereira, V., & Fosso Wamba, S. (2024). Empowering small farmers for sustainable agriculture: A human resource approach to SDG-driven training and innovation. *International Journal of Manpower*. Advance online publication. DOI: 10.1108/IJM-11-2023-0655

Pimentel, D., Hunter, M. S., LaGro, J. A., Efroymson, R. A., Landers, J. C., Mervis, F. T., McCarthy, C. A., & Boyd, A. E. (1989). Benefits and risks of genetic engineering in agriculture. *Bioscience*, *39*(9), 606–614. DOI: 10.2307/1311090

Prasad, R., Bagde, U. S., & Varma, A. (2012). An overview of intellectual property rights in relation to agricultural biotechnology. *African Journal of Biotechnology*, *11*(73), 13476–13752. DOI: 10.5897/AJB12.262

Prasanna, P. A., Rao, L. V., Prasad, S. H., Amtul, W., Meera, S. N., Nirmala, B., & Syamaladevi, D. P. (2020). Intellectual property rights protection for plant varieties in India: Status, emerging issues, and challenges. *Agricultural Economics Research Review*, *32*(2), 259. DOI: 10.5958/0974-0279.2019.00037.5

Prip, C. (2024). Farmers' Rights in the Plant Treaty: Interrelations and recent interactions with other international regimes and processes. *Development Studies Research, 11*(1), 2357095. DOI: 10.1080/21665095.2024.2357095

Pultrone, C. (2012). An overview of contract farming: Legal issues and challenges. *Uniform Law Review-Revue De Droit Uniforme, 17*(1-2), 263–289. DOI: 10.1093/ulr/17.1-2.263

Ramankutty, N., Mehrabi, Z., Waha, K., Jarvis, L., Kremen, C., Herrero, M., & Rieseberg, L. H. (2018). Trends in global agricultural land use: Implications for environmental health and food security. *Annual Review of Plant Biology, 69*(1), 789–815. DOI: 10.1146/annurev-arplant-042817-040256 PMID: 29489395

Rapinski, M., Raymond, R., Davy, D., Herrmann, T., Bedell, J. P., Ka, A., Odonne, G., Chanteloup, L., Lopez, P. J., Foulquier, É., da Silva, E. F., El Deghel, N., Boëtsch, G., Coxam, V., Joliet, F., Guihard-Costa, A.-M., Tibère, L., Nazare, J.-A., & Duboz, P. (2023). Local food systems under global influence: The case of food, health and environment in five socio-ecosystems. *Sustainability (Basel), 15*(3), 2376. DOI: 10.3390/su15032376

Reid, J. (2009). Biopiracy: The struggle for traditional knowledge rights. *Am. Indian L. Rev., 34*, 77.

Robinson, G. M., & Carson, D. A. (2015). The globalisation of agriculture: Introducing the Handbook. In *Handbook on the globalisation of agriculture* (pp. 1–28). Edward Elgar Publishing., DOI: 10.4337/9780857939838.00006

Rodda, D. (2020). Biopiracy Analysis of Existing International and Indian Legal Framework. Issue 2 *Int'l JL Mgmt. &. Human., 3*, 227.

Ruzzante, S., Labarta, R., & Bilton, A. (2021). Adoption of agricultural technology in the developing world: A meta-analysis of empirical literature. *World Development, 146*, 105599. DOI: 10.1016/j.worlddev.2021.105599

Sagoff, M. (2001). Biotechnology and agriculture: The common wisdom and its critics. *Indiana Journal of Global Legal Studies, 9*, 13.

Sawhney, N. (2003). *Cooperative innovation in the commons: rethinking distributed collaboration and intellectual property for sustainable design innovation* (Doctoral dissertation, Massachusetts Institute of Technology).

Sell, S. K. (2009). Corporations, seeds, and intellectual property rights governance.

Shandilya, U., & Khanduja, V. (2020, October). Intelligent farming system with weather forecast support and crop prediction. In *2020 5th International Conference on Computing, Communication and Security (ICCCS)* (pp. 1-6). IEEE. DOI: DOI: 10.1109/ICCCS49678.2020.9277437

Sharma, I. P., Kanta, C., Dwivedi, T., & Rani, R. (2020). Indigenous agricultural practices: A supreme key to maintaining biodiversity. *Microbiological advancements for higher altitude agro-ecosystems & sustainability,* 91-112.

Shiva, V. (1996). Agricultural biodiversity, intellectual property rights and farmers' rights. *Economic and Political Weekly*, 1621–1631.

Singh, K., & Kolar, P. (Eds.). (2024). *Digital Agricultural Ecosystem: Revolutionary Advancements in Agriculture*. John Wiley & Sons., DOI: 10.1002/9781394242962

Smidt, H. J., & Jokonya, O. (2022). Factors affecting digital technology adoption by small-scale farmers in agriculture value chains (AVCs) in South Africa. *Information Technology for Development, 28*(3), 558–584. DOI: 10.1080/02681102.2021.1975256

Smith, T. (2009). Going to Seed: Using Monsanto as a Case Study to Examine the Patent and Antitrust Implications of the Sale and Use of Genetically Modified Seed. *Alabama Law Review, 61*, 629.

Stein, H. (2005). Intellectual property and genetically modified seeds: The United States, trade, and the developing world. *Northwestern Journal of Technology and Intellectual Property, 3*(2), 151.

Subeesh, A., & Mehta, C. R. (2021). Automation and digitization of agriculture using artificial intelligence and the internet of things. *Artificial Intelligence in Agriculture, 5*, 278–291. DOI: 10.1016/j.aiia.2021.11.004

Tantalaki, N., Souravlas, S., & Roumeliotis, M. (2019). Data-driven decision making in precision agriculture: The rise of big data in agricultural systems. *Journal of Agricultural & Food Information, 20*(4), 344–380. DOI: 10.1080/10496505.2019.1638264

Thrupp, L. A., & Giling, R. (1996). *New partnerships for sustainable agriculture* (pp. viii+-136). Washington, DC: World Resources Institute.

Touch, V., Tan, D. K., Cook, B. R., Liu, L. D., Cross, R., Tran, T. A., Utomo, A., Yous, S., Grunbuhel, C., & Cowie, A. (2024). Smallholder farmers' challenges and opportunities: Implications for agricultural production, environment and food security. *Journal of Environmental Management, 370*, 122536. DOI: 10.1016/j.jenvman.2024.122536 PMID: 39299125

Verkey, E. (2007). Shielding farmers' rights. *Journal of Intellectual Property Law & Practice*, *2*(12), 825–831. DOI: 10.1093/jiplp/jpm189

Winter, L. Cultivating Farmers' Rights: Reconciling Food Security, Indigenous Agriculture, and TRIPS' (2010). *Vanderbilt Journal of Transnational Law, 43*, 223.

KEY TERMS AND DEFINITIONS

Asexual Propagation: Also known as vegetative propagation, this method involves creating new plants from existing plant parts, such as stems, roots, or leaves. Techniques include cuttings, layering, and division. Asexual propagation produces genetically identical plants (clones).

Grafting: A horticultural technique whereby tissues from one plant are joined with those of another plant, allowing them to grow together. This method is commonly used to combine desirable traits, such as disease resistance and fruit quality.

Hardening Off: The process of gradually acclimatizing young plants to outdoor conditions after they have been grown in a controlled environment (e.g., a greenhouse). This helps reduce transplant shock and improves survival rates.

Plant Propagation: The process of creating new plants from seeds, cuttings, bulbs, or other plant parts. This technique is essential for the reproduction and cultivation of plants in agriculture, horticulture, and landscaping.

Seed Germination: The biological process by which a seed develops into a new plant. Germination is influenced by factors such as moisture, temperature, light, and oxygen availability. Successful germination is crucial for establishing healthy plants.

Sexual Propagation: A method of plant propagation that involves the use of seeds, which results from the fertilization of ovules by pollen. This method leads to genetic variation among offspring, contributing to biodiversity.

Chapter 4
Environmental Factors Influencing Propagation

Noor Al-Huda Ahmed Mohammed Taher
https://orcid.org/0009-0007-3316-9063
College of Science, University of Mosul, Iraq

Hiba F. A. Shihab
https://orcid.org/0000-0002-9336-4887
University of Mosul, Iraq

ABSTRACT

Propagation, the process by which plants reproduce and spread, is profoundly affected by various environmental factors. Key elements such as light, temperature, water availability, soil conditions, and air movement play a vital role in determining the success and efficiency of both natural and artificial propagation methods. Light intensity and duration influence photosynthesis and germination, while temperature regulates enzymatic activity and growth rates. Adequate water and humidity support seed germination and root development, and soil composition affects nutrient availability and microbial interactions. Additionally, wind and biotic factors, including herbivores, pollinators, and beneficial soil organisms, significantly impact propagation outcomes. Understanding and managing these environmental variables is essential for improving propagation success in agriculture, horticulture, and conservation practices. This chapter explores the complex interplay of these factors, offering insights into optimizing propagation strategies for different plant species and ecological conditions.

DOI: 10.4018/979-8-3373-5626-6.ch004

INTRODUCTION

Environmental factors are among the most important elements affecting plant reproduction and continuity in natural and agricultural environments. These factors are complex and intertwined, affecting the biological processes that enable plants to reproduce effectively, whether through sexual or asexual reproduction. Among the primary factors affecting plant reproduction are: light, which plays a key role in photosynthesis and flowering, affecting flower and seed production (Wu et al., 2025); temperature, which determines growth and maturation periods and influences the success of pollination and fertilization n (Mehmood et al., 2025); humidity, which influences tissue growth and the ease of pollen transfer (Pacini & Dolferus, 2019); and soil, which provides essential nutrients and affects root health and its ability to support plant growth and reproduction. In addition, chemical factors such as oxygen and carbon dioxide levels, and biological factors such as the presence of pollinating or pathogenic insects and animals, play an important role in the reproduction process(Verma et al., 2023).

Understanding the influence of these factors helps explain differences in reproductive patterns among different species and contributes to the development of more effective agricultural methods that ensure high productivity and improve plant resilience to changing environmental conditions. In light of ongoing climate change, it has become necessary to study and evaluate the impact of these factors on an ongoing basis, to ensure the sustainability of plant diversity and increase agricultural production in a manner that is compatible with the environment and maintains its balance (Balliu et al., 2021).

1. EFFECT OF ENVIRONMENTAL FACTORS ON PLANT PROPAGATION

The influence of environmental factors on plant reproduction is a vital topic in botany and agriculture, as these factors play a significant role in determining the success of reproductive processes and the production of new seeds and individuals. The following is a detailed account of the most important effects of environmental factors on plant reproduction:

1.1. Effect of Light on the Propagation Process.

Environmental factors play a major role in plant growth and reproduction, and light is one of the most important of these factors. Light not only provides the energy needed for photosynthesis, but it also controls many vital processes within the plant, including reproduction .By controlling the amount and type of light, humans can directly influence plant productivity and quality, making it a powerful tool in modern agriculture(Chen et al., 2019).

1.2. The importance of light for plants

Light is essential for plants because:

- It provides the energy needed for photosynthesis.
- Regulates internal vital processes such as stomata opening and stem growth.
- It affects the timing of flowering and fruiting, which are key steps in reproduction (Kendrick & Kronenberg, 2012).

1.3. Types of light and their effect on Propagation:

- Light Intensity: Light intensity directly affects a plant's ability to photosynthesize. In conditions with strong light, plants produce more energy, enhancing their ability to reproduce .In conditions with weak light, plant growth may slow, and flower and seed production may decrease(Xu & Hernandez, 2020).
- Light duration (photoperiod): Photoperiod controls the timing of flowering in many plants. Some plants require long days to flower (such as wheat), while others require short days (such as rice).Photoperiod activates hormones that influence the onset of reproduction, such as florescence (Major, 2023).
- Light Type (Light Spectrum): Red and blue light have the greatest impact on plant growth and reproduction .Red light stimulates flowering, while blue light promotes robust leaf and stem growth .Using LED lights that emit specific spectrums can improve reproduction in greenhouses (Ptushenko et al., 2020).

1.4. Effect of light deficiency on reproduction:

- Poor plant growth's
- Delayed or absent flowering
- Reduced number and size of fruits or seeds.
- Some plants fail to reach reproductive maturity (Hart, 2012).

Practical examples:

- Greenhouse cultivation: Artificial lighting is used to stimulate off-season flowering and fruiting.
- Agricultural propagation: Specific light spectrums are used to stimulate plant propagation by cuttings or tissue culture.

2. ROLE OF TEMPERATURE IN GERMINATION AND GROWTH

Temperature is a measure of how hot or cold an environment is, and it directly affects the vital processes within plant cells. Each plant requires a specific temperature range to grow and reproduce normally(Butler et al., 2014).

Temperature is one of the basic factors that directly affects the speed and efficiency of seed germination, as the optimal temperature for germination varies from one plant to another depending on the nature of the environment in which it grows(Watt & Bloomberg, 2012).

Germination is the process by which a seed transitions from a dormant state to an active growth state, allowing the plant to begin developing roots, stems, and leaves. For seeds to germinate, the right environmental conditions must be met, primarily temperature.By understanding the effect of temperature on germination, agricultural production can be improved and crop success ensured (Bahuguna & Jagadish, 2015).

2.1. Role of temperature in germination process

- Stimulating chemical reactions: The right temperatures accelerate chemical reactions within the seed, activating the enzymes necessary to break down nutrients and provide the energy needed for germination (Tesfay et al., 2016).
- Determining germination speed: Seeds germinate faster at optimal temperatures, while low temperatures slow the process, and too high temperatures can kill the embryo inside the seed.
- Determining the germination rate: At appropriate temperatures, the germination rate is high. However, at unsuitable temperatures, the rate may decrease or all seeds may fail to germinate (Gardarin et. al, 2011).

2.2. Suitable temperatures for germination

- Minimum temperature: The lowest temperature at which some seeds can begin to germinate.
- Optimum temperature: The temperature at which the fastest and best germination occurs.

- Maximum temperature: The highest temperature a seed can tolerate before it deteriorates or dies.

These values vary depending on the plant type. For example:

- Wheat germinates at temperatures between 5–30°C, with the optimal temperature being 20–25°C.
- Corn requires 18–35°C, with the optimal temperature being around 30°C.

Temperature plays a crucial role in determining plant growth, survival, and reproduction. Understanding the effects of temperature is important for selecting appropriate crops for the local climate and improving agricultural production (Qiu et al., 2010).

2.3. The importance of temperature for plant growth

- Photosynthesis and Respiration:

Temperature affects the efficiency of photosynthesis, through which plants produce their own food. High temperatures also increase the rate of respiration, which can lead to the useless consumption of stored energy in the plant if it exceeds the optimal level (Dusenge et al., 2019).

- Cell growth and organ development:

Appropriate temperatures contribute to increased cell division and tissue formation, leading to healthy growth of roots, stems, and leaves(Źróbek-Sokolnik, 2011).

2.4. Effect of high and low temperature

- High temperatures: Lead to wilting of plants due to increased water loss through evaporation, and may damage proteins and enzymes within cells (Lohani et al., 2020).
- Low temperatures: Lead to slowing down vital processes, and water freezes inside cells, leading to tissue death (Bhattacharya, 2022).

2.5. The relationship between temperature and plant reproduction

- Sexual Reproduction (by seeds):

- Temperature controls the speed of seed germination. Low temperatures delay or stop germination, while suitable temperatures accelerate the process (Lohani et al., 2020).
- Some plants (such as wheat) require a period of cold to enter the flowering phase, a phenomenon known as vernalization (Rinalducci, et al., 2011).

- Asexual reproduction (such as offshoots and cuttings):
  - The success of this type of reproduction depends on the availability of moderate and stable temperatures.
  - Extremely high temperatures can cause the death of plant tissue used for propagation.

2.6. Effects of high and low temperatures on plant reproduction

- High temperatures: May disrupt the fertilization process. And cause flowers to wilt and fall before fruiting or affect pollen growth and movement(Lohani et al., 2020).
- Low temperatures: May prevent flowers from fully opening. Slow or stop the fertilization process. Reduce the activity of enzymes responsible for cell division(Bhattacharya, 2022).

Real-life examples

- Wheat and barley: They require moderate temperatures during flowering to develop healthy seeds.
- Tomatoes: Fruit set is affected if temperatures rise above 35°C, as the flowers fail to fertilize.
- Strawberries: They grow and reproduce at relatively low temperatures, but their productivity decreases in high temperatures.

3. EFFECT OF HUMIDITY AND WATER AVAILABILITY ON PLANT PROPAGATION

Environmental factors are among the primary determinants of plant growth and reproduction, with humidity and water availability being at the forefront, as they play a vital role in various physiological and biochemical processes within the plant(Mishra, 2023). Reproduction—whether sexual or asexual—is one of the most important stages in the plant life cycle and is greatly influenced by the amount of water available and the degree of humidity in the surrounding environment (Tang et al., 2021). In light of recent climate change and recurring droughts, studying the relationship between water and plant reproduction has become a scientific and agricultural necessity, without sufficient water, plants cannot complete their life cycle. Hence, the importance of water resource management and irrigation techniques to ensure agricultural production.

3.1. The importance of water in plant life

Water is the main component of plant cells, representing approximately 80–90% of their living mass. Its most prominent roles include:

- A general solvent for transporting nutrients and minerals within the plant.
- A mediator of vital reactions, including photosynthesis.
- A thermoregulator within plant tissues.
- A key factor in cell turgor, essential for supporting herbaceous plants (Lange et. Al, 2012).

3.2. Effect of humidity and water availability on sexual reproduction

1. Germination: Seeds need to absorb sufficient water to begin the germination process. Water stimulates enzymes that convert the nutrients stored in the seed into energy for growth (Kauth & Biber, 2015).
2. Flower formation: Lack of water delays or prevents flowering, while high humidity encourages plants to flower on time (Qian et. al, 2025).
3. Pollen and ovule production: Water is essential for the formation of reproductive cells. Drought can lead to fewer or weaker pollen grains (Qian et. al, 2025).
4. Pollination and Fertilization: In some plants, pollination depends on the availability of water for gamete transfer (Qian et. al, 2025).
5. Fruit and Seed Formation: Lack of moisture leads to flower drop or the production of deformed and weak fruit (Rodrigues et al., 2018).

5.3. Effect of Humidity and Water on Asexual (Vegetative) Propagation

1. Cuttings and Seedlings: Successful propagation by cuttings requires sufficient moisture to stimulate root formation (Awotedu et al., 2021).
2. Layering and Grafting: Vegetative propagation methods require a humid environment to ensure the success of the process (Awotedu et al., 2021).

5.4. The relationship between relative humidity and plant reproduction

High relative humidity reduces the rate of transpiration and improves flowering quality and fruit setting.

Effects of Low and High Humidity:

- Low relative humidity Leads to wilting of flowers, reduced fertilization rates, and increased flower and fruit drop.
- High relative humidity, May encourage the spread of fungal diseases, but at moderate levels, it supports reproduction and growth (Ferrante & Mariani, 2018).

5.5. Effect of Water Stress

When the amount of water available to the plant is less than its needs, the plant is exposed to water stress, which leads to reproductive disorders (Johari-Pireivatlou, 2010).

1. Its impact on sexual reproduction:
    - Water stress reduces the number of flowers formed.
    - It causes a decrease in pollen production or reduced pollen viability.
    - It hinders the fertilization process and the formation of seeds and fruits.
    - It increases the shedding of unfertilized flowers (Qianet al., 2025).
2. Its effect on vegetative (asexual) reproduction:
    - Reduces the ability of plant tissues to take root and grow.
    - Delays the formation of reproductive organs such as corms and tubers.
    - Leads to the drying out of parts used for propagation, such as cuttings or suckers (Atzori & Caparrotta, 2023).

4. EFFECT OF SOIL PROPERTIES ON REPRODUCTIVE SUCCESS

Soil is one of the essential factors in the agricultural environment, playing a vital role in plant growth and reproduction. It is the medium that provides plants with the support, nutrients, water, and air necessary for their vital processes. Soil properties vary from place to place, directly affecting the ability of plants to grow and reproduce. They play a fundamental role in the success of plant propagation, whether by seed or vegetative propagation. Understanding and properly managing these characteristics can enhance crop yields and improve a plant's ability to reproduce naturally. Therefore, soil analysis is an integral part of any successful agricultural plan (Jones, 2012).

4.1. Definition of Soil and Its Basic Properties

Soil is a complex mixture of mineral and organic matter, water, and air, and is a dynamic environment that supports plant life (Hillel, 2013). Soil properties include the following:

- Soil Physical Composition:

  1. Soil texture (texture): This refers to the proportion of sand, silt, and clay in the soil. Sandy soil is well-drained but poor in nutrients, while clay soil retains water but is poorly aerated.
  2. Aeration and water-holding capacity: Plants require a balance between air and water in the soil. Soil that retains excessive amounts of water or whose pores are closed impedes root growth and thus limits reproduction (Phogat et al., 2015).

- Soil Chemical Composition:

  1. Soil pH: Soil pH affects the availability of nutrients to plants. Too acidic or alkaline soil can hinder the absorption of essential nutrients.
  2. Soil nutrient content: such as nitrogen, phosphorus, potassium, calcium, and others. These elements are essential for plant growth, development, and reproduction (Rayment & Lyons, 2011) .

- Soil Biological Properties:
  - Microorganisms: such as bacteria and fungi, which contribute to the decomposition of organic matter and the provision of nutrients to plants.
  - Organic Content: Organic matter helps improve soil structure, increase its water-holding capacity, and enhance plant nutrition (Delgado & Gómez, 2024).

4.2. Effect of Soil Properties on Plant Reproduction

1. Effect of Texture and Permeability: Plants that reproduce by seeds require well-aerated and well-drained soil to ensure successful germination. For example, sandy soil facilitates root dispersal, but may not retain enough water for seed

germination. Clay soil, on the other hand, may hinder root growth due to water accumulation and poor aeration(Hosseini et al., 2022).

2. Effect of Soil Fertility: Vegetative propagation (such as tubers, cuttings, and rhizomes) depends on the availability of nutrients. A deficiency in nitrogen or phosphorus leads to poor reproductive organ development, reducing the plant's chances of reproducing (Javed et al., 2022).
3. Effect of Soil pH: The acid-base balance in the soil plays a significant role in plant reproduction. Some plants reproduce best in acidic soil (such as azaleas), while others prefer alkaline soil. An imbalance in pH impedes nutrient uptake and negatively impacts flowering, fruit set, and seed formation (Gentili et al., 2018).
4. The influence of microorganisms: Microorganisms in the soil contribute to improving soil fertility by decomposing organic matter, providing the necessary elements for growth, and producing substances that stimulate flowering and reproduction. Some fungi, such as mycorrhizae, also increase phosphorus uptake, which supports plant reproduction(Chaparro et al., 2012).
5. Soil moisture and temperature: The influence of soil characteristics is not independent of climate. Very dry or cold soil hinders germination, while moist and warm soil facilitates reproduction, especially in perennial plants(Onwukan & Mang, 2018).

4.3. Practices to Improve Soil and Encourage Plant Propagation

a) Improve drainage and aeration by adding sand or organic matter.
b) Add organic and mineral fertilizers to increase fertility.
c) Periodically analyze the soil to adjust pH and adjust if necessary.
d) Use leguminous crops to improve soil nitrogen content.
e) Manage irrigation to maintain adequate moisture without waterlogging.

5. IMPORTANCE OF AIR MOVEMENT AND WIND IN POLLEN DISPERSAL

Wind is one of the most important environmental factors affecting plant growth and reproduction. It not only affects the climatic conditions surrounding the plant, but also plays a direct role in transporting pollen grains and creating favorable conditions for pollination and fertilization, especially in plants that rely on air pollination (Rojo et al., 2015) . Although its accuracy is limited compared to biological pollination, its effectiveness enhances the persistence and spread of plant species in diverse environments(Timerman & Barrett, 2021). Understanding this effect is important when planning agriculture or managing natural ecosystems.

5.1. The Concept of Wind Pollination

Wind pollination is the transfer of pollen through the air without the need for a biological medium such as insects or birds. This type of pollination most often occurs in plants with small, colorless flowers that do not secrete nectar to attract pollinators, such as annuals, grasses, and trees like pine and oak (Faegri & Van Der Pijl, 2013).

5.2. The Role of Wind in Pollen Transport

1. Long-range transport: Wind carries light, dry pollen grains over long distances, allowing plants to be pollinated by distant individuals, thus enhancing genetic diversity.
2. Distribution efficiency: Wind is an effective and widespread means of transporting millions of pollen grains in a short period of time, especially in open areas or areas with sparse vegetation.
3. Timing and synchronization: Peak pollen production coincides with the wind seasons in some environments, increasing the chances of successful aerial pollination (Sofiev et al., 2012).

5.3. Characteristics of Pollen Grains Prepared for Air Transport

Pollen grains that rely on wind for transport have several characteristics:

- Light weight and small size: to enable them to float in the air.
- Smooth or thinly covered surface: to reduce resistance during movement.
- Abundant production: because many grains will not reach their destination, plants produce large quantities (Grega et al., 2013) .

5.4. The Importance of Wind Pollination in the Ecosystem

- Continuing the natural reproduction of wild plants without relying on living organisms.
- Increasing genetic diversity through pollination between genetically different plants.
- Contributing to food production through the pollination of agricultural plants (Songnuan, 2013).

6. EFFECT OF ALTITUDE AND GEOGRAPHIC LOCATION ON PLANT PROPAGATION

Geographic location is one of the vital environmental factors that influence plant growth and reproduction. This concept includes the Earth's position in terms of latitude and longitude, its elevation above sea level, its proximity to bodies of water or deserts, and the direction of mountain slopes. These factors lead to significant variation in climate, soil, and biodiversity, which is reflected in plant characteristics, particularly processes related to sexual and asexual reproduction. Elevation above sea level is an important environmental factor that directly or indirectly affects plant growth and reproduction. Differences in elevation lead to significant changes in climatic factors such as temperature, humidity, solar radiation intensity, wind speed, and air pressure. These environmental changes are reflected in the physiological processes of plants(Sampaio et al., 2016) .

6.1. Defining Geographic Location and Its Environmental Importance

1. Definition of Geographic Location: It is the location of a specific area on the Earth's surface, determined by geographic coordinates (latitude and longitude), in addition to other factors such as proximity to seas or mountains (Mota-Vargas & Rojas-Soto, 2012).
2. Importance of Geographic Location first: Determines the prevailing climate (tropical, subtropical, temperate, polar). Second: affects the distribution of plant species. Third: determines the length of growing seasons. And last controls the abundance of factors that facilitate reproduction, such as light, temperature, and humidity (Wang et al., 2023).

6.2. Geographical Factors Influencing Plant Reproduction

1. Latitude:Areas close to the equator enjoy high temperatures and consistent day and night lengths, which increases the chances of reproduction throughout the year. The farther from the equator, the shorter the growing seasons and the less reproductive activity, especially in cold regions (Åkerström et al., 2010).
2. Longitude: It does not directly affect climate, but may be linked to changes in timing and growing seasons, especially in large areas (Liu et al., 2015).
3. Altitude: The altitude of a region affects temperature, air pressure, and humidity. Plants at high elevations adapt through short life cycles and self-reproduction (Krishnamoorthy et al., 2016) .

4. Proximity to Water Bodies: Areas close to seas and oceans are characterized by higher humidity and a more stable climate. Humidity increases the availability of water necessary for reproduction, especially in flowering plants (Daubenmire, 2012).
5. Terrain (mountains, valleys, and slopes): Southern slopes (in the Northern Hemisphere) receive more light and heat, which stimulates flowering and pollination. Valleys may retain moisture and provide favorable environments for plant reproduction (Daubenmire, 2012).

6.3. The Direct Effects of Geographic Location on Plant Reproduction

1. Flowering times are affected by seasonal changes resulting from geographic location. Plants in temperate regions bloom in the spring, while in tropical regions they may bloom multiple times a year (Opstad et al., 2011) .
2. Availability of Pollinators; Some locations support a high diversity of pollinating insects and birds. Dry or high-altitude areas have fewer pollinators, reducing the chances of sexual pollination (Moeller et al., 2012).
3. Pollination Mechanisms; In windy or open areas, plants prefer wind pollination. In dense forests, plants rely on insects or self-pollination (Moeller et al., 2012).

6.4. Environmental Factors Associated with Altitude and Their Effect on Plant Reproduction

1. Temperature
    - Decreasing temperatures with altitude lead to a shorter growing season and slower plant life processes.
    - Low temperatures negatively affect flower maturation and seed formation.
    - Some plants have developed thermal adaptation mechanisms, such as delaying flowering until cold snaps pass (De Frenne et al., 2013).
2. Light
    - The higher the altitude, the greater the intensity of solar radiation and the amount of ultraviolet radiation.
    - Some mountain plants have developed brightly colored flowers to attract pollinators in an environment with low biodiversity .
    - Increased light can promote flowering in some species, but it can also damage reproductive tissues in cases of intense sunlight (Gratani, 2014).
3. Humidity

- Altitude is often associated with lower atmospheric humidity and increased water loss through evaporation.
- These conditions affect pollination and flower and seed formation.
- Plants at higher altitudes adapt by producing small flowers or being drought-resistant (Pellissier et al., 2010) .

4. Pollinators
   - In high-altitude environments, the number of insects and pollinators decreases due to lower temperatures and scarcity of food.
   - Therefore, some mountain plants rely on self-pollination or wind pollination instead of insects (Zirbel et al., 2017).
5. Wind
   - High-altitude areas are often exposed to strong winds, which:
   - Facilitate pollen transfer in wind-pollinated plants.
   - Damage to flowers or reduce pollination efficiency in other plants (Dai et al., 2015).

6.5. The Effect of Altitude on Different Plant Species

3. Lowland plants

Rely on long growing seasons, require warm temperatures and active pollinators, and may not be able to reproduce at higher elevations due to changing conditions.

4. Highland plants

Have short life cycles and rapid flowering, use more self-pollination, and sometimes exhibit asexual reproduction to compensate for the lack of opportunities for sexual reproduction (Krömer et al., 2013).

6.6. Plant Adaptation to Altitude

- Changes in flowering time: Mountain plants bloom at specific times to avoid cold snaps.
- Modified flower structure: Smaller flowers closer to the ground to protect them from winds.
- Asexual reproduction: Plants propagate by asexual means such as rhizomes or aerial roots.

Elevation is a complex environmental factor that influences plant reproduction by altering a number of climatic and biological factors. Its impact varies depending on the plant species, its degree of adaptation, and the surrounding environmental conditions. Understanding this relationship is important in the fields of agriculture, biodiversity conservation, and the rehabilitation of degraded ecosystems, especially in light of global climate change (Kim & Donohue, 2013).

6.7. Studies and Real-Life Examples

- In the Alps, a study showed that some plant species have adjusted their flowering dates in response to rising temperatures caused by climate change.
- In the Himalayas, plants at elevations above 4,000 meters often rely on self-pollination or vegetative reproduction.
- Desert plants (such as those in the Arabian Desert). Reproduce during limited rainy seasons, relying on asexual reproduction or the production of drought-resistant seeds.
- Tropical plants (such as the Amazon rainforest). Reproduce year-round due to stable temperatures and humidity, and rely heavily on insects and birds for pollination.
- Polar plants (such as the tundra). Reproduce very short reproductive seasons, relying on vegetative reproduction or rapidly growing seeds.

The impact of geographic changes (due to climate change or human intervention); Climate change leads to changes in flowering timing and growing season. Urbanization and urban expansion alter local environmental composition, affecting plant diversity and reproduction. Land use changes may lead to the extinction of plant species or reduce their opportunities for reproduction.

Geographic location profoundly and comprehensively influences plant reproductive processes, determining the prevailing climate and environmental conditions. Understanding this influence helps develop appropriate strategies for agriculture and the conservation of plant diversity in various environments. It is also gaining increasing importance in light of accelerating climate change and modern environmental challenges.

7. ROLE OF LIVING ORGANISMS IN FACILITATING OR IMPEDING PROPAGATION

Living organisms play a crucial role in either facilitating or hindering plant reproduction, through pollination, seed transfer, competition, and disease. The balance of these interactions determines the success of plant reproduction and growth in the environment.

7.1. The role of living organisms in facilitating plant reproduction

Living organisms play a pivotal role in facilitating plant reproduction, whether through pollination, seed dispersal, or improving the environment surrounding plant roots. This role includes interactions between plants and a variety of organisms, such as insects, birds, mammals, and soil microorganisms. These symbiotic or mutualistic relationships are among the most important drivers of biodiversity conservation and ecosystem sustainability (Frankel & Galun, 2012).

Organisms Influencing Plant Sexual Reproduction

1. Insects (such as bees, butterflies, and beetles)
    - They transfer pollen from one flower to another as they search for nectar.
    - Bees are among the most important pollinators in agriculture, contributing to the pollination of a large number of economic crops, such as apples, almonds, and sunflowers.
    - Butterflies prefer colorful, fragrant flowers.
2. Birds (such as hummingbirds and nectar-eating birds)
    - They transfer pollen as they feed on nectar.
    - They help pollinate plants in high-altitude or tropical regions.
    - Some plants have evolved bright red colors to attract these birds.
3. Small Mammals (such as bats and rodents)
    - Some bat species play a role in pollinating nocturnal flowers.
    - Rodents may transport seeds to distant locations while storing food.
4. Humans
    - They interfere with manual plant pollination, especially in commercial agriculture.
    - Cultivates and improves plant species, and develops methods to increase pollination and production.

The Role of Microorganisms in Soil

1. Beneficial bacteria (such as rhizobia): Live in the roots of legumes and fix nitrogen from the air, improving plant nutrition and reproductive capacity.
2. Root fungi (mycorrhizae): Enter into a symbiotic relationship with plant roots, helping them absorb water and minerals. Promote plant health and make them more capable of flowering and seed formation.

Plant Adaptations to Attract Pollinators

- Scent: Flowers release distinctive scents to attract insects and birds.
- Color: Flowers develop specific colors to attract certain pollinator species.
- Shape and Structure: Some flowers have deep tubes for birds, or smooth petals for insects.

The Importance of These Interactions in Agriculture and the Environment

- Improving agricultural yields: Bio-pollination increases the quantity and quality of produce.
- Maintaining plant diversity: The diversity of pollinating organisms contributes to the survival of plant species.
- Combating desertification: Effectively reproducing plants cover the soil and protect it from erosion.

Threats to this relationship

- Pesticide use: kills pollinating insects and destroys the plant-pollinator relationship.
- Climate change: alters flowering timing and affects the timing of pollinator emergence.
- Habitat loss: The reduction of forests and meadows reduces the diversity of organisms essential for plant reproduction.

7.2. Effect of living organisms in hindering plant reproduction

Living organisms play a pivotal role in ecosystems, both facilitating and hindering vital processes, most notably plant reproduction. While some organisms contribute to pollen transfer and seed dispersal, others play the opposite role, impeding this process in various ways, negatively impacting the survival and spread of plants.

Organisms that hinder plant reproduction

Organisms that negatively impact plant reproduction include:

1. Seed Predators: Some animals, such as birds, rodents, and insects, feed on seeds before they germinate. This reduces the chances of seed germination and the growth of new plants. Examples: squirrels, wild mice, and some birds, such as crows.
2. Herbivorous Insects: Some types of insects feed on flowers, destroying the reproductive organs of plants. Some larvae also feed on seeds inside fruits. Examples: beetles, fruit worms, and locusts.

3. Grazing Animals: Livestock and wild animals damage parts of the plant, such as flowers or fruits. Overgrazing limits the plant's ability to reproduce sexually and asexually. Examples: goats, sheep, deer.
4. Pathogenic Microorganisms: Fungi, bacteria, and viruses cause plant diseases, affecting flowers or fruits. They can hinder fertilization or kill seeds before they ripen. Examples: fungi that cause flower or fruit rot.
5. Parasitic Plants: These plants absorb water and nutrients from their host plants, impairing their ability to flower and reproduce. Examples: dodder, hollyhock.

Mechanisms of Obstruction and Their Environmental Impact

1. Reducing the number of seeds available for germination. Altering the geographic distribution of some plant species.
2. Impeding pollination or reducing the quality of the seeds produced. Leading to a decline in sexual reproduction rates.
3. Diseases cause plants to die before fruiting . Some diseases render seeds unfit for germination.
4. Affecting the diversity of plant species. Weakening the biological structure of the soil, affecting the ability of seeds to grow (McIntyre, 2021).

Field Examples

- In tropical forests: Some rodents eat large seeds, preventing them from germinating.
- In dry pastures: Overgrazing causes soil erosion and prevents plant regrowth.
- In agriculture: Pests such as the fall armyworm affect crops like corn and prevent fruiting.

Environmental Solutions and Management

1. Biological Control: Using predators to control harmful insects.
2. Fencing vegetated areas to protect them from grazing.
3. Using disease-resistant plant varieties.
4. Regular monitoring of animals that threaten ecosystem balance.

8. INTERACTION BETWEEN ENVIRONMENTAL FACTORS AND THEIR COMBINED EFFECT ON PROPAGATION

Plant reproduction is a vital process that ensures the continuation and renewal of plant species. A plant's ability to reproduce, whether sexually or asexually, depends on a host of environmental factors that interact in complex ways. The most prominent of these factors are light, temperature, humidity, soil, water, wind, and biotic factors such as other organisms. This complex environmental interaction can enhance or hinder the reproductive process and determine the geographic distribution of plants and their success in their habitats (Gudynaitė-Franckevičienė & Pliūra 2021).

8.1. Light and its interaction with other factors

Light is a crucial factor in stimulating plant physiological processes such as photosynthesis, but it is also a driving factor in reproduction, especially flowering.

The Interaction of Light and Temperature

1. Integration in Determining the Timing of Reproduction

Plants determine the timing of reproduction based on both light and temperature parameters. For example, adequate light may be available, but if temperatures are unsuitable, the plant will not begin to reproduce.

2. The Balance of Light and Temperature for Flower Development

Good lighting with moderate temperatures encourages flowering and fruit formation. Strong light accompanied by excessively high temperatures can stress the plant and delay reproduction.

3. Biochemical Response to the Interaction

Light and temperature together affect the levels of plant hormones such as:

- Gibberellins: Promote flowering and bud development.
- Auxins: Regulate the growth of floral organs.
- Florigen: A hormone that stimulates flowering and is directly affected by the interaction between light and temperature.

Practical Examples

- Strawberry: Blooms only when temperatures are relatively low and a specific photoperiod is present. The interaction between these factors determines the number of flowers and the quality of the fruit.
- Winter Wheat: Requires exposure to a cold period (thermal induction), followed by a long photoperiod to initiate flowering.
- Sunflower: Requires bright light, but if this is accompanied by excessively high temperatures, flower fertility is reduced and seed formation is impaired.

The Interaction Between Light and Humidity

The Interrelationship increased light intensity leads to increased transpiration, making the availability of humidity essential to prevent drought. High humidity reduces water loss despite strong light .Insufficient light combined with high humidity can lead to fungal diseases affecting flowers and fruits .The Combined Effect on Reproduction ; Reproduction occurs actively in bright, humid environments. In bright, dry environments, the plant is forced to close its stomata. In dark, humid environments, the energy available for growth decreases, and rot may occur (Driesen et al., 2020) .

Application Examples from Agriculture

- Tomatoes: Require bright light and moderate humidity.
- Wheat: Its flowering timing is affected by light intensity and humidity.
- Lilies: Grow and flower prolifically in bright light and moderate relative humidity.

8.2. Temperature and its interaction with environmental factors

Temperature plays a fundamental role in regulating growth and reproduction, as it affects seed germination, flowering, and fruit ripening.

Environmental Factors Interacting with Temperature

1. Light and temperature together affect photosynthesis, affecting the energy production necessary for flower and fruit formation .Some plants require prolonged light and certain temperatures to initiate reproduction (such as long-day plants) (Legris et al., 2017) .
2. High temperatures combined with low humidity can cause water stress, hindering flowering or leading to flower and fruit drop .Conversely, a balance between humidity and temperature creates an ideal environment for successful pollination and fruiting (Ferrante & Mariani, 2018).

3. Soil temperature affects root activity and nutrient uptake. The interaction of air temperature with soil temperature affects the plant's water and mineral balance, which is essential for reproduction.

4.High-temperature winds increase evaporation rates, causing a lack of moisture necessary for flowering .Sometimes, winds contribute to pollen transport, but if accompanied by excessive heat, they can dry out and spoil the pollen (Shilo et al., 2015).

Effect of Interaction on Plant Reproduction

1. Positive Effects: A balanced environment in terms of temperature, light, and humidity encourages abundant flowering and fruit set .The interaction between appropriate temperatures and ideal humidity contributes to enhanced pollination and fertilization processes.
2. Negative Effects: Excessive heat combined with dry weather leads to failure of pollination and may cause plant sterility .An imbalance between temperature and light may lead to early or late flowering, which affects agricultural production.

Examples of interactions in some crops

- Wheat: Requires relatively low temperatures during the vegetative growth phase.The interaction between low temperature and sufficient light is important for successful flowering and ornamental development.
- Tomato: Is significantly affected by high temperatures, which may lead to flower drop.If sufficient humidity is available, the negative effects of high temperatures can be mitigated.
- Corn: Requires relatively high temperatures, but with sufficient humidity. Heat stress combined with low humidity leads to poor pollen production and reduced fruit set.

8.3. Moisture, Water, and Their Interaction with the Soil

Water is an essential element for all stages of plant reproduction, from seed germination to flower and fruit development.

- Moisture-Soil Interaction: Soil quality determines its ability to retain moisture. Sandy soil, for example, dries quickly, which can reduce reproductive success in arid regions.
- Moisture-Heat Interaction: Excessive evaporation at high temperatures reduces water availability to the plant, which can lead to reduced flower fertility and flower drop before pollination (Seneviratne et al., 2010).

8.4. Wind and its Interactive Role in Reproduction

Wind directly affects pollen transfer (especially in wind-pollinated plants) and indirectly through water evaporation from the soil and plants.

- Interaction with Humidity: Dry winds increase water loss from leaves, which can negatively impact delicate flowers and reduce the chances of fertilization.
- Interaction with Living Organisms: Wind affects the activity of pollinating insects, as it may prevent them from flying in windy conditions, hindering the pollination process (Saunders, 2018).

8.5. Soil Properties and Their Interaction with Environmental Factors

Soil physical and chemical properties (such as fertility, aeration, and drainage) affect plant growth and reproduction.

- Interaction with Water: Clay soil may retain water, which harms roots and affects flower formation.
- Interaction with Microorganisms: Beneficial organisms in the soil (such as bacteria and fungi) contribute to improving the absorption of nutrients necessary for flower and fruit formation (Li et al., 2023).

8.6. Living Organisms and Their Environmental Interactions

Influential organisms include insects, birds, animals, microorganisms, and even other plants.

- Interaction with Light and Weather: The activity of pollinating insects is linked to temperature and light availability, and therefore, weather changes affect the effectiveness of these organisms in transmitting pollen (Evstropov et al., 2019).
- Interaction with Other Plants: Some plants secrete chemicals that inhibit the growth of nearby plants or hinder their reproduction, a phenomenon known as allelopathy.

Environmental factors, primarily light and temperature, are the cornerstone of a plant's successful life cycle, particularly during the reproductive stage. These factors control the timing of flowering, the formation of reproductive organs, seed maturation, and even the quality and quantity of production (Lowry, 2013).

9. METHODS OF PLANT ADAPTATION TO DIFFERENT ENVIRONMENTAL FACTORS

Plants are highly adaptable to diverse environments, from arid deserts to humid tropical forests, thanks to a series of morphological, physiological, and behavioral adaptations. These adaptations enable them to survive, grow, and reproduce despite harsh environmental conditions that may include drought, poor soil, limited light, high or low temperatures, and variable humidity.

9.1. Plant Adaptation to Temperature

Temperature is an important environmental factor that affects plant growth. Plants have developed several methods to adapt to different temperatures.

The Effects of Temperature on Plants

1. High Temperatures lead to increased evaporation and transpiration rates, resulting in water loss. Cause damage to cell membranes and inhibit enzymes. Negatively impact photosynthesis and lead to the production of toxic substances such as peroxides (Crawford et al., 2012).
2. Low Temperatures, cause water to freeze inside cells, leading to cell damage. Slows down enzyme activity. May cause the plant to enter a dormant state (Körner, 2016).

Plant Adaptations to High Temperatures

1. Anatomical Mechanisms
2. Small or atrophied leaves, to reduce evaporation surface area. Hairs or a waxy layer, to reduce water loss and protect against solar radiation. Deep roots, to absorb water from lower soil layers. Leaves become spines, as in cacti (Bano et al., 2019).
3. Physiological Mechanisms

Closing stomata at noon to reduce water loss. Increasing the production of thermo protective compounds to protect intracellular proteins. Self-evaporation by secreting certain fluids that cool the leaf surface (Yamori et al., 2014).

4. Behavioral or Cyclical Mechanisms

Flowers open at night in desert plants to reduce heat stress. Rapid growth in temperate seasons and cessation of growth during hot periods (Zandalinas et al., 2018).

Plant Adaptations to Low Temperatures

1. Anatomical Mechanisms

Accumulation of hairs on the surface to insulate against cold air. Thick or leathery leaves reduce heat loss (Schweingruber et al., 2012).

2. Physiological Mechanisms

Production of antifreeze compounds, such as sugars and alcohols, which lower the freezing point of cell fluid. Increasing the concentration of salts in the cell sap. Stabilizing the cell membrane to prevent it from breaking during freezing (Körner, 2016).

3. Behavioral or Cyclical Mechanisms

Winter dormancy, where the plant enters a dormant phase to reduce its vital activity. Timing flowering to coincide with the warming of early spring (Fjellheim et al., 2014).

9.2. Plant Adaptations to Humidity

Humidity is one of the primary environmental factors that directly influences plant growth and reproduction. It refers to the amount of water vapor in the air or the amount of water available in the soil. Plants vary in their needs and responses to humidity, prompting them to develop diverse adaptation mechanisms that help them survive and grow in different environments, whether very humid or extremely dry.

Plant Adaptations to Humid Environments

Broad, thin leaves: to increase the rate of transpiration, such as ferns. The presence of aerial tissue (aerenchyma), to facilitate gas exchange in roots that are submerged in water. Shallow or rootless roots, as they do not need to penetrate the soil to obtain water. The formation of floating organs, as in lotus and water hyacinth plants, to help them remain above the water surface. The development of ventilation mechanisms: such as the presence of ventilating holes in the stems and leaves (Du et al., 2019).

Plant Adaptations to Dry Environments

Reducing the size of leaves or converting them to spines, as in cacti, to reduce water loss. Presence of a thick waxy layer on the surface of leaves to prevent evaporation. Deep or extensive roots, to obtain water from deep or wide areas of the soil.

Storing water in tissues: in stems or leaves, as in succulents. Opening stomata at night (CAM system), to reduce daily transpiration (Al-Tawaha et al., 2017).

Plant Adaptations to Temperate Environments

Medium-sized leaves: adapted to moderate humidity levels. Medium-deep roots: to absorb water without the need for extensive adaptation. Balance between transpiration and absorption: to avoid water stress.

The Importance of Plant Adaptation to Humidity

1. Ensuring survival in changing environments.
2. Maximizing the use of available water resources.
3. Contributing to biodiversity and geographic distribution.
4. Improving agricultural production in different regions (Du et al., 2019).

9.3. Plant adaptation mechanisms to light intensity

Light is a critical environmental factor that influences plant life processes, most notably photosynthesis. Light intensity varies across different environments, prompting plants to develop specific responses and adaptations for survival and growth, whether in strong light conditions (open, sunny environments) or weak light conditions (shades and dense forests). This report explains the adaptive mechanisms plants use to cope with changes in light intensity.

Adapting to Strong Light

1. Small, thick leaves: Reduce water loss due to evaporation.
2. Waxy or hairy layer: Reflects some excess light and protects against desiccation.
3. Increased chlorophyll: (a) content relative to (b) Maximizes chlorophyll absorption.
4. Concentrate chloroplasts in appropriate locations within cells: Avoid damage from excessive radiation.
5. Orient leaves at specific angles: Reduce direct light intensity (Zhang et al., 2024).

Adapting to Low Light

1. Wide, thin leaves: to increase the surface area exposed to light.
2. Increased chlorophyll (b) content: to take advantage of scattered, low light.
3. Increased number of chloroplasts: to improve photosynthesis.
4. Increased transpiration rate to enhance water and nutrient uptake.
5. Plant growth toward the light (phototropism) (Zhang et al., 2024).

The Importance of Adapting to Light Intensity

1. Maximize the use of available light sources.
2. Prevent damage from excessive radiation.
3. Ensure continued photosynthesis in shaded conditions.
4. Maintain a balance between light absorption and water loss (Zandalinas et al., 2018).

The Role of Light in the Geographic Distribution of Plants

Light intensity controls the distribution of plant species on Earth's surface. Some plants only grow in bright light, such as desert plants, while others prefer shade, such as rainforest plants (Moeslund et al., 2019). Therefore, clear classifications of plants emerge based on their response to light:

- Light-loving plants (Heliophytes).
- Shade-loving plants (Sciophytes).
- Light- and shade-tolerant plants (Intermediate).

9.4. Plant adaptation mechanisms to winds

Wind is one of the most important environmental factors affecting plants, both directly through its physical impact on plant tissues, and indirectly through its influence on evaporation, temperature, relative humidity, and pollen and seed transport. Plants vary in their ability to cope with winds and have developed multiple adaptation mechanisms that enable them to survive in environments exposed to strong or persistent winds, such as coasts and desert plains (Li & Gong, 2011).

5. Morphological (structural) adaptations

Small or needle-shaped leaves, reduce wind resistance and evaporation (such as pine). Short, flexible stems, to avoid breakage and excessive bending (such as in desert plants). Dwarf or trailing growth, the plant sticks to the ground and is less exposed to wind (such as in plants in mountainous regions). Balanced branch distribution, reduces stress on any one part of the plant (Gardiner et al., 2016).

6. Physiological Adaptations

Closing stomata during periods of strong winds; to reduce water loss through transpiration. Thicker leaves and a waxy coating; to reduce evaporation caused by air movement. Improving root anchorage in the soil using deep or fibrous roots to stabilize against tug forces (Calfapietra et al., 2015).

7. Reproductive Adaptations

Using wind as a means of pollen transport, as in wind-borne plants such as corn and grasses. Lightness of seeds or the presence of wings or hairs. These enable them to be carried by the wind over long distances (such as the teeth in dandelions). Environmental Benefits of Plant Adaptations to Wind .Continuing plant life in stormy environments. Utilizing wind as a means of propagation and dispersal (Nathan et al., 2011).

8. Contributing to soil stabilization and reducing erosion.
9. Providing shelter and protection for other plants.

9.5. Plant Adaptation to Soil and Salinity

1. In poor soils: Symbiotic relationships with fungi (mycorrhizae) to increase nutrient uptake and formation of root nodules containing nitrogen-fixing bacteria (as in legumes).
2. In saline soils (halophytes): Excretion of excess salts through special glands in the leaves and storage of salts in cell vacuoles to reduce their impact on vital processes (Zhao et al., 2020).

REFERENCES

Al-Tawaha, A. R., Turk, M. A., Abu-Zaitoon, Y. M., Aladaileh, S. H., Al-Rawashdeh, I. M., Alnaimat, S., & Wedyan, M. (2017). Plants adaptation to drought environment. *Bulgarian Journal of Agricultural Science*, *23*(3), 381–388.

Atzori, G., & Caparrotta, S. (2023). Different salt stress response in Kalanchoe plants during vegetative and asexual propagation. *South African Journal of Botany*, *154*, 282–288. DOI: 10.1016/j.sajb.2023.01.027

Awotedu, B. F., Omolola, T. O., Akala, A. O., Awotedu, O. L., & Olaoti-Laaro, S. O. (2021). Vegetative propagation: A unique technique of improving plants growth. *World News of Natural Sciences*, *35*, 83–101.

Bahuguna, R. N., & Jagadish, K. S. (2015). Temperature regulation of plant phenological development. *Environmental and Experimental Botany*, *111*, 83–90. DOI: 10.1016/j.envexpbot.2014.10.007

Balliu, A., Zheng, Y., Sallaku, G., Fernández, J. A., Gruda, N. S., & Tuzel, Y. (2021). Environmental and cultivation factors affect the morphology, architecture and performance of root systems in soilless grown plants. *Horticulturae*, *7*(8), 243. DOI: 10.3390/horticulturae7080243

Bano, C., Amist, N., & Singh, N. B. (2019). Morphological and anatomical modifications of plants for environmental stresses. *Molecular plant abiotic stress: Biology and biotechnology*, 29-44.

Bhattacharya, A. (2022). Effect of low-temperature stress on germination, growth, and phenology of plants: A review. Physiological processes in plants under low temperature stress, 1-106.

Butler, T. J., Celen, A. E., Webb, S. L., Krstic, D., & Interrante, S. M. (2014). Temperature affects the germination of forage legume seeds. *Crop Science*, *54*(6), 2846–2853. DOI: 10.2135/cropsci2014.01.0063

Calfapietra, C., Peñuelas, J., & Niinemets, Ü. (2015). Urban plant physiology: Adaptation-mitigation strategies under permanent stress. *Trends in Plant Science*, *20*(2), 72–75. DOI: 10.1016/j.tplants.2014.11.001 PMID: 25476199

Chaparro, J. M., Sheflin, A. M., Manter, D. K., & Vivanco, J. M. (2012). Manipulating the soil microbiome to increase soil health and plant fertility. *Biology and Fertility of Soils*, *48*(5), 489–499. DOI: 10.1007/s00374-012-0691-4

Chen, Y. M., Huang, J. Z., Hou, T. W., & Pan, I. C. (2019). Effects of light intensity and plant growth regulators on callus proliferation and shoot regeneration in the ornamental succulent Haworthia. *Botanical Studies*, *60*(1), 10. DOI: 10.1186/s40529-019-0257-y PMID: 31267253

Crawford, A. J., McLachlan, D. H., Hetherington, A. M., & Franklin, K. A. (2012). High temperature exposure increases plant cooling capacity. *Current Biology*, *22*(10), R396–R397. DOI: 10.1016/j.cub.2012.03.044 PMID: 22625853

Dai, K., Bergot, A., Liang, C., Xiang, W. N., & Huang, Z. (2015). Environmental issues associated with wind energy–A review. *Renewable Energy*, *75*, 911–921. DOI: 10.1016/j.renene.2014.10.074

Daubenmire, R. (2012). *Plant geography: with special reference to North America*. Elsevier.

De Frenne, P., Graae, B. J., Rodríguez-Sánchez, F., Kolb, A., Chabrerie, O., Decocq, G., De Kort, H., De Schrijver, A., Diekmann, M., Eriksson, O., Gruwez, R., Hermy, M., Lenoir, J., Plue, J., Coomes, D. A., & Verheyen, K. (2013). Latitudinal gradients as natural laboratories to infer species' responses to temperature. *Journal of Ecology*, *101*(3), 784–795. DOI: 10.1111/1365-2745.12074

Delgado, A., & Gómez, J. A. (2024). The soil: Physical, chemical, and biological properties. In *Principles of agronomy for sustainable agriculture* (pp. 15–30). Springer International Publishing. DOI: 10.1007/978-3-031-69150-8_2

Driesen, E., Van den Ende, W., De Proft, M., & Saeys, W. (2020). Influence of environmental factors light, CO2, temperature, and relative humidity on stomatal opening and development: A review. *Agronomy (Basel)*, *10*(12), 1975. DOI: 10.3390/agronomy10121975

Du, Q., Liu, T., Jiao, X., Song, X., Zhang, J., & Li, J. (2019). Leaf anatomical adaptations have central roles in photosynthetic acclimation to humidity. *Journal of Experimental Botany*, *70*(18), 4949–4962. DOI: 10.1093/jxb/erz238 PMID: 31145790

Dusenge, M. E., Duarte, A. G., & Way, D. A. (2019). Plant carbon metabolism and climate change: Elevated CO 2 and temperature impacts on photosynthesis, photorespiration and respiration. *The New Phytologist*, *221*(1), 32–49. DOI: 10.1111/nph.15283 PMID: 29983005

Evstropov, V., Trushkova, E., & Nikhayeva, A. (2019). Recent research questions of ecological aspects of the interaction of the organism and the environment. *Научный альманах стран Причерноморья*, (3 (19)), 41-51.

Faegri, K., & Van Der Pijl, L. (2013). Principles of pollination ecology.

Ferrante, A., & Mariani, L. (2018). Agronomic management for enhancing plant tolerance to abiotic stresses: High and low values of temperature, light intensity, and relative humidity. *Horticulturae*, *4*(3), 21. DOI: 10.3390/horticulturae4030021

Ferrante, A., & Mariani, L. (2018). Agronomic management for enhancing plant tolerance to abiotic stresses: High and low values of temperature, light intensity, and relative humidity. *Horticulturae*, *4*(3), 21. DOI: 10.3390/horticulturae4030021

Fjellheim, S., Boden, S., & Trevaskis, B. (2014). The role of seasonal flowering responses in adaptation of grasses to temperate climates. *Frontiers in Plant Science*, *5*, 431. DOI: 10.3389/fpls.2014.00431 PMID: 25221560

Frankel, R., & Galun, E. (2012). *Pollination mechanisms, reproduction and plant breeding* (Vol. 2). Springer Science & Business Media.

Gardarin, A., Dürr, C., & Colbach, N. (2011). Prediction of germination rates of weed species: Relationships between germination speed parameters and species traits. *Ecological Modelling*, *222*(3), 626–636. DOI: 10.1016/j.ecolmodel.2010.10.005

Gardiner, B., Berry, P., & Moulia, B. (2016). Wind impacts on plant growth, mechanics and damage. *Plant Science*, *245*, 94–118. DOI: 10.1016/j.plantsci.2016.01.006 PMID: 26940495

Gentili, R., Ambrosini, R., Montagnani, C., Caronni, S., & Citterio, S. (2018). Effect of soil pH on the growth, reproductive investment and pollen allergenicity of Ambrosia artemisiifolia L. *Frontiers in Plant Science*, *9*, 1335. DOI: 10.3389/fpls.2018.01335 PMID: 30294333

Gratani, L. (2014). Plant phenotypic plasticity in response to environmental factors. *Advances in Botany*, *2014*(1), 208747. DOI: 10.1155/2014/208747

Grega, L., Anderson, S., Cheetham, M., Clemente, M., Colletti, A., Moy, W., Talarico, D., Thatcher, S. L., & Osborn, J. M. (2013). Aerodynamic characteristics of saccate pollen grains. *International Journal of Plant Sciences*, *174*(3), 499–510. DOI: 10.1086/668694

Gudynaitė-Franckevičienė, V., & Pliūra, A. (2021). The Impact of different environmental conditions during vegetative propagation on growth, survival, and biochemical characteristics in Populus hybrids in clonal field trial. *Forests*, *12*(7), 892. DOI: 10.3390/f12070892

Hart, J. W. (2012). *Light and plant growth* (Vol. 1). Springer Science & Business Media.

Hillel, D. (2013). *Introduction to soil physics*. Academic press.

Hosseini, N., & Rezanejad, F., & ZamaniBahramabadi, E. (2022). Effects of soil texture, irrigation intervals, and cultivar on some nut qualities and different types of fruit blankness in pistachio (Pistacia vera L.). *International Journal of Horticultural Science and Technology*, *9*(1), 41–53.

Javed, A., Ali, E., Afzal, K. B., Osman, A., & Riaz, S. (2022). Soil fertility: Factors affecting soil fertility, and biodiversity responsible for soil fertility. *International Journal of Plant. Animal and Environmental Sciences*, *12*(1), 21–33.

Johari-Pireivatlou, M. (2010). Effect of soil water stress on yield and proline content of four wheat lines. *African Journal of Biotechnology*, *9*(1).

Jones, J. R. (2012). *Plant nutrition and soil fertility manual*. CRC press. DOI: 10.1201/b11577

Kauth, P. J., & Biber, P. D. (2015). Moisture content, temperature, and relative humidity influence seed storage and subsequent survival and germination of Vallisneria americana seeds. *Aquatic Botany*, *120*, 297–303. DOI: 10.1016/j.aquabot.2014.09.009

Kendrick, R. E., & Kronenberg, G. H. (Eds.). (2012). *Photomorphogenesis in plants*. Springer Science & Business Media.

Kim, E., & Donohue, K. (2013). Local adaptation and plasticity of Erysimum capitatum to altitude: Its implications for responses to climate change. *Journal of Ecology*, *101*(3), 796–805. DOI: 10.1111/1365-2745.12077

Körner, C. (2016). Plant adaptation to cold climates. [-Faculty.]. *F1000 Research*, *5*, F1000. DOI: 10.12688/f1000research.9107.1 PMID: 27990251

Körner, C. (2016). Plant adaptation to cold climates. [-Faculty.]. *F1000 Research*, *5*, F1000. DOI: 10.12688/f1000research.9107.1 PMID: 27990251

Krishnamoorthy, M., Palanisamy, K., Francis, A. P., & Gireesan, K. (2016). Impact of environmental factors and altitude on growth and reproductive characteristics of teak (Tectona grandis Linn. f.) in Southern India. *Journal of Forest and Environmental Science*, *32*(4), 353–366. DOI: 10.7747/JFES.2016.32.4.353

Krömer, T., Acebey, A., Kluge, J., & Kessler, M. (2013). Effects of altitude and climate in determining elevational plant species richness patterns: A case study from Los Tuxtlas, Mexico. Flora-Morphology, Distribution. *Flora (Jena)*, *208*(3), 197–210. DOI: 10.1016/j.flora.2013.03.003

Lange, O. L., Kappen, L., & Schulze, E. D. (Eds.). (2012). *Water and plant life: problems and modern approaches* (Vol. 19). Springer Science & Business Media.

Legris, M., Nieto, C., Sellaro, R., Prat, S., & Casal, J. J. (2017). Perception and signalling of light and temperature cues in plants. *The Plant Journal*, *90*(4), 683–697. DOI: 10.1111/tpj.13467 PMID: 28008680

Li, H., Chang, L., Wei, Y., & Li, Y. (2023). Interacting effects of land use type, soil attributes, and environmental factors on aggregate stability. *Land (Basel)*, *12*(7), 1286. DOI: 10.3390/land12071286

Li, Z. G., & Gong, M. (2011). Mechanical stimulation-induced cross-adaptation in plants: An overview. *Journal of Plant Biology*, *54*(6), 358–364. DOI: 10.1007/s12374-011-9178-3

Liu, W., Liu, J., Yin, D., & Zhao, X. (2015). Influence of ecological factors on the production of active substances in the anti-cancer plant Sinopodophyllum hexandrum (Royle) TS Ying. *PLoS One*, *10*(4), e0122981. DOI: 10.1371/journal.pone.0122981 PMID: 25874701

Lohani, N., Singh, M. B., & Bhalla, P. L. (2020). High temperature susceptibility of sexual reproduction in crop plants. *Journal of Experimental Botany*, *71*(2), 555–568. DOI: 10.1093/jxb/erz426 PMID: 31560053

Lohani, N., Singh, M. B., & Bhalla, P. L. (2020). High temperature susceptibility of sexual reproduction in crop plants. *Journal of Experimental Botany*, *71*(2), 555–568. DOI: 10.1093/jxb/erz426 PMID: 31560053

Lowry, W. P. (2013). *Weather and life: An introduction to biometeorology*. Elsevier.

Major, D. J. (2023). Light and photoperiod. In CRC Handbook of plant science in agriculture (pp. 195-216). CRC Press. Major, D. J. (2023). Light and photoperiod. In CRC Handbook of plant science in agriculture (pp. 195-216). CRC Press.

McIntyre, I. (2021). *Environmental blockades: Obstructive direct action and the history of the environmental movement*. Routledge. DOI: 10.4324/9781003037781

Mehmood, M., Tanveer, N. A., Joyia, F. A., Ullah, I., & Mohamed, H. I. (2025). Effect of high temperature on pollen grains and yield in economically important crops: A review. *Planta*, *261*(6), 141. DOI: 10.1007/s00425-025-04714-0 PMID: 40374974

Mishra, R. K. (2023). Fresh water availability and its global challenge. *British Journal of Multidisciplinary and Advanced Studies*, *4*(3), 1–78. DOI: 10.37745/bjmas.2022.0208

Moeller, D. A., Geber, M. A., Eckhart, V. M., & Tiffin, P. (2012). Reduced pollinator service and elevated pollen limitation at the geographic range limit of an annual plant. *Ecology*, *93*(5), 1036–1048. DOI: 10.1890/11-1462.1 PMID: 22764490

Moeslund, J. E., Zlinszky, A., Ejrnæs, R., Brunbjerg, A. K., Bøcher, P. K., Svenning, J. C., & Normand, S. (2019). Light detection and ranging explains diversity of plants, fungi, lichens, and bryophytes across multiple habitats and large geographic extent. *Ecological Applications*, *29*(5), e01907. DOI: 10.1002/eap.1907 PMID: 31002436

Mota-Vargas, C., & Rojas-Soto, O. R. (2012). The importance of defining the geographic distribution of species for conservation: The case of the Bearded Wood-Partridge. *Journal for Nature Conservation*, *20*(1), 10–17. DOI: 10.1016/j.jnc.2011.07.002

Nathan, R., Katul, G. G., Bohrer, G., Kuparinen, A., Soons, M. B., Thompson, S. E., Trakhtenbrot, A., & Horn, H. S. (2011). Mechanistic models of seed dispersal by wind. *Theoretical Ecology*, *4*(2), 113–132. DOI: 10.1007/s12080-011-0115-3

Onwuka, B., & Mang, B. J. A. P. A. R. (2018). Effects of soil temperature on some soil properties and plant growth. *Advances in Plants & Agriculture Research*, *8*(1), 34–37. DOI: 10.15406/apar.2018.08.00288

Opstad, N., Sønsteby, A., Myrheim, U., & Heide, O. M. (2011). Seasonal timing of floral initiation in strawberry: Effects of cultivar and geographic location. *Scientia Horticulturae*, *129*(1), 127–134. DOI: 10.1016/j.scienta.2011.03.022

Pacini, E., & Dolferus, R. (2019). Pollen developmental arrest: Maintaining pollen fertility in a world with a changing climate. *Frontiers in Plant Science*, *10*, 679. DOI: 10.3389/fpls.2019.00679 PMID: 31178886

Pellissier, L., Fournier, B., Guisan, A., & Vittoz, P. (2010). Plant traits co-vary with altitude in grasslands and forests in the European Alps. *Plant Ecology*, *211*(2), 351–365. DOI: 10.1007/s11258-010-9794-x

Phogat, V. K., Tomar, V. S., & Dahiya, R. I. T. A. (2015). Soil physical properties. Soil science: An introduction, 135-171.

Ptushenko, O. S., Ptushenko, V. V., & Solovchenko, A. E. (2020). Spectrum of light as a determinant of plant functioning: A historical perspective. *Life (Chicago, Ill.)*, *10*(3), 25. DOI: 10.3390/life10030025 PMID: 32192016

Qian, D., Wang, M., Niu, Y., Yang, Y., & Xiang, Y. (2025). Sexual reproduction in plants under high temperature and drought stress. *Cell Reports*, *44*(3), 115390. DOI: 10.1016/j.celrep.2025.115390 PMID: 40056418

Qian, D., Wang, M., Niu, Y., Yang, Y., & Xiang, Y. (2025). Sexual reproduction in plants under high temperature and drought stress. *Cell Reports*, *44*(3), 115390. DOI: 10.1016/j.celrep.2025.115390 PMID: 40056418

Qiu, J., Bai, Y., Fu, Y. B., & Wilmshurst, J. F. (2010). Spatial variation in temperature thresholds during seed germination of remnant Festuca hallii populations across the Canadian prairie. *Environmental and Experimental Botany*, *67*(3), 479–486. DOI: 10.1016/j.envexpbot.2009.09.002

Rayment, G. E., & Lyons, D. J. (2011). *Soil chemical methods: Australasia* (Vol. 3). CSIRO publishing.

Rinalducci, S., Egidi, M. G., Mahfoozi, S., Godehkahriz, S. J., & Zolla, L. (2011). The influence of temperature on plant development in a vernalization-requiring winter wheat: A 2-DE based proteomic investigation. *Journal of Proteomics*, *74*(5), 643–659. DOI: 10.1016/j.jprot.2011.02.005 PMID: 21320650

Rodrigues, B. R. A., Nietsche, S., Mercadante-Simões, M. O., Pereira, M. C. T., & Ribeiro, L. M. (2018). Climatic seasonality influences the development of pollen grains and fruiting in Annona squamosa. *Environmental and Experimental Botany*, *150*, 240–248. DOI: 10.1016/j.envexpbot.2018.03.025

Rojo, J., Rapp, A., Lara, B., Fernández-González, F., & Pérez-Badia, R. (2015). Effect of land uses and wind direction on the contribution of local sources to airborne pollen. *The Science of the Total Environment*, *538*, 672–682. DOI: 10.1016/j.scitotenv.2015.08.074 PMID: 26327635

Sampaio, B. L., Edrada-Ebel, R., & Da Costa, F. B. (2016). Effect of the environment on the secondary metabolic profile of Tithonia diversifolia: A model for environmental metabolomics of plants. *Scientific Reports*, *6*(1), 29265. DOI: 10.1038/srep29265 PMID: 27383265

Saunders, M. E. (2018). Insect pollinators collect pollen from wind-pollinated plants: Implications for pollination ecology and sustainable agriculture. *Insect Conservation and Diversity*, *11*(1), 13–31. DOI: 10.1111/icad.12243

Schweingruber, F. H., Börner, A., & Schulze, E. D. (2012). Anatomical adaptations to environmental conditions. In *Atlas of Stem Anatomy in Herbs, Shrubs and Trees* (Vol. 2, pp. 355–370). Springer Berlin Heidelberg.

Seneviratne, S. I., Corti, T., Davin, E. L., Hirschi, M., Jaeger, E. B., Lehner, I., Orlowsky, B., & Teuling, A. J. (2010). Investigating soil moisture–climate interactions in a changing climate: A review. *Earth-Science Reviews*, *99*(3-4), 125–161. DOI: 10.1016/j.earscirev.2010.02.004

Shilo, E., Ziv, B., Shamir, E., & Rimmer, A. (2015). Evaporation from Lake Kinneret, Israel, during hot summer days. *Journal of Hydrology (Amsterdam)*, *528*, 264–275. DOI: 10.1016/j.jhydrol.2015.06.042

Sofiev, M., Belmonte, J., Gehrig, R., Izquierdo, R., Smith, M., Dahl, Å., & Siljamo, P. (2012). Airborne pollen transport. Allergenic pollen: A review of the production, release, distribution and health impacts, 127-159.

Songnuan, W. (2013). Wind-pollination and the roles of pollen allergenic proteins. *Asian Pacific Journal of Allergy and Immunology*, *31*(4), 261. PMID: 24383968

Tang, M., Zhao, W., Xing, M., Zhao, J., Jiang, Z., You, J., Ni, B., Ni, Y., Liu, C., Li, J., & Chen, X. (2021). Resource allocation strategies among vegetative growth, sexual reproduction, asexual reproduction and defense during growing season of Aconitum kusnezoffii Reichb. *The Plant Journal*, *105*(4), 957–977. DOI: 10.1111/tpj.15080 PMID: 33180340

Tesfay, S. Z., Modi, A. T., & Mohammed, F. (2016). The effect of temperature in moringa seed phytochemical compounds and carbohydrate mobilization. *South African Journal of Botany*, *102*, 190–196. DOI: 10.1016/j.sajb.2015.07.003

Timerman, D., & Barrett, S. C. (2021). The biomechanics of pollen release: New perspectives on the evolution of wind pollination in angiosperms. *Biological Reviews of the Cambridge Philosophical Society*, *96*(5), 2146–2163. DOI: 10.1111/brv.12745 PMID: 34076950

Verma, R. C., Waseem, M. A., Sharma, N., Bharathi, K., Singh, S., Anto Rashwin, A., & Singh, B. V. (2023). The role of insects in ecosystems, an in-depth review of entomological research. *International Journal of Environment and Climate Change*, *13*(10), 4340–4348. DOI: 10.9734/ijecc/2023/v13i103110

Wang, X., Morin, X., Zhang, J., Chen, G., Mao, L., Chen, Y., Song, Z., Du, Y., & Ma, K. (2023). Geographical patterns and determinants in plant reproductive phenology duration. *Frontiers in Plant Science*, *14*, 1199316. DOI: 10.3389/fpls.2023.1199316 PMID: 37396633

Watt, M. S., & Bloomberg, M. (2012). Key features of the seed germination response to high temperatures. *The New Phytologist*, *196*(2), 332–336. DOI: 10.1111/j.1469-8137.2012.04280.x PMID: 22882304

Wu, W., Chen, L., Liang, R., Huang, S., Li, X., Huang, B., Luo, H., Zhang, M., Wang, X., & Zhu, H. (2025). The role of light in regulating plant growth, development and sugar metabolism: A review. *Frontiers in Plant Science*, *15*, 1507628. DOI: 10.3389/fpls.2024.1507628 PMID: 39840366

Xu, X., & Hernandez, R. (2020). The effect of light intensity on vegetative propagation efficacy, growth, and morphology of "Albion" strawberry plants in a precision indoor propagation system. *Applied Sciences (Basel, Switzerland), 10*(3), 1044. DOI: 10.3390/app10031044

Yamori, W., Hikosaka, K., & Way, D. A. (2014). Temperature response of photosynthesis in C3, C4, and CAM plants: Temperature acclimation and temperature adaptation. *Photosynthesis Research, 119*(1), 101–117. DOI: 10.1007/s11120-013-9874-6 PMID: 23801171

Zandalinas, S. I., Mittler, R., Balfagón, D., Arbona, V., & Gómez-Cadenas, A. (2018). Plant adaptations to the combination of drought and high temperatures. *Physiologia Plantarum, 162*(1), 2–12. DOI: 10.1111/ppl.12540 PMID: 28042678

Zandalinas, S. I., Mittler, R., Balfagón, D., Arbona, V., & Gómez-Cadenas, A. (2018). Plant adaptations to the combination of drought and high temperatures. *Physiologia Plantarum, 162*(1), 2–12. DOI: 10.1111/ppl.12540 PMID: 28042678

Zhang, M., Ming, Y., Wang, H. B., & Jin, H. L. (2024). Strategies for adaptation to high light in plants. *aBIOTECH, 5*(3), 381–393. DOI: 10.1007/s42994-024-00164-6 PMID: 39279858

Zhao, C., Zhang, H., Song, C., Zhu, J. K., & Shabala, S. (2020). Mechanisms of plant responses and adaptation to soil salinity. *Innovation (Cambridge (Mass.)), 1*(1), 100017. DOI: 10.1016/j.xinn.2020.100017 PMID: 34557705

Zirbel, C. R., Bassett, T., Grman, E., & Brudvig, L. A. (2017). Plant functional traits and environmental conditions shape community assembly and ecosystem functioning during restoration. *Journal of Applied Ecology, 54*(4), 1070–1079. DOI: 10.1111/1365-2664.12885

Źróbek-Sokolnik, A. (2011). Temperature stress and responses of plants. In *Environmental adaptations and stress tolerance of plants in the era of climate change* (pp. 113–134). Springer New York.

Chapter 5
Scientific Foundations of Plant Propagation

Azza Adel
https://orcid.org/0009-0000-9075-8912
Ministry of Higher Education and Scientific Research, Iraq & University of Mosul, Iraq

ABSTRACT

The scientific foundations of plant propagation integrate classical knowledge with cutting-edge biotechnology. Sexual propagation ensures diversity and adaptability, while vegetative methods maintain clonal fidelity. Tissue culture and totipotency provide tools for mass production and conservation, guided by hormonal and genetic regulation. Anatomical, histological, and physiological processes form the basis of successful regeneration, while environmental factors determine practical outcomes. As global challenges intensify, the refinement of propagation science will remain central to agricultural innovation, biodiversity conservation, and sustainable development.

1. INTRODUCTION

The ability to propagate plants efficiently has always been a cornerstone of agricultural and horticultural progress (Wang et al., 2025). From ancient seed sowing to modern in vitro culture techniques, propagation methods have evolved alongside advances in plant biology (Taji et al., 2024). The scientific foundations of propagation rest upon interdisciplinary insights from seed physiology, cellular totipotency, hormonal signaling, anatomical structures, genetic regulation, and environmental interactions (Barrero et al., 2025). Understanding these foundations not only facilitates efficient plant multiplication but also supports breeding programs,

DOI: 10.4018/979-8-3373-5626-6.ch005

conservation of endangered species, and adaptation strategies under global climate challenges (Jain et al., 2023).

2. SEXUAL PROPAGATION THROUGH SEEDS

Sexual propagation is the natural and most widespread method of plant reproduction. It ensures genetic variability, which is crucial for adaptation and crop improvement (Salgotra and Chauhan, 2023). Sexual propagation is the fundamental biological process through which plants reproduce by the formation of seeds, following the fusion of male and female gametes (Pereira and Coimbra, 2019). This process begins with pollination, where pollen grains are transferred from the anther (the male part of a flower) to the stigma (the female part) (Bhatla and Shakya, 2023). Once a compatible pollen grain lands on the stigma, it germinates, sending a pollen tube down the style to the ovary (Mehmood et al., 2025). Here, fertilization occurs: one sperm cell fuses with the egg cell to form a zygote, which will develop into the embryo of the seed (Khalimova, 2024), while another sperm cell fuses with polar nuclei to form the endosperm, a nutrient-rich tissue that will nourish the developing embryo (Jin et al., 2025). The ovary itself matures into a fruit, which protects the seeds and often aids in their dispersal (Sripathy and Groot, 2023). The primary advantage of this method is the introduction of extensive genetic variability, as the resulting offspring inherit a unique combination of traits from both parent plants (Joshi et al., 2023). This genetic diversity is paramount for natural selection, allowing plant species to adapt to changing environmental conditions, resist diseases, and evolve over time (Gibson, 2022). In agriculture and horticulture, this variability is harnessed through plant breeding programs to develop new cultivars with improved yield, hardiness, flavor, or other desirable characteristics, making sexual propagation the cornerstone of both natural ecosystems and global food security (Weiss and Gruda, 2025).

3.SEED DISPERSAL AND ESTABLISHMENT

The survival, colonization, and genetic diversity of plant species are fundamentally dependent on effective seed dispersal and subsequent successful establishment (Wu et al., 2023). Dispersal is the critical process of transporting seeds away from the parent plant (Jordano, 2017), a strategy that mitigates several risks; it reduces competition for resources like light, water, and nutrients between the parent and its offspring (Yu et al.,2024), minimizes the spread of species-specific pests and diseases that may inhabit the soil beneath the parent, and enables the species to colonize new and potentially more favorable habitats (Fuller et al.,2023), which is crucial for range expansion especially in changing climates (Osland et al., 2023). Plants have evolved a remarkable array of dispersal mechanisms, broadly categorized as abiotic and biotic (Morris and Blackwood, 2024). Abiotic methods include wind dispersal (anemochory), where seeds possess wings or parachutes for buoyancy (Poschlod et al., 2025), water dispersal (hydrochory), utilizing currents for transport, and mechanical dispersal (autochory), where seed pods dry and explode, propelling the seeds outward (Volovnik, 2025). Biotic methods involve animals, either through ingestion and later excretion (endozoochory), where seeds are adapted to survive digestive tracts (Green and Wilkinson, 2024), external attachment to fur or feathers (epizoochory) via hooks and barbs, or caching by animals like squirrels and jays, which often forget their stores (Gómez et al., 2019). However, dispersal alone is insufficient; the ultimate goal is establishment, or germination and growth into a viable seedling (Ruxton and Schaefer, 2012). This requires the seed to land in a location with the correct soil composition, moisture, temperature, and light exposure(Shaw et al., 2020). Many seeds also exhibit dormancy, a timed delay that ensures they only germinate when seasonal conditions are optimal, thereby synchronizing their most vulnerable growth stage with favorable environmental windows to maximize their chances of survival and maturation (Absakine et al., 2024).

4. SEED PHYSIOLOGY AND DORMANCY MECHANISMS

Seed physiology encompasses the biological processes that govern seed development, dormancy, germination, and seedling establishment, with dormancy representing a critical adaptive mechanism for survival (Nautiyal et al., 2023) . Dormancy is a state of arrested metabolic activity that prevents a viable seed from germinating even when environmental conditions, such as water, oxygen, and suitable temperatures, are ostensibly favorable (Bhatla and Kathpalia, 2023). This evolutionary strategy ensures that germination is temporally delayed until conditions are optimal for seedling survival, such as the correct season (Reed et al., 2022), thereby avoiding germination during brief, unseasonal warm spells in autumn that would be fatal with the onset of winter (Bhattacharya, 2022). The mechanisms of dormancy are complex and varied, often classified into two primary types: physical and physiological (Lamont and Pausas, 2023). Physical dormancy (PY) is caused by a hard, impermeable seed coat or fruit wall that prevents water uptake and/or gas exchange (Wen et al., 2024); this type of dormancy is broken by natural abrasion, microbial action, fire, or the scarifying action of an animal's digestive tract. In contrast (Awasthi, 2023), physiological dormancy (PD) is caused by a physiological inhibiting mechanism within the embryo itself (Corbineau, 2022), often regulated by a balance between growth promoting hormones (like gibberellins) and growth inhibiting hormones (like abscisic acid, or ABA) (Shah et al., 2023). Breaking physiological dormancy typically requires specific environmental cues such as a period of cold, moist stratification (mimicking winter), exposure to light of a specific wavelength, or diurnal temperature fluctuations (Lamont and Pausas, 2023). Some seeds exhibit combinational dormancy, possessing both physical and physiological barriers (Arpitha and Navya, 2025). Understanding these intricate physiological mechanisms is paramount in agriculture, horticulture, and ecological restoration, as it allows for the artificial breaking of dormancy to achieve synchronized germination, propagate desirable species, and conserve those threatened by environmental change (COST, 2025).

Physiological Processes in Propagation

Physiological processes are the fundamental engine driving successful plant propagation, governing every stage from the initial development of a propagule to its establishment as an independent, photosynthesizing organism (Roddy et al., 2020). In sexual propagation these processes commence with the critical hormonal and enzymatic events of fertilization, leading to seed formation (Kandar and Pal, 2024). The subsequent physiology of the seed is dominated by two key states, dormancy and germination (Klupczyńska and Pawłowski, 2021). Dormancy, often regulated by the ratio of growth inhibitors like abscisic acid (ABA) to growth promoters like gibberellins (GA), is a protective mechanism that prevents premature germination (Sabir et al., 2025). Its breakage requires specific environmental cues such as moisture, stratification, or light exposure that trigger an internal hormonal shift, metabolically activating the seed (Malviya and Gayen, 2024). Upon germination, the imbibition of water rehydrates tissues and reignites metabolic activity, mobilizing stored reserves in the endosperm or cotyledons through the action of enzymes like amylases and proteases (Ali and Elozeiri, 2017). This provides the energy and building blocks for the radicle to emerge and establish a root system, followed by the plumule's growth towards light (Lewsey et al., 2025). In asexual propagation, the physiological principles are equally complex (Tang et al., 2024). The regeneration of tissues from a cutting or graft relies on totipotency the ability of a single cell to dedifferentiate and regenerate an entire plant (Malabadi et al., 2025). This process is driven by plant growth regulators, particularly auxins, which are often applied externally to stimulate adventitious root formation at the base of a stem cutting (Lakehal and Bellini, 2019). Auxins promote cell division and differentiation at the cambium, leading to callus formation and ultimately, organized root primordial (Ma et al., 2025). Simultaneously, the cutting must manage water stress through transpiration control, often requiring high humidity environments, while reactivating photosynthetic pathways to become autotrophic (Jiang et al., 2025). Thus, whether through seeds or cuttings, successful propagation is a meticulously orchestrated physiological symphony of water uptake, hormone signaling, enzyme activation, and energy mobilization, all aimed at achieving genetic continuity and growth (Lokare and Pandya, 2024).

Vegetative (Asexual) Propagation

Vegetative propagation, also known as asexual propagation, is a method of plant reproduction that utilizes vegetative parts of a plant such as stems, leaves, roots, or buds to generate new individuals that are genetically identical clones of the parent plant (Chauhan et al., 2021). This process preserves the exact genotype of desirable plants, ensuring that valuable traits such as superior fruit quality, flower characteristics, disease resistance, or growth habit are maintained without the genetic variability introduced by sexual reproduction (Raghavan et al., 2025). A wide range of methods are employed to achieve this. Cuttings are among the most common techniques (Grigoriev et al., 2019), involving the removal and rooting of a portion of a stem, leaf, or root in a suitable growing medium; success depends on factors like humidity, temperature, and the application of rooting hormones to stimulate adventitious root development (Pandey and Husen, 2022). Layering encourages roots to develop on a stem while it is still attached to the parent plant, as seen in methods like simple layering, air layering, or tip layering, after which the rooted section is severed to become an independent plant (Okocha and Ogbu, 2021). Division involves physically splitting a plant into multiple sections, each with its own roots and shoots, a method particularly suited for herbaceous perennials, bulbs, and rhizomes (Chauhan et al., 2021). Grafting and budding are advanced techniques where tissues from one plant (the scion or bud) are joined to the root system of another (the rootstock) (Nazir et al., 2023), combining desirable traits from both, such as disease resistance from the rootstock and high quality fruit from the scion (Shivran et al., 2022). Other methods include the use of specialized structures like runners, suckers, tubers, and bulbs (Vibha et al., 2024). Vegetative propagation allows for the rapid multiplication of plants that are difficult to grow from seed, the preservation of unique genetic hybrids, and the production of uniform crops in agriculture and horticulture (Ak et al., 2024). However, a significant drawback is the potential for monoculture, where genetic uniformity can make entire populations susceptible to pests or diseases, and the accumulation of viral infections over generations (Gibson, 2022).

Plant Tissue Culture and *In Vitro* Propagation

Plant tissue culture is a sophisticated form of asexual propagation that exploits the fundamental biological principle of totipotency the inherent capacity of a single plant cell to dedifferentiate (Malabadi et al., 2025), proliferate, and regenerate into an entire, fully functional plant when provided with the appropriate chemical and physical environment (Pandey and Chaudhary, 2025) . This process occurs under sterile (*in vitro*) conditions on a defined nutrient culture medium, allowing for the precise manipulation of growth (Phillips and Garda, 2019). The foundation of this technology lies in the principles of totipotency, where mature, differentiated plant cells can be induced to revert to a meristematic state, regaining the potential for unlimited division and differentiation (Wang et al., 2025). This reprogramming is primarily driven by a carefully balanced ratio of plant growth regulators, particularly auxins and cytokinins (Thilakarathne et al., 2025). The practical application of this principle is realized through micropropagation, a multi-stage process (Nowakowska et al., 2023). It begins with the:-

1. **establishment of sterile explants** (small tissue samples) on a medium designed to initiate growth (Beyl, 2018).
2. **multiplication stage**, where explants are transferred to a new medium; a high cytokinin-to-auxin ratio typically promotes prolific shoot proliferation, while a high auxin-to-cytokinin ratio can induce callus formation (Pandey and Chaudhary, 2025).
3. **rooting** is induced by transferring shoots to a medium enriched with auxins, which stimulates the development of adventitious roots (Druege et al., 2019).
4. **acclimatization**, where the fragile plantlets are gradually weaned from the high-humidity, heterotrophic in vitro conditions to the variable, autotrophic environment of a greenhouse or field (Shiwani et al., 2022).

The versatility of tissue culture has led to significant biotechnological extensions (Twaij et al., 2020). These include somatic embryogenesis for efficient large scale clonal propagation (Egertsdotter et al., 2019), the creation of synthetic seeds by encapsulating somatic embryos for easy storage and handling, cryopreservation for the long term conservation of valuable germplasm in liquid nitrogen (Ghosh et al., 2025), and serving as the essential platform for genetic transformation, where foreign genes are introduced into plant cells which are then regenerated into whole transgenic plants (Wang et al., 2025). Thus, tissue culture is not merely a propagation tool but a cornerstone of modern plant biotechnology, enabling rapid clonal multiplication, genetic improvement, and the preservation of plant biodiversity (Sirangi, 2025).

Biotechnological Extensions

Plant tissue culture transcends its role as a mere propagation technique (Shilpa et al., 2025), serving as an indispensable foundational platform for a suite of advanced biotechnological applications that are revolutionizing plant science and industry (Verma, 2025). One of the most significant extensions is somatic embryogenesis (De-la-Peña et al., 2015), a sophisticated process wherein individual somatic cells are reprogrammed through specific hormonal and stress treatments to develop into bipolar somatic embryos (Xu et al., 2023). These structures possess both a shoot and a root pole, mimicking their zygotic counterparts but without any genetic recombination (Leljak-Levanić et al., 2015), making this an exceptionally efficient and scalable method for large scale clonal propagation in liquid cultures and bioreactors (Aronen et al., 2025). This technology directly enables the production of synthetic seeds, a groundbreaking innovation where these somatic embryos or alternatively (Ghosh et al., 2025), apical or axillary buds are encapsulated within a protective, nutrient fortified hydrogel matrix, typically composed of compounds like sodium alginate (Nongdam, 2016) . This artificial coating acts as an artificial endosperm and seed coat, providing essential nourishment and physical protection (Pandey et al., 2025), thereby allowing for the easy handling, mechanized sowing, long-term storage, and direct delivery of clonal plants to agricultural fields (Shukla et al., 2025), circumventing the need for traditional seed production and offering a powerful solution for propagating sterile hybrids, rare species, or genetically modified plants (Tailor and Joshi, 2024). Furthermore, tissue culture provides the essential biological material for cryopreservation, the premier technology for the long term, secure conservation of plant biodiversity (Correia et al., 2024). In this process, explants like meristems, embryos, or calli are treated with cryoprotectants and stored at the ultra-low temperature of liquid nitrogen (-196°C), a state where all cellular metabolic and biochemical activities are virtually halted (Majumder et al., 2023). This suspended animation allows for the theoretically indefinite storage of valuable germplasm including endangered species, elite breeding lines, and unique genetic variants without risk of genetic erosion, somaclonal variation, or loss of viability that plagues traditional *in vitro* collections (Kumari et al., 2025), thus forming a critical global strategy for safeguarding genetic resources for future generations in cryobanks (Boes et al., 2023). Collectively, these biotechnological extensions demonstrate how tissue culture is the cornerstone of modern plant biotechnology, enabling not just multiplication, but also genetic preservation, enhancement, and innovation (Nazarov et al., 2024).

Genetic Transformation Platform

Plant tissue culture serves as the fundamental and indispensable enabler of genetic transformation (Efferth, 2019), forming the critical bridge between gene discovery and the creation of stably transformed plants (Kumlehn et al., 2018). The entire process is wholly dependent on the *in vitro* environment to manipulate, select, and regenerate genetically modified cells (Grath and Dai, 2019). The two most common methods, *Agrobacterium tumefaciens* mediated transformation and the direct DNA delivery method of the gene gun, both rely entirely on tissue culture principles (Rahman et al., 2024) . *Agrobacterium*, a natural genetic engineer, is used to transfer a segment of DNA (T-DNA) from its plasmid into the plant genome (Shreni Agrawal, 2022), but this requires a target tissue that is both susceptible to infection and capable of subsequent regeneration, such as leaf discs, cotyledons, or embryogenic callus (Long et al., 2022). Biolistics, by contrast, physically bombards cells with DNA-coated microscopic particles (e.g., gold or tungsten) (Tripathi and Shukla, 2024), forcing foreign DNA through the cell wall and membrane into the nucleus; this method is particularly vital for transforming species or tissues recalcitrant to *Agrobacterium* infection (Mangena, 2022). Following the initial gene transfer, the truly crucial role of tissue culture begins (Pamies et al., 2022) . The vast majority of cells in the target tissue are not transformed (Altpeter et al., 2016). Therefore, the next step is the selection of transformed cells (Rubin and Rubin, 2018). This is achieved by incorporating a selectable marker gene (e.g., for antibiotic or herbicide resistance) alongside the gene of interest into the transferred DNA (Dormatey et al., 2021). When the explants are placed on a culture medium containing the corresponding antibiotic or herbicide, only the rare, successfully transformed cells can survive and proliferate, while all non transformed cells die (Parzymies and Dabski, 2015). The final and most complex step is the regeneration of intact, fertile plants from these selected transformed cells (Bennur et al., 2025). This is accomplished through the precise hormonal manipulation of the culture medium to induce either organogenesis or somatic embryogenesis from a single transformed cell (Neves et al., 2021). This ensures that the entire resulting plant is clonally derived from that cell and therefore uniformly carries and expresses the transgene (Wang et al., 2025). Without the controlled conditions of tissue culture providing sterile explants, a means for selection, and a pathway for whole plant regeneration genetic engineering in plants would be impossible (Long et al., 2022). Thus, tissue culture is not merely a supporting technique but the very foundation upon which modern plant biotechnology is built (Trigiano and Gray, 2016).

Plant Hormones in Propagation

The success of both sexual and asexual plant propagation is fundamentally governed by the precise regulation of plant hormones (phytohormones) (Tang et al., 2024), which act as chemical messengers to control cellular division, differentiation, and development (Nair et al., 2019). The orchestrated balance and interaction of these hormones determine the physiological responses critical for germination, rooting, and shoot formation (Jhanji et al., 2024).

Auxins (e.g., IAA, IBA, NAA) Auxins are the primary hormones responsible for adventitious root initiation and are consequently indispensable in cutting propagation (Liu et al., 2024). They stimulate the differentiation of vascular tissues and promote cell elongation (Velasquez et al., 2016). Indole-3-acetic acid (IAA) is the naturally occurring form, but it is easily broken down by light and enzymes (Duca and Glick, 2020). Therefore, synthetic analogs like Indole-3-butyric acid (IBA) and 1-Naphthaleneacetic acid (NAA) are widely used in commercial rooting powders and solutions due to their stability and efficacy (Pincelli-Souza et al., 2024). Auxins exert their effects by redirecting nutrients and energy to the site of application inducing root primordia formation (Vanneste et al., 2025). In tissue culture, a high auxin-to-cytokinin ratio is essential for the rooting phase of micropropagation (Adu Donyina et al., 2025).

Cytokinins (e.g., Kinetin, BAP, Zeatin) Cytokinins are pivotal for cell division and shoot proliferation (Svolacchia and Sabatini, 2023). They work antagonistically to auxins; where a high cytokinin-to-auxin ratio promotes shoot formation from callus or lateral buds by overcoming apical dominance (Bennur et al., 2025). In micropropagation cytokinins like Benzylaminopurine (BAP) or Kinetin are added to the culture medium to stimulate the development of multiple shoots from a single explant (Martins et al., 2022), leading to rapid multiplication (Li et al., 2022). They also play a key role in delaying senescence (aging) by promoting nutrient mobilization (Zhou and Yang, 2023), which helps maintain the vitality of plant tissues during the propagation process (Awotedu et al., 2021).

Gibberellins (e.g., GA_3)

Gibberellins are crucial for breaking seed dormancy (Yang et al., 2020) and promoting seed germination (Al-Tai and Mohammed, 2024). They achieve this by stimulating the production of hydrolytic enzymes like amylase (Farooq et al., 2021), which break down stored reserves in the endosperm to provide energy for the growing embryo (Yan et al., 2014). Additionally, gibberellins are potent promoters of internode elongation, which is vital for the rapid growth of young seedlings (Castro-Camba et al., 2022). In some species they can also be used to promote flowering, though this is less common in standard propagation protocols (Yasemin and Beruto, 2024).

Abscisic Acid (ABA) Abscisic Acid acts as a growth inhibitor and is the primary hormone responsible for maintaining seed and bud dormancy (Sabir et al., 2025). It counteracts the effects of gibberellins and promotes tolerance to various environmental stresses, such as drought, by closing stomata (Abbas et al., 2023). In propagation a decline in ABA levels is often necessary for germination to proceed or for buds to break dormancy (Sabir et al., 2025). Managing ABA is therefore critical in strategies like stratification, where cold treatment reduces ABA sensitivity to allow for subsequent growth (Abley et al., 2024).

Ethylene

Ethylene is a gaseous hormone with a diverse range of effects on plant development (Dubois et al., 2018). It can either promote or inhibit adventitious root formation depending on the species and concentration (Baharudin and Osman, 2023). It is also involved in leaf abscission (leaf drop), fruit ripening, and the induction of callus formation (Tipu and Sherif, 2024). In tissue culture ethylene accumulation in culture vessels can be detrimental often inhibiting growth and development (Pasternak and Steinmacher, 2024), which is why ventilation or ethylene absorbents are sometimes necessary (Kumar et al., 2024).

Genetic Basis of Propagation

The fundamental processes of plant propagation, whether sexual or asexual (Duke et al., 2018), are governed by intricate genetic principles that determine both the variability and the fidelity of the resulting plants (Biswas and Kumar, 2023). In sexual propagation, Mendelian inheritance plays a crucial role dictating the segregation and expression of traits in offspring seeds (Muthazhagu Kuppuraj et al., 2025). This genetic recombination during meiosis ensures diversity (Johnston, 2024), influencing essential characteristics such as seed vigor, disease resistance, and adaptive potential, though it can also lead to unpredictable trait combinations (Reed et al., 2022). Many critical propagation traits including germination rate depth of dormancy and rooting ability are not controlled by single genes but are instead polygenic (Ma and Li, 2024), influenced by the complex interaction of multiple genes and environmental factors, making them a key focus for selective breeding programs (Pathak et al., 2025). In contrast, asexual propagation aims to preserve genetic identity, producing clones that are genetically identical to the parent plant (Paul, 2024). However, maintaining clonal fidelity is paramount, as somaclonal variations unintended genetic mutations can arise during *in vitro* culture (Majumder et al., 2025). To ensure genetic integrity advanced molecular markers such as Simple Sequence Repeats (SSRs), Amplified Fragment Length Polymorphisms (AFLPs), and Single Nucleotide Polymorphisms (SNPs) are employed to authenticate clones (Jajoriya et al., 2025), detect off-types, and confirm genetic stability across propagation cycles (Kovincic et al., 2023). Beyond monitoring, modern biotechnological tools have revolutionized propagation practices (Hayat et al., 2025). CRISPR-Cas genome editing enables precise targeted modifications to improve propagation efficiency (Li et al., 2024), such as editing genes responsible for dormancy to achieve faster germination Rachappanavar, 2025), enhancing rooting capabilities (Boudadi et al., 2025), or introducing traits for abiotic stress tolerance (Gilliham et al., 2017). This powerful technology allows for the development of elite plant varieties with optimized propagation characteristics without introducing foreign DNA (Sahin et al., 2024), streamlining both traditional breeding and clonal propagation systems (Banadka et al., 2025). Thus, the genetic basis of propagation encompasses a spectrum from classical inheritance to cutting edge molecular biology (Sun et al., 2024), providing the foundation for enhancing both the diversity and uniformity of plants to meet agricultural, horticultural, and conservation goals (Ochar and Kim, 2023).

Environmental Factors Influencing Propagation

The success of plant propagation, whether through sexual or asexual means, is profoundly influenced by a complex interplay of environmental factors that directly regulate physiological processes at every stage (Guo et al., 2025).

Light serves as both an energy source and a developmental signal (Wu et al., 2025); its intensity, quality, and photoperiod dictate photomorphogenic responses such as seed germination, chlorophyll synthesis, and root initiation (Razzaq and Du, 2024). For instance, many seeds require specific light wavelengths to break dormancy (Yang et al., 2020), while cuttings often root more successfully under controlled light conditions that optimize photosynthesis without causing desiccation (De Almeida et al., 2017).

Temperature acts as a master regulator of metabolic and enzymatic activity (Arcus et al., 2016), influencing seed dormancy release through stratification, determining germination rates, and affecting cell division and differentiation in tissue culture (Stoian-Dod et al., 2023). Each species has optimal temperature ranges for propagation, and deviations can lead to poor rooting, microbial contamination, or developmental abnormalities (Abdalla et al., 2022).

Humidity is critical for maintaining tissue turgor and preventing desiccation (Ievinsh, 2023), particularly in cuttings and during the acclimatization of *in vitro* derived plantlets (Grzelak et al., 2024). High humidity reduces transpirational water loss in vulnerable tissues lacking fully developed cuticles or functional root systems (Tafolla-Arellano et al., 2018), though it must be carefully balanced to avoid pathogenic growth. Beyond these physical factors (Nega, 2014)

soil microbiota, particularly beneficial organisms like mycorrhizal fungi and plant growth promoting rhizobacteria (Li et al., 2023), form symbiotic relationships that significantly enhance propagation success by improving nutrient and water uptake (Radi et al., 2025), producing growth stimulating phytohormones (Sultan et al., 2024) and providing systemic resistance against pathogens (Ali et al., 2024).

Response to climate change, understanding and selecting for abiotic stress tolerance including resilience to drought, salinity, and temperature extremes has become a central focus of propagation research (Kapazoglou et al., 2025). Modern techniques increasingly aim to develop propagules pre adapted to harsh conditions (Clements and Jones, 2021), leveraging both traditional hardening practices and biotechnological tools to ensure survival and growth in challenging environments (Parikh et al., 2025). Thus, effective propagation requires not only the control of these external conditions but also an integrated approach that harnesses biological interactions and anticipates future climatic challenges (Salse et al., 2024).

CONCLUSION

The scientific foundations of plant propagation integrate classical knowledge with cutting-edge biotechnology. Sexual propagation ensures diversity and adaptability, while vegetative methods maintain clonal fidelity. Tissue culture and totipotency provide tools for mass production and conservation, guided by hormonal and genetic regulation. Anatomical, histological, and physiological processes form the basis of successful regeneration, while environmental factors determine practical outcomes. As global challenges intensify, the refinement of propagation science will remain central to agricultural innovation, biodiversity conservation, and sustainable development.

REFERENCES

Abbas, K., Li, J., Gong, B., Lu, Y., Wu, X., Lü, G., & Gao, H. (2023). Drought stress tolerance in vegetables: The functional role of structural features, key gene pathways, and exogenous hormones. *International Journal of Molecular Sciences, 24*(18), 13876. DOI: 10.3390/ijms241813876 PMID: 37762179

Abdalla, N., El-Ramady, H., Seliem, M. K., El-Mahrouk, M. E., Taha, N., Bayoumi, Y., Shalaby, T. A., & Dobránszki, J. (2022). An academic and technical overview on plant micropropagation challenges. *Horticulturae*, *8*(8), 677. DOI: 10.3390/horticulturae8080677

Abley, K., Goswami, R., and Locke, J. C. (2024). Bet-hedging and variability in plant development: seed germination and beyond. *Philosophical Transactions of the Royal Society B, 379*(1900), 20230048.

Absakine, S. I., Tchimbi, B., and Ousman, A. H. (2024). The Difficulty of Chadian Forest Tree Seed Germination: Overcoming Physical Dormancy. *African Journal of Advanced Pure and Applied Sciences (AJAPAS)*, 98-109.

Adu Donyina, G., Szarvas, A., Opoku, V. A., Miko, E., Tar, M., Czóbel, S., & Monostori, T. (2025). Enhancing sweet potato production: A comprehensive analysis of the role of auxins and cytokinins in micropropagation. *Planta*, *261*(4), 1–14. DOI: 10.1007/s00425-025-04650-z PMID: 40035890

Ak, B. E., Hatipoğlu, İ. H., & Hatipoğlu, M. (2024). Modern Methods in Ornamental Plant Propagation: Improving Resistance and Sustainability. *Endangered Species*, *1*(2), 3.

Al-Tai, A.A., Mohammed, A.A.H. (2024). Germination of lavender (*Lavandula angustifolia*) plant seeds and production of its seedlings callus cultures. *Malaysian Journal of Microbiology*. 20 (Special issue 6): 36 - 39.

Ali, A. S., and Elozeiri, A. A. (2017). Metabolic processes during seed germination. *Advances in seed biology, 2017*, 141-166.

Ali, J., Mukarram, M., Ojo, J., Dawam, N., Riyazuddin, R., Ghramh, H. A., Khan, K. A., Chen, R., Kurjak, D., & Bayram, A. (2024). Harnessing phytohormones: Advancing plant growth and defence strategies for sustainable agriculture. *Physiologia Plantarum*, *176*(3), e14307. DOI: 10.1111/ppl.14307 PMID: 38705723

Altpeter, F., Springer, N. M., Bartley, L. E., Blechl, A. E., Brutnell, T. P., Citovsky, V., Conrad, L., Gelvin, S. B., Jackson, D., Kausch, A. P., Lemaux, P. G., Medford, J. I., Orozo-Cardenas, M., Tricoli, D., VanEck, J., Voytas, D. F., Walbot, V., Wang, K., Zhang, Z. J., & Stewart, C. N.Jr. (2016). Advancing crop transformation in the era of genome editing. *The Plant Cell*, *28*(7), 1510–1520. DOI: 10.1105/tpc.16.00196 PMID: 27335450

Arcus, V. L., Prentice, E. J., Hobbs, J. K., Mulholland, A. J., Van der Kamp, M. W., Pudney, C. R., Parker, E. J., & Schipper, L. A. (2016). On the temperature dependence of enzyme-catalyzed rates. *Biochemistry*, *55*(12), 1681–1688. DOI: 10.1021/acs.biochem.5b01094 PMID: 26881922

Aronen, T., Varis, S., & Tikkinen, M. (2025). Somatic embryogenesis: concept, principles, and applications. In *Forest Microbiology* (pp. 373–388). Academic press. DOI: 10.1016/B978-0-443-21903-0.00022-9

Arpitha, K., & Navya, V. (2025). SEED DORMANCY AND IT'S MECHANISM. *Seed Biology and Biotechnology*, 18.

Awasthi, P. (2023). Review on dormancy, causes, uses, and measures of overcoming it. *Science*, *7*(2), 24–27.

Awotedu, B. F., Omolola, T. O., Akala, A. O., Awotedu, O. L., & Olaoti-Laaro, S. O. (2021). Vegetative propagation: A unique technique of improving plants growth. *World News of Natural Sciences*, *35*, 83–101.

Baharudin, N. F., & Osman, N. I. (2023). Plant development, stress responses, and secondary metabolism under ethylene regulation. *Plant Stress*, *7*, 100146. DOI: 10.1016/j.stress.2023.100146

Banadka, A., Toppo, V., Kashyap, K., Nandhini, M., Akshita, B., Sudheer, W. N., & Nagella, P. (2025). Revolutionizing Ashwagandha Breeding: From Traditional to Modern Breeding Techniques for Indian Ginseng, Withania somnifera (L.) Dunal. In *Biodiversity and Genetic Improvement of Medicinal and Aromatic Plants II* (pp. 1–34). Springer. DOI: 10.1007/978-3-031-81857-8_1

Barrero, J. M., Buitink, J., Chen, S. C., Da Silva, E. A. A., Hay, F. R., Kawakami, N., Leprince, O., Ojiewo, C. O., Pawłowski, T. A., Pérez, H. E., & Vijay, D. (2025). Celebrating the 25th Anniversary of the International Society for Seed Science. *Seed Science Research*, *35*(1), 1–8. DOI: 10.1017/S0960258525000029

Bennur, P. L., O'Brien, M., Fernando, S. C., & Doblin, M. S. (2025). Improving transformation and regeneration efficiency in medicinal plants: Insights from other recalcitrant species. *Journal of Experimental Botany*, *76*(1), 52–75. DOI: 10.1093/jxb/erae189 PMID: 38652155

Beyl, C. A. (2018). Getting started with tissue culture—media preparation, sterile technique, and laboratory equipment. In *Plant tissue culture concepts and laboratory exercises* (pp. 21–38). Routledge.

Bhatla, S. C., & Kathpalia, R. (2023). Seed dormancy and germination. In *Plant physiology, development and metabolism* (pp. 625–640). Springer Nature Singapore. DOI: 10.1007/978-981-99-5736-1_28

Bhatla, S. C., & Shakya, R. (2023). Pollination, fertilization, and seed development. In *Plant physiology, development and metabolism* (pp. 583–605). Springer Nature Singapore. DOI: 10.1007/978-981-99-5736-1_26

Bhattacharya, A. (2022). Effect of low-temperature stress on germination, growth, and phenology of plants: A review. *Physiological processes in plants under low temperature stress*, 1-106.

Biswas, P., and Kumar, N. (2023). Application of molecular markers for the assessment of genetic fidelity of in vitro raised plants: Current status and future prospects. *Molecular Marker Techniques: A Potential Approach of Crop Improvement*, 233-256.

Boes, J., Boettcher, P., & Honkatukia, M. (2023). *Innovations in cryoconservation of animal genetic resources: practical guide*. Food and Agriculture Org.

Boudadi, I., El Merzougui, S., Lachguer, K., Lagram, K., El Boullani, R., & Serghini, M. A. (2025). Enhancing in vitro rooting and acclimatization of Spathiphyllum wallisii through explant size, NAA, and substrate selection. [PCTOC]. *Plant Cell, Tissue and Organ Culture*, *160*(2), 38. DOI: 10.1007/s11240-025-02987-7

Castro-Camba, R., Sánchez, C., Vidal, N., & Vielba, J. M. (2022). Plant development and crop yield: The role of gibberellins. *Plants*, *11*(19), 2650. DOI: 10.3390/plants11192650 PMID: 36235516

Chauhan, A., Sharma, D., Kumar, R., Shiwani, K., & Sharma, N. (2021). *Methods of propagation in vegetable crops. Recent trends in propagation of forest and horticultural crops*. Taran Publication.

Chauhan, A., Sharma, D., Kumar, R., Shiwani, K., & Sharma, N. (2021). *Methods of propagation in vegetable crops. Recent trends in propagation of forest and horticultural crops*. Taran Publication.

Clements, D. R., & Jones, V. L. (2021). Ten ways that weed evolution defies human management efforts amidst a changing climate. *Agronomy (Basel)*, *11*(2), 284. DOI: 10.3390/agronomy11020284

Corbineau, F. (2022). Oxygen, a key signalling factor in the control of seed germination and dormancy. *Seed Science Research*, *32*(3), 126–136. DOI: 10.1017/S096025852200006X

Correia, S., Matos, M., & Leal, F. (2024). Advances in blueberry (Vaccinium spp.) in vitro culture: A review. *Horticulturae*, *10*(6), 533. DOI: 10.3390/horticulturae10060533

COST, R. (2025). Integrative approaches to enhance reproductive resilience of crops for climate-proof agriculture. *Plant Stress*, •••, 15.

De Almeida, M. R., Aumond, M.Jr, Da Costa, C. T., Schwambach, J., Ruedell, C. M., Correa, L. R., & Fett-Neto, A. G. (2017). Environmental control of adventitious rooting in Eucalyptus and Populus cuttings. *Trees (Berlin)*, *31*(5), 1377–1390. DOI: 10.1007/s00468-017-1550-6

De-la-Peña, C., Nic-Can, G. I., Galaz-Ávalos, R. M., Avilez-Montalvo, R., & Loyola-Vargas, V. M. (2015). The role of chromatin modifications in somatic embryogenesis in plants. *Frontiers in Plant Science*, *6*, 635. DOI: 10.3389/fpls.2015.00635 PMID: 26347757

Dormatey, R., Sun, C., Ali, K., Fiaz, S., Xu, D., Calderón-Urrea, A., Bi, Z., Zhang, J., & Bai, J. (2021). ptxD/Phi as alternative selectable marker system for genetic transformation for bio-safety concerns: A review. *PeerJ*, *9*, e11809. DOI: 10.7717/peerj.11809 PMID: 34395075

Druege, U., Hilo, A., Pérez-Pérez, J. M., Klopotek, Y., Acosta, M., Shahinnia, F., Zerche, S., Franken, P., & Hajirezaei, M. R. (2019). Molecular and physiological control of adventitious rooting in cuttings: Phytohormone action meets resource allocation. *Annals of Botany*, *123*(6), 929–949. DOI: 10.1093/aob/mcy234 PMID: 30759178

Dubois, M., Van den Broeck, L., & Inzé, D. (2018). The pivotal role of ethylene in plant growth. *Trends in Plant Science*, *23*(4), 311–323. DOI: 10.1016/j.tplants.2018.01.003 PMID: 29428350

Duca, D. R., & Glick, B. R. (2020). Indole-3-acetic acid biosynthesis and its regulation in plant-associated bacteria. *Applied Microbiology and Biotechnology*, *104*(20), 8607–8619. DOI: 10.1007/s00253-020-10869-5 PMID: 32875364

Duke, S. O., Kigel, J., & Koller, D. (2018). Asexual reproduction of weeds. In *Weed physiology* (pp. 65–100). CRC Press. DOI: 10.1201/9781351077743-3

Efferth, T. (2019). Biotechnology applications of plant callus cultures. *Engineering (Beijing)*, *5*(1), 50–59. DOI: 10.1016/j.eng.2018.11.006

Egertsdotter, U., Ahmad, I., & Clapham, D. (2019). Automation and scale up of somatic embryogenesis for commercial plant production, with emphasis on conifers. *Frontiers in Plant Science*, *10*, 109. DOI: 10.3389/fpls.2019.00109 PMID: 30833951

Farooq, M. A., Ali, S., Hassan, A., Tahir, H. M., Mumtaz, S., & Mumtaz, S. (2021). Biosynthesis and industrial applications of α-amylase: A review. *Archives of Microbiology*, *203*(4), 1281–1292. DOI: 10.1007/s00203-020-02128-y PMID: 33481073

Fuller, E., Germaine, K. J., & Rathore, D. S. (2023). The good, the bad, and the useable microbes within the common Alder (Alnus glutinosa) microbiome—Potential bio-agents to combat Alder dieback. *Microorganisms*, *11*(9), 2187. DOI: 10.3390/microorganisms11092187 PMID: 37764031

Ghosh, T., Samai, I., Debnath, S., Singh, A. P., & Atta, K. (2025). Synthetic Seeds: Innovation in Plant Propagation and Conservation for Sustainable Agriculture. In *Advances in Seed Quality Evaluation and Improvement* (pp. 431–446). Springer Nature Singapore. DOI: 10.1007/978-981-96-5178-8_17

Ghosh, T., Samai, I., Debnath, S., Singh, A. P., & Atta, K. (2025). Synthetic Seeds: Innovation in Plant Propagation and Conservation for Sustainable Agriculture. In *Advances in Seed Quality Evaluation and Improvement* (pp. 431–446). Springer Nature Singapore. DOI: 10.1007/978-981-96-5178-8_17

Gibson, A. K. (2022). Genetic diversity and disease: The past, present, and future of an old idea. *Evolution; International Journal of Organic Evolution*, *76*(S1, s1), 20–36. DOI: 10.1111/evo.14395 PMID: 34796478

Gilliham, M., Able, J. A., & Roy, S. J. (2017). Translating knowledge about abiotic stress tolerance to breeding programmes. *The Plant Journal*, *90*(5), 898–917. DOI: 10.1111/tpj.13456 PMID: 27987327

Gómez, J. M., Schupp, E. W., & Jordano, P. (2019). Synzoochory: The ecological and evolutionary relevance of a dual interaction. *Biological Reviews of the Cambridge Philosophical Society*, *94*(3), 874–902. DOI: 10.1111/brv.12481 PMID: 30467946

Grath, A., & Dai, G. (2019). Direct cell reprogramming for tissue engineering and regenerative medicine. *Journal of Biological Engineering*, *13*(1), 14. DOI: 10.1186/s13036-019-0144-9 PMID: 30805026

Green, A. J., & Wilkinson, D. M. (2024). Darwin's digestion myth: Historical and modern perspectives on our understanding of seed dispersal by waterbirds. *Seeds (New York, N.Y.), 3*(4), 505–527. DOI: 10.3390/seeds3040034

Grigoriev, S. N., Fedorov, S. V., & Hamdy, K. (2019). Materials, properties, manufacturing methods and cutting performance of innovative ceramic cutting tools– a review. *Manufacturing Review, 6*, 19. DOI: 10.1051/mfreview/2019016

Grzelak, M., Pacholczak, A., & Nowakowska, K. (2024). Challenges and insights in the acclimatization step of micropropagated woody plants. [PCTOC]. *Plant Cell, Tissue and Organ Culture, 159*(3), 72. DOI: 10.1007/s11240-024-02923-1

Guo, Y. W., Liu, Y., Huang, P. C., Rong, M., Wei, W., Xu, Y. H., & Wei, J. H. (2025). Adaptive Changes and Genetic Mechanisms in Organisms Under Controlled Conditions: A Review. *International Journal of Molecular Sciences, 26*(5), 2130. DOI: 10.3390/ijms26052130 PMID: 40076752

Hayat, M. K., Khan, T. Y., Sial, T. N., Khan, H. A., Ashraf, Z. U., Sheraz, M., & Belizaire, P. J. (2025). The Biotech And Precision Agriculture Revolution: Feeding the World Sustainably. *Policy Research Journal, 3*(3), 293–316.

Ievinsh, G. (2023). Water content of plant tissues: So simple that almost forgotten? *Plants, 12*(6), 1238. DOI: 10.3390/plants12061238 PMID: 36986926

Jain, S., Singh, H., Rathod, M., Meena, R., Deshmukh, R. N., Mohapatra, A., Sharma, R., & Anushi, . (2023). Preserving for the future: The critical role of germplasm conservation in fruit crop resilience. *Int J Environ Clim Change, 13*(11), 4651–4661. DOI: 10.9734/ijecc/2023/v13i113644

Jajoriya, R., Sharma, R., Meena, V. K., Vadithya, A. S., Singh, K., Kumar, P., & Nathawat, T. S. (2025). EMERGENCE OF MOLECULAR MARKER FROM RFLP TO SNP: A REVIEW. *Plant Archives, 25*(1), 2062–2074.

Jhanji, S., Goyal, E., Chumber, M., & Kaur, G. (2024). Exploring fine tuning between phytohormones and ROS signaling cascade in regulation of seed dormancy, germination and seedling development. *Plant Physiology and Biochemistry, 207*, 108352. DOI: 10.1016/j.plaphy.2024.108352 PMID: 38266558

Jiang, Z., Li, Y. P., Gai, P. Z., Gao, J., & Xu, L. (2025). Exogenously applied ABA alleviates dysplasia of maize (Zea mays L.) ear under drought stress by altering photosynthesis and sucrose transport. *Plant Signaling & Behavior, 20*(1), 2462497. DOI: 10.1080/15592324.2025.2462497 PMID: 39907639

Jin, L., Sharpe, A. G., Ji, Y., Hewavithana, T., Sharpe, A. G., and Jin, L. (2025). Yuanyuan Ji1, 2, Thulani Hewavithana1, Andrew G. Sharpe2* and Lingling Jin1. *Omics in Seed Development: Challenges and Opportunities for Improving of Seed Quality and Yield in Model and Crop Plants.*

Johnston, S. E. (2024). Understanding the genetic basis of variation in meiotic recombination: Past, present, and future. *Molecular Biology and Evolution, 41*(7), msae112. DOI: 10.1093/molbev/msae112 PMID: 38959451

Jordano, P. (2017). What is long-distance dispersal? And a taxonomy of dispersal events. *Journal of Ecology, 105*(1), 75–84. DOI: 10.1111/1365-2745.12690

Joshi, B. K., Ghimire, K. H., Neupane, S. P., Gauchan, D., & Mengistu, D. K. (2023). Approaches and advantages of increased crop genetic diversity in the fields. *Diversity, 15*(5), 603. DOI: 10.3390/d15050603

Kandar, C. C., & Pal, D. (2024). Relation Between Seed Life Cycle and Cell Proliferation. Metabolic Changes in Seed Germination. In *Seeds: anti-proliferative storehouse for bioactive secondary metabolites* (pp. 49–79). Springer Nature Singapore. DOI: 10.1007/978-981-97-3014-8_2

Kapazoglou, A., Tani, E., Papasotiropoulos, V., Letsiou, S., Gerakari, M., Abraham, E., & Bebeli, P. J. (2025). Enhancing abiotic stress resilience in Mediterranean woody perennial fruit crops: Genetic, epigenetic, and microbial molecular perspectives in the face of climate change. *International Journal of Molecular Sciences, 26*(7), 3160. DOI: 10.3390/ijms26073160 PMID: 40243927

Khalimova, Y. S. (2024). Features of Sperm Development: Spermatogenesis and Fertilization. *American Journal of Bioscience and Clinical Integrity, 1*(11), 90–98.

Klupczyńska, E. A., & Pawłowski, T. A. (2021). Regulation of seed dormancy and germination mechanisms in a changing environment. *International Journal of Molecular Sciences, 22*(3), 1357. DOI: 10.3390/ijms22031357 PMID: 33572974

Kovincic, A., Markovic, K., Ristic, D., Babic, V., Petrovic, T., Zivanovic, T., & Kravic, N. (2023). Efficiency of biological typing methods in maize hybrid genetic purity estimation. *Genes, 14*(6), 1195. DOI: 10.3390/genes14061195 PMID: 37372375

Kumar, S., Kumar, R., Bibwe, B. R., Nath, P., Singh, R. K., Mandhania, S., Pal, A., Soni, R., & Kumar, A. (2024). Postharvest handling of ethylene with oxidative and absorptive means. *Journal of Food Science and Technology, 61*(5), 813–832. DOI: 10.1007/s13197-023-05777-1 PMID: 38487289

Kumari, A., Sharma, L., Yadav, Y., Karma, A., Singh, J., & Arya, S. C. (2025). Biotechnological Strategies for Seed Germplasm Maintenance and Plant Genetic Resource Conservation. In *Advances in Seed Quality Evaluation and Improvement* (pp. 379–403). Springer Nature Singapore. DOI: 10.1007/978-981-96-5178-8_15

Kumlehn, J., Pietralla, J., Hensel, G., Pacher, M., & Puchta, H. (2018). The CRISPR/Cas revolution continues: From efficient gene editing for crop breeding to plant synthetic biology. *Journal of Integrative Plant Biology*, *60*(12), 1127–1153. DOI: 10.1111/jipb.12734 PMID: 30387552

Lakehal, A., & Bellini, C. (2019). Control of adventitious root formation: Insights into synergistic and antagonistic hormonal interactions. *Physiologia Plantarum*, *165*(1), 90–100. DOI: 10.1111/ppl.12823 PMID: 30159890

Lamont, B. B., & Pausas, J. G. (2023). Seed dormancy revisited: Dormancy-release pathways and environmental interactions. *Functional Ecology*, *37*(4), 1106–1125. DOI: 10.1111/1365-2435.14269

Leljak-Levanić, D., Mihaljević, S., & Bauer, N. (2015). Somatic and zygotic embryos share common developmental features at the onset of plant embryogenesis. *Acta Physiologiae Plantarum*, *37*(7), 127. DOI: 10.1007/s11738-015-1875-y

Lewsey, M. G., Bassel, G. W., & Whelan, J. (2025). Dynamic and spatial control of cellular activity during seed germination. *Current Opinion in Plant Biology*, *86*, 102754. DOI: 10.1016/j.pbi.2025.102754 PMID: 40555155

Li, B., Sun, C., Li, J., & Gao, C. (2024). Targeted genome-modification tools and their advanced applications in crop breeding. *Nature Reviews. Genetics*, *25*(9), 603–622. DOI: 10.1038/s41576-024-00720-2 PMID: 38658741

Li, W., Li, W. B., Xing, L. J., & Guo, S. X. (2023). Effect of arbuscular mycorrhizal fungi (AMF) and plant growth-promoting rhizobacteria (PGPR) on microorganism of phenanthrene and pyrene contaminated soils. *International Journal of Phytoremediation*, *25*(2), 240–251. DOI: 10.1080/15226514.2022.2071832 PMID: 35549569

Li, Y. Y., Hao, Z. G., Miao, S., Zhang, X., Li, J. Q., Guo, S. X., & Lee, Y. I. (2022). Profiles of cytokinins metabolic genes and endogenous cytokinins dynamics during shoot multiplication in vitro of Phalaenopsis. *International Journal of Molecular Sciences*, *23*(7), 3755. DOI: 10.3390/ijms23073755 PMID: 35409120

Liu, S., Li, X., Xu, L., & Zhang, G. (2024). Hormone functions in adventitious root formation during cutting propagation of woody plants. *Journal of Plant Research*, •••, 1–8. DOI: 10.1007/s10265-024-01602-8 PMID: 39656343

Lokare, P. D., & Pandya, J. B. (2024). *Entrepreneurial Botany*. Book Saga Publications. DOI: 10.60148/978-8197603839

Long, Y., Yang, Y., Pan, G., & Shen, Y. (2022). New insights into tissue culture plant-regeneration mechanisms. *Frontiers in Plant Science, 13*, 926752. DOI: 10.3389/fpls.2022.926752 PMID: 35845646

Ma, J., Wang, Y., Chen, S., & Xu, T. (2025). The Central Role of Auxin in Orchestrating Apical Stem Cells in Plants. *Plant, Cell & Environment, 48*(7), 5053–5072. DOI: 10.1111/pce.15464 PMID: 40152539

Ma, Y., & Li, Y. (2024). The research progress on genomic selection and breeding for drought resistance in wheat: From genetic analysis to smart breeding applications. *Geographical Research Bulletin, 3*, 625–646.

Majumder, A., Roychowdhury, D., and Ray, S. (2023). Biotechnological approaches for ex situ conservation of medicinal plants. *Medicinal Plants: Biodiversity, Biotechnology and Conservation*, 729-800.

Majumder, S., Igamberdiev, A. U., & Debnath, S. C. (2025). Somaclonal Variation and Clonal Fidelity in Commercial Micropropagation: Challenges and Perspectives. *Agronomy (Basel), 15*(6), 1489. DOI: 10.3390/agronomy15061489

Malabadi, R. B., Chalannavar, R. K., & Kolkar, K. P.Ravindra B. MalabadiRaju K. ChalannavarKiran P. Kolkar. (2025). Plant cell totipotency: Plant tissue culture applications-An updated review. *World Journal of Advanced Engineering Technology and Sciences, 16*(02), 112–135. DOI: 10.30574/wjaets.2025.16.2.1262

Malviya, R., & Gayen, D. (2024). Seed Deterioration: Unraveling the Role of Phytohormones on Seed Germination Under Aging Condition. *Journal of Plant Growth Regulation*, •••, 1–17.

Mangena, P. (2022). Plant transformation via Agrobacterium tumefaciens: culture conditions, recalcitrance and advances in soybean.

Martins, J. P. R., Wawrzyniak, M. K., Ley-López, J. M., Kalemba, E. M., Mendes, M. M., & Chmielarz, P. (2022). 6-Benzylaminopurine and kinetin modulations during in vitro propagation *of Quercus robur* (L.): An assessment of anatomical, biochemical, and physiological profiling of shoots. [PCTOC]. *Plant Cell, Tissue and Organ Culture, 151*(1), 149–164. DOI: 10.1007/s11240-022-02339-9

Mehmood, M., Tanveer, N. A., Joyia, F. A., Ullah, I., & Mohamed, H. I. (2025). Effect of high temperature on pollen grains and yield in economically important crops: A review. *Planta, 261*(6), 141. DOI: 10.1007/s00425-025-04714-0 PMID: 40374974

Morris, S. J., & Blackwood, C. B. (2024). The ecology of soil biota and their function. In *Soil microbiology, ecology and biochemistry* (pp. 275–302). Elsevier. DOI: 10.1016/B978-0-12-822941-5.00010-7

Muthazhagu Kuppuraj, S. A., Chokkalingam, Y., Jothiganapathy, K., Vedachalam, V., Bisht, D. S., Cholin, S., & Saminadane, T. (2025). Maternal effect on the inheritance of pericarp colour and grain dimension in rice (*Oryza sativa* L.). *Journal of Genetics*, *104*(1), 5. DOI: 10.1007/s12041-024-01490-w PMID: 39959938

Nair, A., Chauhan, P., Saha, B., & Kubatzky, K. F. (2019). Conceptual evolution of cell signaling. *International Journal of Molecular Sciences*, *20*(13), 3292. DOI: 10.3390/ijms20133292 PMID: 31277491

Nautiyal, P. C., Sivasubramaniam, K., & Dadlani, M. (2023). Seed dormancy and regulation of germination. *Seed Science and Technology*, *52*, 39–66.

Nazarov, A., Kulikova, E., & Molokova, E. (2024). Economic security through technological advancements in agriculture: A pathway to sustainable agro-industrial growth. In *BIO Web of Conferences* (Vol. 121, p. 02012). EDP Sciences. DOI: 10.1051/bioconf/202412102012

Nazir, N., Yaseen, I., Been, T. J., Khalil, A., Iqbal, U., Mir, M. M., & Bhat, S. (2023). Improved Propagation Techniques in Temperate Nuts. In *Temperate Nuts* (pp. 113–148). Springer Nature Singapore. DOI: 10.1007/978-981-19-9497-5_5

Nega, A. (2014). Review on concepts in biological control of plant pathogens. *Journal of Biology, Agriculture and Healthcare*, *4*(27), 33–54.

Neves, M., Correia, S., Cavaleiro, C., & Canhoto, J. (2021). Modulation of organogenesis and somatic embryogenesis by ethylene: An overview. *Plants*, *10*(6), 1208. DOI: 10.3390/plants10061208 PMID: 34198660

Nongdam, P. (2016). Development of synthetic seed technology in plants and its applications: A review. *International Journal of Current Science*, *19*(4), 86–101.

Nowakowska, K., Nongdam, P., Samsurizal, N. A., & Pacholczak, A. (2023). An efficient micropropagation protocol for the endangered European shrub February daphne (Daphne mezereum L.) and Identification of bacteria in culture. *Agriculture*, *13*(9), 1692. DOI: 10.3390/agriculture13091692

Ochar, K., & Kim, S. H. (2023). Conservation and global distribution of onion (Allium cepa L.) germplasm for agricultural sustainability. *Plants*, *12*(18), 3294. DOI: 10.3390/plants12183294 PMID: 37765458

Okocha, O. I., and Ogbu, J. U. (2021). Horticultural crops production. *Agricultural technology for colleges, 151.*

Osland, M. J., Chivoiu, B., Feher, L. C., Dale, L. L., Lieurance, D., Daniel, W. M., & Spencer, J. E. (2023). Plant migration due to winter climate change: Range expansion of tropical invasive plants in response to warming winters. *Biological Invasions*, *25*(9), 2813–2830. DOI: 10.1007/s10530-023-03075-7

Pamies, D., Leist, M., Coecke, S., Bowe, G., Allen, D., Gstraunthaler, G., ... and Stacey, G. (2022). Guidance document on good cell and tissue culture practice 2.0 (GCCP 2.0).

Pandey, D. K., & Chaudhary, B. (Eds.). (2025). *Regeneration Potential in Plants.* CRC Press. DOI: 10.1201/9781032702391

Pandey, S., & Husen, A. (2022). Role of plant stem or shoot cutting positions and hormone treatments in adventitious root formation. In *Environmental, Physiological and Chemical Controls of Adventitious Rooting in Cuttings* (pp. 367-379). Academic Press. DOI: 10.1016/B978-0-323-90636-4.00009-X

Pandey, V. K., Shafi, Z., Choudhary, P., Pathak, A., & Rustagi, S. (2025). Navigating the Functional and Bioactive Landscape of Sodium Alginate-Based Edible Coatings for Modernized Fruit Preservation. *Journal of Food Safety*, *45*(3), e70020. DOI: 10.1111/jfs.70020

Parikh, H. S., Dave, G., Singh, P. K., & Tiwari, A. (2025). Exploring the significance of xenic and axenic cultures for diatoms: Methodological insights, scope, and biotechnological implications. *Systems Microbiology and Biomanufacturing*, *5*(3), 1–21. DOI: 10.1007/s43393-025-00363-2

Parzymies, M., & Dabski, M. (2015). Methods of eliminating contaminations in tissue cultures on the example of selected ornamental plant species. *BioTechnologia. Journal of Biotechnology Computational Biology and Bionanotechnology*, *96*(1).

Pasternak, T. P., & Steinmacher, D. (2024). Plant growth regulation in cell and tissue culture in vitro. *Plants*, *13*(2), 327. DOI: 10.3390/plants13020327 PMID: 38276784

Pathak, S., Singh, J., Kumar, K., & Bisht, V. (2025). Breeding for the Future: A Comprehensive Review of Recurrent Selection. *Journal of Scientific Research and Reports*, *31*(7), 694–708. DOI: 10.9734/jsrr/2025/v31i73289

Paul, M. (2024). BREEDING TECHNIQUES FOR ASEXUALLY PROPAGATED CROPS. *Fundamentals of plant breeding*, 154.

Pereira, A. M., & Coimbra, S. (2019). Advances in plant reproduction: From gametes to seeds. *Journal of Experimental Botany*, *70*(11), 2933–2936. DOI: 10.1093/jxb/erz227 PMID: 31152172

Phillips, G. C., & Garda, M. (2019). Plant tissue culture media and practices: An overview. *In Vitro Cellular & Developmental Biology. Plant*, *55*(3), 242–257. DOI: 10.1007/s11627-019-09983-5

Pincelli-Souza, R. P., Tang, Q., Miller, B. M., & Cohen, J. D. (2024). Horticultural potential of chemical biology to improve adventitious rooting. *Horticulture Advances*, *2*(1), 12. DOI: 10.1007/s44281-024-00034-7

Poschlod, P., Mašková, T., Chen, S. C., Phartyal, S., Rosbakh, S., Silveira, F. A., ... and Vandelook, F. (2025). A handbook for standardised measurements of regenerative plant functional traits.

Rachappanavar, V. (2025). Utilizing CRISPR-based genetic modification for precise control of seed dormancy: Progress, obstacles, and potential directions. *Molecular Biology Reports*, *52*(1), 204. DOI: 10.1007/s11033-025-10285-w PMID: 39907946

Radi, H., Koufan, M., Belkoura, I., Koussa, T., & Mazri, M. A. (2025). In Vitro Mycorrhization for Plant Propagation and Enhanced Resilience to Environmental Stress: A Review. *Plants*, *14*(14), 2097. DOI: 10.3390/plants14142097 PMID: 40733334

Raghavan, M., Pradhan, V., Singh, S., Jain, S., Kumari, N., Kaur, R., Kumar, G., & Kaur, R. (2025). Advances in Genetic Plant Breeding for Horticultural Crop Improvement. *Plant Cell Biotechnology and Molecular Biology*, *26*(7-8), 432–444. DOI: 10.56557/pcbmb/2025/v26i7-89576

Rahman, S. U., Khan, M. O., Ullah, R., Ahmad, F., & Raza, G. (2024). Agrobacterium-mediated transformation for the development of transgenic crops; present and future prospects. *Molecular Biotechnology*, *66*(8), 1836–1852. DOI: 10.1007/s12033-023-00826-8 PMID: 37573566

Razzaq, K., & Du, J. (2024). Phytohormonal regulation of plant development in response to fluctuating light conditions. *Journal of Plant Growth Regulation*, •••, 1–34.

Reed, R. C., Bradford, K. J., & Khanday, I. (2022). Seed germination and vigor: Ensuring crop sustainability in a changing climate. *Heredity*, *128*(6), 450–459. DOI: 10.1038/s41437-022-00497-2 PMID: 35013549

Roddy, A. B., Théroux-Rancourt, G., Abbo, T., Benedetti, J. W., Brodersen, C. R., Castro, M., Castro, S., Gilbride, A. B., Jensen, B., Jiang, G.-F., Perkins, J. A., Perkins, S. D., Loureiro, J., Syed, Z., Thompson, R. A., Kuebbing, S. E., & Simonin, K. A. (2020). The scaling of genome size and cell size limits maximum rates of photosynthesis with implications for ecological strategies. *International Journal of Plant Sciences*, *181*(1), 75–87. DOI: 10.1086/706186

Rubin, H., & Rubin, A. L. (2018). Phenotypic selection as the biological mode of epigenetic conversion and reversion in cell transformation. *Proceedings of the National Academy of Sciences of the United States of America*, *115*(4), E725–E732. DOI: 10.1073/pnas.1717299115 PMID: 29311337

Ruxton, G. D., & Schaefer, H. M. (2012). The conservation physiology of seed dispersal. *Philosophical Transactions of the Royal Society of London. Series B, Biological Sciences*, *367*(1596), 1708–1718. DOI: 10.1098/rstb.2012.0001 PMID: 22566677

Sabir, I. A., Hu, X., Khan, I., & Qin, Y. (2025). Regulatory Mechanisms of Bud Dormancy: Environmental, Hormonal, and Genetic Perspectives. *International Journal of Molecular Sciences*, *26*(6), 2517. DOI: 10.3390/ijms26062517 PMID: 40141161

Sahin, E. C., Aydin, Y., & Uncuoglu, A. A. (2024). Molecular Marker Applications in the Selection of Elite Genotypes for Plant Stress Tolerance and Genetic Fidelity. *OBM Genetics*, *8*(3), 1–25. DOI: 10.21926/obm.genet.2403260

Salgotra, R. K., & Chauhan, B. S. (2023). Genetic diversity, conservation, and utilization of plant genetic resources. *Genes*, *14*(1), 174. DOI: 10.3390/genes14010174 PMID: 36672915

Salse, J., Barnard, R. L., Veneault-Fourrey, C., & Rouached, H. (2024). Strategies for breeding crops for future environments. *Trends in Plant Science*, *29*(3), 303–318. DOI: 10.1016/j.tplants.2023.08.007 PMID: 37833181

Shah, S. H., Islam, S., Mohammad, F., & Siddiqui, M. H. (2023). Gibberellic acid: A versatile regulator of plant growth, development and stress responses. *Journal of Plant Growth Regulation*, *42*(12), 7352–7373. DOI: 10.1007/s00344-023-11035-7

Shaw, N., Barak, R. S., Campbell, R. E., Kirmer, A., Pedrini, S., Dixon, K., & Frischie, S. (2020). Seed use in the field: Delivering seeds for restoration success. *Restoration Ecology*, *28*(S3), S276–S285. DOI: 10.1111/rec.13210

Shilpa, P., Subbalakshmi, G., & Anuradha, M. (2025). In Vitro Production of Colors and Pigments. In *In Vitro Production of Plant Secondary Metabolites: Theory and Practice* (pp. 233–251). Springer Nature Singapore. DOI: 10.1007/978-981-97-8808-8_18

Shivran, M., Sharma, N., Dubey, A. K., Singh, S. K., Sharma, N., Sharma, R. M., Singh, N., & Singh, R. (2022). Scion–rootstock relationship: Molecular mechanism and quality fruit production. *Agriculture, 12*(12), 2036. DOI: 10.3390/agriculture12122036

Shiwani, K., Sharma, D., & Kumar, A. (2022). Improvement of plant survival and expediting acclimatization process. In *Commercial scale tissue culture for horticulture and Plantation crops* (pp. 277–291). Springer Nature Singapore. DOI: 10.1007/978-981-19-0055-6_12

Shreni Agrawal, E. R. (2022). A review: Agrobacterium-mediated gene transformation to increase plant productivity. *J Phytopharm, 11*(2), 111–117. DOI: 10.31254/phyto.2022.11211

Shukla, G., Pradhan, S., Debberma, S., Singh, M., Sharma, P., Roy, B., & Shukla, P. S. (2025). Revolutionizing Tree Propagation by Exploiting the Power of Synthetic Seed Technology. In *Advances in Seed Quality Evaluation and Improvement* (pp. 447–464). Springer Nature Singapore. DOI: 10.1007/978-981-96-5178-8_18

Sirangi, S. K. (2025). *Tissue Culture and Conservation Strategies for Indian Flora.* Deep Science Publishing. DOI: 10.70593/978-93-7185-376-7

Sripathy, K. V., & Groot, S. P. (2023). Seed development and maturation. In *Seed science and technology: Biology, production, quality* (pp. 17–38). Springer Nature Singapore. DOI: 10.1007/978-981-19-5888-5_2

Stoian-Dod, R. L., Dan, C., Morar, I. M., Sestras, A. F., Truta, A. M., Roman, G., & Sestras, R. E. (2023). Seed germination within genus Rosa: The complexity of the process and influencing factors. *Horticulturae, 9*(8), 914. DOI: 10.3390/horticulturae9080914

Sultan, S. J., Qaddawi, Z. T., & Mohammed, A. A. H. (2024). Molecular diagnosis of bacteria isolated from *Trifolium repens* root nodules. *Journal of Applied and Natural Science, 16*(3), 1153–1157. DOI: 10.31018/jans.v16i3.5404

Sun, L., Lai, M., Ghouri, F., Nawaz, M. A., Ali, F., Baloch, F. S., Nadeem, M. A., Aasim, M., & Shahid, M. Q. (2024). Modern plant breeding techniques in crop improvement and genetic diversity: From molecular markers and gene editing to artificial intelligence—a critical review. *Plants*, *13*(19), 2676. DOI: 10.3390/plants13192676 PMID: 39409546

Svolacchia, N., & Sabatini, S. (2023). Cytokinins. *Current Biology*, *33*(1), R10–R13. DOI: 10.1016/j.cub.2022.11.022 PMID: 36626853

Tafolla-Arellano, J. C., Báez-Sañudo, R., & Tiznado-Hernández, M. E. (2018). The cuticle as a key factor in the quality of horticultural crops. *Scientia Horticulturae*, *232*, 145–152. DOI: 10.1016/j.scienta.2018.01.005

Tailor, A., & Joshi, A. (2024). The Role of Synthetic Seeds in Germplasm Conservation of Forest Tree Species. In *Biotechnological Approaches for Sustaining Forest Trees and Their Products* (pp. 103–150). Springer Nature Singapore. DOI: 10.1007/978-981-97-4363-6_5

Taji, A., Kumar, P., & Lakshmanan, P. (2024). *In vitro plant breeding*. CRC Press. DOI: 10.1201/9781003578529

Tang, L., Syed, A. U. A., Otho, A. R., Junejo, A. R., Tunio, M. H., Hao, L., Asghar Ali, M. N. H., Brohi, S. A., Otho, S. A., & Channa, J. A. (2024). Intelligent Rapid Asexual Propagation Technology—A Novel Aeroponics Propagation Approach. *Agronomy (Basel)*, *14*(10), 2289. DOI: 10.3390/agronomy14102289

Thilakarathne, A. S., Liu, F., & Zou, Z. (2025). Plant Signaling Hormones and Transcription Factors: Key Regulators of Plant Responses to Growth, Development, and Stress. *Plants*, *14*(7), 1070. DOI: 10.3390/plants14071070 PMID: 40219138

Tipu, M. M., & Sherif, S. M. (2024). Ethylene and its crosstalk with hormonal pathways in fruit ripening: Mechanisms, modulation, and commercial exploitation. *Frontiers in Plant Science*, *15*, 1475496. DOI: 10.3389/fpls.2024.1475496 PMID: 39574438

Trigiano, R. N., & Gray, D. J. (2016). *Plant tissue culture, development, and biotechnology*. CRC Press. DOI: 10.1201/9781439896143

Tripathi, A., & Shukla, S. (2024). Methods of genetic transformation: Major emphasis to crop plants. *Journal of Microbiology, Biotechnology and Food Sciences*, *13*(4), e10276–e10276.

Twaij, B. M., Jazar, Z. H., & Hasan, M. N. (2020). Trends in the use of tissue culture, applications and future aspects. *International Journal of Plant Biology*, *11*(1), 8385. DOI: 10.4081/pb.2020.8385

Vanneste, S., Pei, Y., & Friml, J. (2025). Mechanisms of auxin action in plant growth and development. *Nature Reviews. Molecular Cell Biology*, *26*(9), 1–19. DOI: 10.1038/s41580-025-00851-2 PMID: 40389696

Velasquez, S. M., Barbez, E., Kleine-Vehn, J., & Estevez, J. M. (2016). Auxin and cellular elongation. *Plant Physiology*, *170*(3), 1206–1215. DOI: 10.1104/pp.15.01863 PMID: 26787325

Verma, K. (2025). *Plant Biotechnology and Genetic Advances*. Educohack Press.

Vibha, B., Oza, D. N., and Hathi, H. S. (2024). Specialized Propagation Structure for Ornamental Plants.

Volovnik, S. V. (2025). Phytochory–the dispersal of animals by terrestrial and aquatic plants. *Journal of Natural History*, *59*(21-24), 1469–1539. DOI: 10.1080/00222933.2025.2475536

Wang, P., Si, H., Li, C., Xu, Z., Guo, H., Jin, S., & Cheng, H. (2025). Plant genetic transformation: Achievements, current status and future prospects. *Plant Biotechnology Journal*, *23*(6), 2034–2058. DOI: 10.1111/pbi.70028 PMID: 40052992

Weiss, J., & Gruda, N. S. (2025). Novel breeding techniques and strategies for enhancing greenhouse vegetable product quality. *Agronomy (Basel)*, *15*(1), 207. DOI: 10.3390/agronomy15010207

Wen, Z., Lu, X., Wen, J., Wang, Z., & Chai, M. (2024). Physical seed dormancy in legumes: Molecular advances and perspectives. *Plants*, *13*(11), 1473. DOI: 10.3390/plants13111473 PMID: 38891282

Wu, W., Chen, L., Liang, R., Huang, S., Li, X., Huang, B., Luo, H., Zhang, M., Wang, X., & Zhu, H. (2025). The role of light in regulating plant growth, development and sugar metabolism: A review. *Frontiers in Plant Science*, *15*, 1507628. DOI: 10.3389/fpls.2024.1507628 PMID: 39840366

Wu, Z. Y., Milne, R. I., Liu, J., Nathan, R., Corlett, R. T., & Li, D. Z. (2023). The establishment of plants following long-distance dispersal. *Trends in Ecology & Evolution*, *38*(3), 289–300. DOI: 10.1016/j.tree.2022.11.003 PMID: 36456382

Xu, X., Zhang, C., Xu, X., Cai, R., Guan, Q., Chen, X., Chen, Y., Zhang, Z., Xu-Han, X., Lin, Y., & Lai, Z. (2023). Riboflavin mediates m6A modification targeted by miR408, promoting early somatic embryogenesis in longan. *Plant Physiology*, *192*(3), 1799–1820. DOI: 10.1093/plphys/kiad139 PMID: 36930572

Yan, D., Duermeyer, L., Leoveanu, C., & Nambara, E. (2014). The functions of the endosperm during seed germination. *Plant & Cell Physiology*, *55*(9), 1521–1533. DOI: 10.1093/pcp/pcu089 PMID: 24964910

Yang, L., Liu, S., & Lin, R. (2020). The role of light in regulating seed dormancy and germination. *Journal of Integrative Plant Biology*, *62*(9), 1310–1326. DOI: 10.1111/jipb.13001 PMID: 32729981

Yasemin, S., & Beruto, M. (2024). A review on flower bulb micropropagation: Challenges and opportunities. *Horticulturae*, *10*(3), 284. DOI: 10.3390/horticulturae10030284

Yu, H., Jia, H., Shen, N., Gang, D., Yuan, W., Yang, Y., Hu, C., & Qu, J. (2024). Can "risk-sharing" mechanisms help clonal aquatic plants mitigate the stress of nanoplastics? *Environmental Science & Technology*, *58*(6), 2984–2997. DOI: 10.1021/acs.est.3c09436 PMID: 38306608

Zhou, M., & Yang, J. (2023). Delaying or promoting? Manipulation of leaf senescence to improve crop yield and quality. *Planta*, *258*(3), 48. DOI: 10.1007/s00425-023-04204-1 PMID: 37477756

Chapter 6
Sexual Propagation Seeds and Seedling Production

Fadiyya Nooruldeen Saeed
https://orcid.org/0000-0002-0860-6838
University of Mosul, Iraq

Burcu Yilmaz Çıtak
University of Selçuk, Turkey

ABSTRACT

Sexual propagation through seeds and seedlings remains a cornerstone of agricultural and horticultural production. It ensures genetic diversity, enhances adaptability to changing environmental conditions, and contributes to sustainable crop improvement. The success of this approach depends on the integration of proper seed quality management, advanced cultivation techniques, and effective disease control strategies. By improving seed systems, optimizing germination conditions, and maintaining seedling health, agricultural productivity and biodiversity conservation can be secured. Therefore, sexual propagation continues to play a decisive role in food security and the future of sustainable agriculture.

INTRODUCTION

Plant propagation is defined as the process of plant reproduction, regeneration, and multiplication, allowing plants to produce new offspring through the production of seeds.

Sexual reproduction is a method of using viable seeds to grow plants. It is a convenient and easy way to propagate crops such as fruits, medicinal plants, ornamentals, and others. It can be faster and more economical than asexual reproduction

DOI: 10.4018/979-8-3373-5626-6.ch006

and may be the only method for some plants. In sexual plant reproduction, the seed gestation period is longer, delaying flowering and fruiting. Crops that do not produce seeds cannot be propagated this way (Awotedu et al., 2021).

Sexual reproduction can also be defined as the process of producing seeds through successful fertilization and fusion of parental gametes. Good seeds are essential for successful vegetable production. Most vegetables are propagated by true seeds, with the exception of a few vegetables (Chauhan et al., 2021).

Seed and seedling propagation remain the primary methods for producing food crops, fiber plants, and other essential raw materials. Therefore, seed propagation is the foundation of agricultural and horticultural production. This technique not only enhances genetic diversity but also contributes to improved plant quality compared to vegetative propagation. Furthermore, plants grown from seeds often have a longer lifespan and are less susceptible to diseases and viruses (Cabahug et al., 2018).

There are differences between seed production and seedling production, and it is important to study each separately. Seeds are primarily associated with sexual reproduction, while seedlings are the later stage toward mature plants. Seedling production relies on established techniques previously developed within sexual reproduction. Seed and seedling production remain important activities within the general framework of sexual reproduction, especially when selectively focusing on the role of seeds in producing new plants.

Table 1. Comparison between Sexual and Vegetative Propagation

Aspect	Sexual Propagation	Vegetative Propagation
Basis	Fusion of gametes leading to seed formation	Use of vegetative organs (stem, root, leaf). (Awotedu et al., 2021)
Genetic diversity	High, generates variability	Low, genetically identical clones. (Maity et al., 2023)
Disease resistance	Generally higher	May transmit pathogens. (Ventura et al., 2019)
Production speed	Slower, requires germination	Faster, direct growth. (Chauhan et al., 2021)

1. SEEDS

Since seeds are one of the most important inputs for plant production, developed countries have encouraged extensive research into this important agricultural input.

As a result of these efforts, these industrialized nations have developed their seed industries and gained a prominent position in the global seed trade.

As a result of developed countries' investments in the seed sector over the years, seed technology has rapidly evolved, and these rapidly changing technologies have begun to spread to developing countries. The introduction of new technologies to farmers has increased agricultural production and quality, enhancing the importance of the seed sector in global trade (Elçi, 2000).

A seed is defined as a mature ovule and consists of three parts: the embryo, the stored food material in the form of the endosperm, and the seed coat. For successful crop production, high-quality seeds are of paramount importance. Even though seeds account for only a minor share of overall crop production costs, their quality plays a crucial role in determining final yield.

High-quality seeds must be carefully sorted, clean, size-graded, viable, and free from diseases and insects. These seeds must be obtained from a reliable and safe source. Seed viability, longevity, and shelf life are among the most important characteristics. Dry seeds for all vegetables must be packed in sealed containers under vacuum. For successful vegetable production, high-quality seeds are essential, as most vegetables are propagated from true seeds (Chauhan et al., 2021).

Climate change has negative impacts on agriculture, including drought, flooding, extreme temperatures, salinity, chemicals, and heavy metals. These factors collectively negatively impact seed production and quality, disrupting plant morphological, physiological, cellular, biochemical, and molecular activities more than any single factor, leading to poor seed production. Furthermore, the exogenous use of plant hormones, stress-reducing compounds, modified agronomic management, and modern breeding strategies, such as the development of resistant varieties, all help combat stress and produce high-quality seeds (Das & Biswas, 2022).

Seeds form the foundation of agricultural productivity. The availability and use of high-quality seeds are essential for achieving optimal yields, especially under the pressures of climate change. Producing such seeds in environments exposed to different abiotic stresses requires careful management, as these stresses negatively influence plants. They can reduce both yield and quality by impacting plant growth from the morphological scale down to molecular processes. Nonetheless, plants have evolved mechanisms to counter these challenges, including stress avoidance, stress escape, and stress tolerance. To survive in adverse environments, plants undergo specific molecular, cellular, and physiological adjustments. Gaining insight into these adaptive responses is vital for advancing plant research and securing global food supplies (Maity et al., 2023)

Seeds vary greatly in size and shape, from the smallest orchid seeds to the largest coconut seeds. The survival time of seeds varies; some, such as willow seeds, are short-lived, while lotus seeds can live for thousands of years. During germination, the embryo divides and differentiates into roots and stems, with nutrients stored in the embryo or endosperm. The state of the seed varies from dormant to non-dormant, and the dormant state is influenced by several factors, including hormones, environment, and genetics.

The germination process begins with water uptake and metabolic activation, followed by the expression of degradative enzymes to release nutrients, with a sensitive response to light and temperature. The radicle appears as the visible beginning of germination, growing slowly in the dark, while the growth of aerial parts accelerates with the availability of light, promoting the seedling's greening and the initiation of photosynthesis (Bhatla & Kathpalia, 2023).

2. SEED PRODUCTION

Seeds are the foundation of nutrition and play a vital role in agricultural productivity. They are also considered an essential asset for the global community. To achieve a hunger-free world, the development of seed systems must be prioritized.

in several countries, seed production is generally categorized into two systems. The formal system involves distinct processes such as developing new cultivars, ensuring genetic purity, certifying seed quality, and distributing them through officially approved channels. By contrast, the informal system encompasses locally regulated seed-related practices, often managed without national legislation. Oversight of the certified formal system is carried out by the National Seed Authority along with National Seed Service Agencies, operating under the Ministry of Agriculture and Forestry. This ministry holds the central regulatory role for the seed sector, particularly for the formal seed system (Konja et al., 2019).

Unlike animals, plants follow a distinct life cycle marked by the alternation of two multicellular phases: the sporophyte and the gametophyte. The reproductive structures, stamens (androecium) for males and the carpel (gynoecium) for females, generate haploid spores through meiosis—microspores and macrospores. These spores then undergo successive divisions to form the male (microgametophyte) and female (macrogametophyte) gametes. The male gametophyte develops inside the anther, while the female gametophyte forms within the ovule. Once mature, mitotic division produces sperm and egg cells, which unite during fertilization to create a zygote. This zygote grows into a sporophyte, initiating the next plant generation (Bhatla & Shakya, 2023).

Most seed studies are conducted in controlled environments. Therefore, it is important to recognize that various field environmental conditions influence the physiological and biochemical characteristics of crops and seed quality. Therefore, it is important to examine changes in seed maturity and quality across different field production environments. This will contribute to supporting decision support systems for harvesting seeds from different crops while maintaining optimal seed quality under different environmental conditions. It will also contribute to expanding the scope of seed production programs within countries (Weerasekara et al., 2021).

3. SEED COLLECTION AND PROCESSING

Seeds are the primary and most important input to agricultural production systems, so the success of all technologies and inputs hinges on seed quality. Cereal seed production, in particular, has declined in recent years due to significant losses during processing and storage. To minimize the effects of deterioration during processing, storage, and seed management, pre-planting seed treatment methods must be implemented.

When harvesting crop seeds, a rotary combine should be used. It produces a coarse pile of husks and grains better than a drum combine. Combines equipped with a rotary separator and threshing device are known to break grains 2.20–13.41 times less. Meanwhile, the amount of whole grains is increased (Li et al., 2018).

Immediately after harvesting, the crop is more likely to spoil due to high humidity. Therefore, post-harvest seed treatment is performed to extend its shelf life and maintain its quality. Seed moisture and temperature are among the most important factors in maintaining its quality.

Grain seed cleaning and quality improvement are achieved through winnowing fans, which remove impurities such as weed seeds, husks, stalk fragments, broken or empty grains, dust, sand, stones, plastic, and even metallic or glass particles. Improper drying makes seeds vulnerable to insect attack and reduces germination capacity. After drying, sorting is essential to separate underdeveloped or shriveled seeds from well-formed ones. Additionally, treatments with fungicides, insecticides, or both are applied to protect seeds against pathogens and storage pests, thereby maintaining their viability (Sharma et al., 2023).

4. SEED GERMINATION

The ability of a seed to germinate at the appropriate time of year is of great importance from an environmental and economic perspective for the plant. The speed of germination and the vigor of the plant will affect the quality of the crop. These traits are influenced by the plant's genetic makeup and the environment.

Seed production and germination are closely linked to the survival and spread of plant species. Seeds protect the embryo and sense environmental information to adjust the timing of germination to suit the appropriate climatic seasons to complete the plant's life cycle. The germination process proceeds from water absorption to the emergence of the radicle through the seed coats. In the natural environment, the spatial pattern of seed dispersal depends on the habitat of the parent plant and the characteristics of the fruit and seed, while the temporal distribution of germination is determined by the interaction between the environment and the plant's genetic makeup, which influences dormancy and germination potential.

Physiological dormancy is the most common and provides seeds with important advantages, such as increased dispersal and reduced competition with the parent plant, as well as inhibiting germination at inopportune times. After ripening, seeds undergo a post-ripening phase, where water content and dormancy gradually decrease. The rate of this decline is influenced by factors such as humidity and temperature during storage.

When conditions are favorable, non-dormant seeds resume the chemical and biological processes related to germination, such as resumption of respiration and energy production, activation of repair mechanisms and protein synthesis, which leads to elongation of the embryonic axis, rupture of the seed coat, emergence of the radicle and growth of the seedling (Carrera-Castaño et al., 2020)

Table 2. Major Factors Affecting Seed Germination

Factor	Influence on Germination
Moisture	Initiates water uptake and metabolic activation. (Bhatla & Kathpalia, 2023)
Temperature	Determines germination rate and percentage. (Li et al., 2018)
Light	Regulates chlorophyll formation and photosynthesis. (Carrera-Castaño et al., 2020)
Aeration	Ensures respiration and prevents seed decay. (Grossnickle & MacDonald, 2018)

5. SEEDLING PRODUCTION TECHNIQUES

Since the early 1900s, forestry experts have emphasized the value of raising seedlings with superior traits. Evaluating seedling quality requires examining both morphological and physiological features. Such high-quality seedlings are crucial in reforestation and restoration efforts. To ensure this, systems should be in place to monitor cultivation practices and plant development, enabling production according to defined standards. Proper nursery techniques are essential for the success of reforestation projects using bare-root or container-grown seedlings.

Figure 1. Seedling Production Process

Seed selection → Seed treatment → Sowing under optimal conditions → Germination and nursery care → Transplanting
(Grossnickle & MacDonald, 2018)

Understanding how plants react to their growing environment is vital, as physiological and developmental responses dictate overall growth performance. Environmental conditions influencing germination—such as temperature, light, carbon dioxide, humidity, and soil moisture—can be continuously tracked with automated sensors. Fertilization and irrigation aspects, including pH and electrical conductivity, may also be measured with portable probes.

Crop progress is assessed by recording morphological and physiological indicators at key growth stages, which can be automated for real-time monitoring. Quality assurance programs for seedlings assess both material and performance traits. Material traits provide snapshot measurements of individual plant subsystems (e.g., morphology, nutrient content, osmotic potential, gas exchange, or root leakage), while performance traits integrate multiple factors and are tested under specific conditions (e.g., root growth potential, stress tolerance, or gas exchange over time). Both categories offer insights into survival rates and field performance (Grossnickle & MacDonald, 2018).

Table 3. Climate Change Challenges in Seed Production

Challenge	Effect on Seeds	Mitigation Strategies	
Drought	Reduced germination and seed vigor	Breeding drought-tolerant varieties. (Maity et al., 2023)	
Salinity	Damage to embryonic tissues	Application of growth regulators and anti-stress agents.(Das & Biswas, 2022)	
Heat stress	Reduced pollination and fertilization	Adjusting planting dates.(Weerasekara et al., 2021)	

6. TRANSPLANTING SEEDLINGS

The potted vegetable seedling transplanter is designed to improve the efficiency and quality of dryland cultivation. It relies on a rotating disc feed mechanism and a five-bar duckbill mechanism. Design parameters were determined by seedling characteristics, such as duckbill diameter, opening and closing angles, and slope, as well as different shaft and bar lengths and cam stroke control.

Tests were conducted on pepper seedlings, and the machine demonstrated a capacity to transplant 50-70 plants per minute with a seedling straightness rate of 91.1% to 93% and a qualified planting depth rate of 92% to 96%. It also achieved a very low leakage rate (0-0.26%), consistent spacing, no seedling damage, and minimal mechanical damage to the soil surface.The machine effectively improves planting quality and meets the production and design requirements of modern agricultural machinery (Jin et al., 2020).

An automatic dry-land vegetable potting machine was developed to enhance mechanization efficiency and automate planting operations. The machine consists of rotary seedling picking mechanisms (planetary with elliptical and non-elliptic gears), a horizontal and longitudinal intermittent seedling transmitter, and eccentric disc planting mechanisms with an integrated power transmission system. Seedlings are automatically fed from 128-hole pots and are transported and received by synchronized mechanisms to ensure precise planting. Experiments were conducted on tomato seedlings, evaluating performance at planting speeds ranging from 45 to 96 plants per minute

The results showed that increasing the planting speed resulted in a slight increase in the variation of plant spacing and the under-planting rate, while slightly decreasing the rates of qualified planting depth, seedling straightness, and qualified planting rate. All indicators met the standards of the Chinese agricultural and mechanical industry. The coordination systems between the transmitter, picker, and planter operate seamlessly, enabling an integrated automated process that includes automatic seedling transfer, planting, and soil covering, enhancing the quality and efficiency of dryland vegetable cultivation (Wang et al., 2018).

7. PEST AND DISEASE MANAGEMENT

Healthy and well-formed propagating material is essential for the economic success of fruit cultivation. It is also essential to ensure the health of seedlings to prevent plant death and the spread of diseases that can increase costs and reduce yields. The best strategy for controlling propagating diseases is to use resistant varieties, although sources of resistance are not always identified, and resistance can be broken by the emergence of new strains of disease. Other important measures include integrated disease management, cultural and biological methods, substrate preparation, irrigation management, grafting, balanced nutrition, and the use of organic matter.

In nurseries, disease-free seeds and cuttings are used, tools, hands, and containers are disinfected, irrigation water quality is maintained, and invasive plants are eliminated. Regular cleaning of nurseries and the provision of appropriate waste disposal facilities is vital. Production history must be recorded, and entry into greenhouses must be monitored to ensure seedling health. Implementing the best integrated management strategies ensures quality production and crop success. Preventive measures to produce healthy seedlings are among the most effective methods for disease control, as control becomes difficult once a disease has appeared in the field or nursery. Pathogens are numerous in seedling production, Sources of seedling infections include contaminated seeds, diseased cuttings, growth substrates, irrigation water, airborne fungi and bacteria, unsanitized equipment, surrounding host plants, soil, workers' hands, and insect carriers. Seedling disorders may be biotic, caused by pathogens like fungi, bacteria, and viruses, or abiotic, resulting from environmental stresses such as excess heat, humidity, or poor cultivation practices. Because many diseases present similar symptoms, accurate identification is often difficult without adequate environmental management. This challenge highlights the need for improved diagnostic tools and modern disease management strategies (Ventura et al., 2019).

In nurseries, root diseases are a major cause of reduced fruit plant productivity. Root diseases arise from soil-borne pathogens that invade and damage the underground organs of plants. The main groups responsible include fungi, bacteria, and nematodes (Iskra, Voigt, & Raab, 2024).

To ensure the production of healthy seedlings, it is necessary to use seeds free of plant pathogens and treat the planting substrates in clean locations, with the necessity of disinfecting hands, tools and containers. It is also necessary to rely on high-quality water for irrigation, eliminating harmful weeds and controlling disease-carrying insects. In addition, it is necessary to secure the nurseries and control the entry of individuals to them and clean them using suitable floors such as concrete, gravel or rubber to facilitate sanitation and find a designated place to dispose of seedlings, substrates and plant waste, and burn them weekly. It is also necessary to maintain a periodic and historical record of seedling production operations, in addition to managing irrigation efficiently, as excess water may cause flooding and affect root respiration and growth. It is also necessary to maintain a balanced nutrition for plants, as nutrients play an important role in plant defense, as they work within metabolism as stimulants, regulators or inhibitors (Ventura et al., 2019).

The health of seedlings and seeds is the cornerstone of successful agricultural production. Seedlings produced from infected seeds are diseased and represent a primary source of pathogens for the transfer of plant diseases to new areas, leading to equipment contamination and increased costs for plant disease control. In modern seed and seedling production, health is of paramount importance, requiring qualified personnel to ensure the quality of propagating material and issue the necessary certifications. Understanding the relationship between microorganisms and seeds, with a focus on seed-associated pathogens and their transmission dynamics and mechanisms, along with developing effective detection and control methods, helps implement the health standards stipulated by legislation, ensuring the production of healthy seedlings (Ventura et al., 2019).

From above we conclude

Sexual propagation through seeds and seedlings remains a cornerstone of agricultural and horticultural production. It ensures genetic diversity, enhances adaptability to changing environmental conditions, and contributes to sustainable crop improvement. The success of this approach depends on the integration of proper seed quality management, advanced cultivation techniques, and effective disease control strategies. By improving seed systems, optimizing germination conditions, and maintaining seedling health, agricultural productivity and biodiversity conservation can be secured. Therefore, sexual propagation continues to play a decisive role in food security and the future of sustainable agriculture.

Figure 2. Determinants of Seed Quality

Environmental conditions (temperature, light, water)

↓

Management practices (sorting, treatment, storage)

↓

Pathogen control (fungicides, resistant varieties)

(Ventura et al., 2019; Maity et al., 2023)

REFERENCES

Awotedu, B. F., Omolola, T. O., Akala, A. O., Awotedu, O. L., & Olaoti-Laaro, S. O. (2021). Vegetative propagation: A unique technique of improving plants growth. *World News of Natural Sciences*, *35*, 83–101.

Bhatla, S. C., & Kathpalia, R. (2023). Seed dormancy and germination. In Bhatla, S. C., & Lal, M. A. (Eds.), *Plant physiology, development and metabolism* (pp. 625–640). Springer Nature Singapore. DOI: 10.1007/978-981-99-5736-1_28

Bhatla, S. C., & Shakya, R. (2023). Pollination, fertilization, and seed development. In Bhatla, S. C., & Lal, M. A. (Eds.), *Plant physiology, development and metabolism* (pp. 583–605). Springer Nature Singapore. DOI: 10.1007/978-981-99-5736-1_26

Cabahug, R. A. M., Nam, S. Y., Lim, K. B., Jeon, J. K., & Hwang, Y. J. (2018). Propagation techniques for ornamental succulents. *Flower Research Journal*, *26*(3), 90–101. DOI: 10.11623/frj.2018.26.3.02

Carrera-Castaño, G., Calleja-Cabrera, J., Pernas, M., Gómez, L., & Oñate-Sánchez, L. (2020). An updated overview on the regulation of seed germination. *Plants*, *9*(6), 703. DOI: 10.3390/plants9060703 PMID: 32492790

Chauhan, A., Sharma, D., Kumar, R., Shiwani, K., & Sharma, N. (2021). Methods of propagation in vegetable crops. In *Recent trends in propagation of forest and horticultural crops* (pp. 270–281). Taran Publication.

Das, R., & Biswas, S. (2022). Influence of abiotic stresses on seed production and quality. In *Seed biology updates*. IntechOpen., DOI: 10.5772/intechopen.106045

Elçi, A. (2000). Türkiye'de tohumculuğun durumu ve gelişimi. In *Türkiye Ziraat Mühendisliği V* (Vol. 2, pp. 859–870). Teknik Kongresi.

Grossnickle, S. C., & MacDonald, J. E. (2018). Seedling quality: History, application, and plant attributes. *Forests*, *9*(5), 283. DOI: 10.3390/f9050283

Iskra, M., Voigt, L., & Raab, M. (2024). Accounting for dynamic cognition–action interaction in decision-making tasks in sports: A scoping review. *Sport, Exercise, and Performance Psychology*. Advance online publication. DOI: 10.1037/spy0000345

Jin, X., Cheng, Q., Zhao, B., Ji, J., & Li, M. (2020). Design and test of 2ZYM-2 potted vegetable seedlings transplanting machine. *International Journal of Agricultural and Biological Engineering*, *13*(1), 101–110. DOI: 10.25165/j.ijabe.20201301.5494

Li, B., Lecourt, J., & Bishop, G. (2018). Advances in non-destructive early assessment of fruit ripeness towards defining optimal time of harvest and yield prediction—A review. *Plants*, *7*(1), 3. DOI: 10.3390/plants7010003 PMID: 29320410

Maity, A., Paul, D., Lamichaney, A., Sarkar, A., Babbar, N., Mandal, N., & Chakrabarty, S. K. (2023). Climate change impacts on seed production and quality: Current knowledge, implications, and mitigation strategies. *Seed Science and Technology*, *51*(1), 65–96. DOI: 10.15258/sst.2023.51.1.07

Sharma, P., Roy, M., Roy, B., & Deka, S. D. (2023). Post harvest management strategies and storage approaches for quality seed production. Emerging issues in agricultural sciences, 2, 110–129.

Tasila Konja, D., Mabe, F. N., & Oteng-Frimpong, R. (2019). Profitability and profit efficiency of certified groundnut seed and conventional groundnut production in Northern Ghana: A comparative analysis. *Cogent Economics & Finance*, *7*(1), 1631525. DOI: 10.1080/23322039.2019.1631525

Ventura, J. A., Lima, I. D. M., Martins, M. V. V., Culik, M. P., & Costa, H. (2019). Impact and management of diseases in the propagation of fruit plants. *Revista Brasileira de Fruticultura*, *41*(4), e-647. DOI: 10.1590/0100-29452019647

Ventura, J. A., Lima, I. D. M., Martins, M. V. V., Culik, M. P., & Costa, H. (2019). Impacto e manejo das doenças na propagação das fruteiras [Impact and management of diseases in the propagation of fruit plants]. *Revista Brasileira de Fruticultura*, *41*, e-647. DOI: 10.1590/0100-29452019647

Wang, Y., He, Z., Wang, J., Wu, C., Yu, G., & Tang, Y. (2018). Experiment on transplanting performance of automatic vegetable pot seedling transplanter for dry land. *Nongye Gongcheng Xuebao (Beijing)*, *34*(3), 19–25. DOI: 10.11975/j.issn.1002-6819.2018.03.003

Weerasekara, I., Sinniah, U. R., Namasivayam, P., Nazli, M. H., Abdurahman, S. A., & Ghazali, M. N. (2021). The influence of seed production environment on seed development and quality of soybean (Glycine max (L.) Merrill). *Agronomy (Basel)*, *11*(7), 1430. DOI: 10.3390/agronomy11071430

Chapter 7
The Impact of Plant Reproduction on Biodiversity

Marwan Y. M. El Fargani
https://orcid.org/0000-0002-5919-7632
Ministry of Higher Education and Scientific Research, Libyan Arab Jamahiriya

Muheddin Mahmud Rteba
https://orcid.org/0000-0003-1804-600X
Libyan Authority for Scientific Research, Libya

Amal Mahmud El Khaddari
https://orcid.org/0000-0001-7701-1726
Ibn Tofil University, Morocco

ABSTRACT

Reproduction in living organisms is an important means for the continuation of life and the transmission of genetic traits between generations and their preservation from disappearance, especially those good genetic traits. The importance of biological diversity is due to the strength of different ecosystems. Biodiversity is represented in three forms: the multiplicity of living species, the diversity of genetic structures, and the multiplicity of ecosystems. Genetic diversity is one of the components of biological diversity in plants through sexual reproduction among individuals of the same species, and this is Genetic crossing between genes produces individuals with multiple genotypes. Asexual reproduction does not achieve genetic diversity, as the individuals resulting from this reproduction carry identical genetic structures to

DOI: 10.4018/979-8-3373-5626-6.ch007

the original, thus not achieving the targeted biodiversity. Different ecosystems also support biodiversity, and the diversity of individuals within communities supports the stability of the biosphere and achieves sustainability on planet Earth.

INTRODUCTION

Biodiversity and plant reproduction are two important and interesting topics in biology. The Biodiversity concept Refers to the diversity of living organisms in an environment, including plants and Animals and microbes are affected and Vegetative propagation on biodiversity and its one of the most important factors affecting.

Plant propagation is an important means of preserving the genetic diversity of plants, as plants can reproduce by seeds, roots, or other parts of the plant, which helps improve their ability to adapt to Environmental changes. Biodiversity and plant reproduction contribute to ecosystem balance, as the diversity of plant species provides diverse food sources for animals and microbes, thus maintaining ecosystem balance. Plants can contribute to improving environmental quality through the process of vegetative reproduction, as they help purify the air and soil, and absorb harmful gases. So it is, Biodiversity and reproduction Vegetarian two-elements Interconnected and integrated in preserving the environment and improving its quality and Support for ecosystem Sustainability. (Johri & Srivastava, 2012)

There is also a close link between Biodiversity and cover Veganism, and each affects the other. Vegetation is an essential part of biodiversity in ecosystems. Where Vegetation contributes to providing diverse and suitable habitats for living organisms such as animals, insects, and microbes. Plants provide food, shelter, and protection for other living organisms. Which Leads to increase diversity of living organisms in an ecosystem.

Vegetation helps maintain the genetic diversity of plants and, consequently, other organisms. The presence of a diverse range of plants in vegetation helps maintain genetic and species diversity.

Vegetation provides many ecological services essential for preserving the environment and biodiversity. Plants contribute to air purification, rationalizing water use, providing oxygen, and stabilizing Soil, as well Vegetation helps maintain the balance of the ecosystem and the complementary relationships between different living organisms. The vegetation cover in promoting environmental sustainability and diversity (Blicharska., et al;2019).

Vegetation can contribute to combating environmental and climate change and providing solutions for adapting to it. Plants play a vital role in improving air and water quality and mitigate the effects of climate change on the environment (S. Ollinger;2003)

Therefore, Vegetation cover and biodiversity interact with each other and promote sustainability and balance in ecosystems. Humans must work to protect and to support Vegetation cover and biodiversity enhancement to ensure sustainable life on the Earth's surface, where it is considered, Vegetation is one of the most important factors affecting biodiversity in the environment. The vegetation cover is plant group That covers the surface of the earth in a certain area, and includes trees, grasses, shrubs and other plants that are found in this area. Vegetation provides food resources for animals and microbes that live within this range, which contributes to the diversity and sustainability of living organisms and vegetation is an important habitat for many animals, as animals use plants as shelter, a source of food, and protection. Vegetation purifies the air from harmful gases and improves soil quality through its work for decomposition of nutrients. Vegetation contributes to the balance of the ecosystem by providing nutrition and protection for various living organisms, it works on stability order Environmental. Vegetation plays a vital role in maintaining biodiversity in the environment. It is considered a fundamental factor in achieving environmental balance and the sustainability of living organisms in nature. It plays a crucial role in achieving sustainability in many environmental, economic and social aspects. Such as balancing the ecosystem, achieving food security, environmental health, human health, and supporting a sustainable economy. So, Biodiversity plays a vital role in achieving sustainability in various aspects of life. It is considered a fundamental factor in achieving a balance between human needs and environmental needs (S. Locki, et al; 2015).

BIODIVERSITY

It is a concept that refers to the presence of a variety of living organisms, plants, and microbes in a specific place, it is important for the stability and health of ecosystems, as every living organism plays an important role in the balance of the ecosystem and contributes to the preservation of biodiversity.

That Preserving biodiversity is essential to maintaining a healthy environment and preserving their sources natural for present and future generations and it is a source of food, medicine, raw materials, and other environmental services such as air and water purification, and animal and plant agricultural breeding. (Rodolfo, Eduardo; 2019)

Biodiversity Faces pressures many such as the degradation of natural resources, climate change, Pollution and resource exploitation natural unsustainably. For these reasons we need for Protect and enhance biodiversity through sustainable and informed actions and decisions to conserve organisms and ecosystems.

The United Nations defines biodiversity as: the diversity of living organisms, including animals, plants, and microbes, and the genetic variations associated with them, including diversity in species, genetic diversity, and diversity in the environments in which they are found. Biodiversity conservation aims to preserve living organisms, sustainably manage natural resources, and maintain the ecological services provided by the environment. Where The United Nations is considered that All countries the members committed to protecting biodiversity and to support manage it sustainably. www.un.org(CBD)

Biodiversity sections

Biodiversity is divided into several types:

1. Genetic diversity encompasses the first type of diversity within individuals. This type of biological diversity is essential for organisms to adapt to environmental changes and maintain genetic diversity.
2. Species diversity: This refers to the number of different species present in an area. The distribution of this type of biodiversity reflects the environment's richness in various living organisms.
3. Ecosystem diversity is the variety of different ecosystems and habitats that contain a variety of organisms and their ecological relationships.
4. Functional diversity is the diversity of biological functions and roles that living organisms play in the environment, such as decomposing organic matter, pollinating plants and recycling materials.
5. Aesthetic diversity: This type of biodiversity relies on the aesthetics and natural diversity of the environment, such as the diversity of landscapes, terrain, and natural patterns. (Blicharska., et al;2019)

Figure 1. Biodiversity

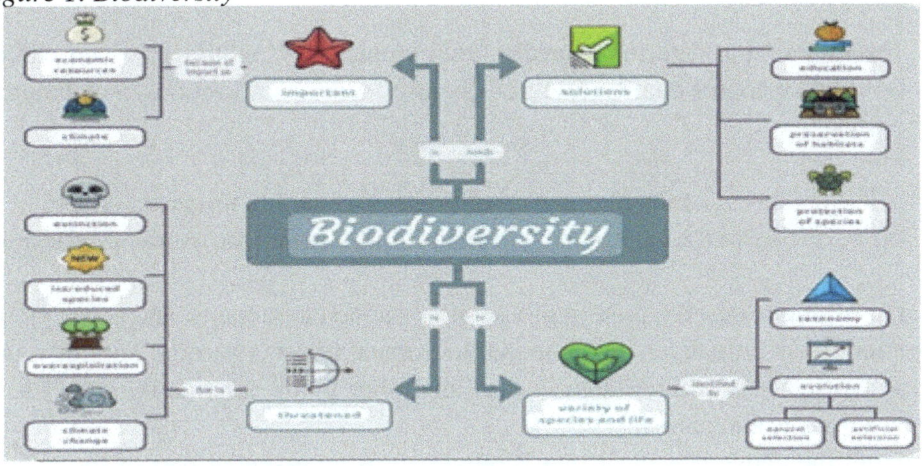

Importance of biodiversity

Biodiversity is one of the most important factors contributing to the sustainability of life on earth. Its importance is due to:

- Maintaining the balance of the ecosystem: Biodiversity is a key factor in maintaining ecosystem balance and it is contributes to regulating the interrelationships between living organisms and their environment.
- Providing food and medicine: Biodiversity is a major source of food and medicine, such as agriculture, fishing, and the pharmaceutical industry depend on the diversity of natural resources.
- Maintaining genetic diversity: Biodiversity is an important source of genetic diversity that contributes to the adaptation of organisms to changes in the environment and resistance to diseases, pests, adverse environmental and climatic conditions.
- Maintaining ecological services: Biodiversity provides important ecological services such as air, water, and soil purification and coastal protection, thus contributing to improved environmental quality and the conservation of wildlife and marine life.
- Economic benefits: Biodiversity contributes to the economy by providing job opportunities in the agricultural, tourism, and industrial sectors that rely on natural resources. (S. Locki, et al;2015)

Factors determining diversity:

The factors that determine species are complex and multifaceted and can be grouped into resource quality, resource quantity, species interactions, and dynamic processes.

- The harshness of the physical conditions to which life forms are adapted.
- The quality of the environment in terms of its size and the nature of its terrain.

For example, deserts, polar regions, snow-capped mountains, and ocean floors have relatively little diversity compared to tropical forests (warmth, humidity, and sufficient light). It is more diverse than any other living community on earth, in terms of plants and animals.

From the above it is clear that Biodiversity is an essential part of the natural and cultural heritage of peoples. As contributes to improving the quality of life and health Public and everyone is required to work on the Protection and governorate on biodiversity to ensure the sustainability of life on Earth. (Hardinsyah et al., 2007)

The concept of sustainable development

Sustainable development means meeting the needs of the present generation without compromising the ability of future generations to meet their needs. It aims to achieve a balance between economic, social and environmental dimensions to ensure sustainable growth and development in the long term. (Racel E.;2015)

Sustainable development Included the following:

- Focus on a sustainable economy: which includes providing job opportunities and achieving economic growth without harming the environment.
- Social justice: means providing equal opportunities for all and guaranteeing basic rights and freedoms for all segments of society.
- Environmental protection: it is preserving natural and vital resources and reducing pollution and negative environmental impacts.

That sustainable development Achieving a balance between the economy, society, and the environment to ensure a decent and sustainable life for current and future generations.

The relationship between sustainable development and biodiversity

Sustainable development and biodiversity have close relationship Between them. Biodiversity is an essential element in achieving sustainable development. It affects directly impacting the environment, economy and the society where humans depend on biodiversity for food, medicine, raw materials, etc. environmental services. Deterioration leads to Biodiversity having negative impacts on the economy, public health and the environment.

Conversely, conserving biodiversity contributes to sustainable development by preserving the environment, providing sustainable economic opportunities, and ensuring sustainability. Conserving natural resources and preserving biodiversity must be an essential part of sustainable development strategies. Efforts must be integrated between conserving biodiversity and achieving sustainable development to ensure the sustainability of life on our planet. (Clark, & Wu, 2016)

Reproduction in living organisms

That the proliferation in Living things a basic and vital process for the survival and continuity of species in the world. It is the reproductive process by which organisms produce new offspring. it is one the most important functions of living organisms to ensure the continuity and diversity of life, Reproduction occurs in different ways that vary among different organisms, such as cell division and sexual reproduction. So, cell division is a reproductive process that occurs in simple organisms such as bacteria and algae, where they divide a cell into two or more cells to produce new cells identical to the parent cell. Sexual reproduction is a reproductive process that occurs in organisms with different sexes, where goatherder cells of different sexes to produce new offspring containing a mixture of genetic traits for parents and children reproduction increases the diversity of living organisms and their adaptation to the surrounding environment. It also contributes to the renewal of generations and ensures the continuity of life. Reproduction is considered one of the most important functions of living organisms for the following reasons:

- Maintaining genetic diversity: Reproduction increases genetic diversity among individuals, making species more resilient, adaptable to changes in the environment, and resistant to diseases and adverse conditions.
- Increase in the number of individuals: Reproduction allows the number of individuals in a species to increase, making it easier for them to survive and spread in the environment.

- o Evolution of organisms: Reproduction allows organisms to evolve and improve from generation to generation, which contributes to improving their characteristics and adapting to the environment.
- o Continuity of life: Without the function of reproduction, there would be no continuity of life on earth, as all living species depend on reproduction for the survival of subsequent generations.

The importance of reproduction is due to it being the cornerstone of the life cycle of living organisms, and it is necessary for the continuity of life on Earth planet.) Aguilar, et al., 2006)

Reproduction in plants

Reproduction in plants is a vital process essential for the survival and reproduction of species. There are several different ways to propagate plants, including:

- Sexual reproduction: Sexual reproduction involves the merging of embryos (male and female gametes) from two different plants to produce a new seed. The seeds contain a different genetic makeup than the original two plants, allowing for increased genetic diversity.

Figure 2. Reproduction in plants

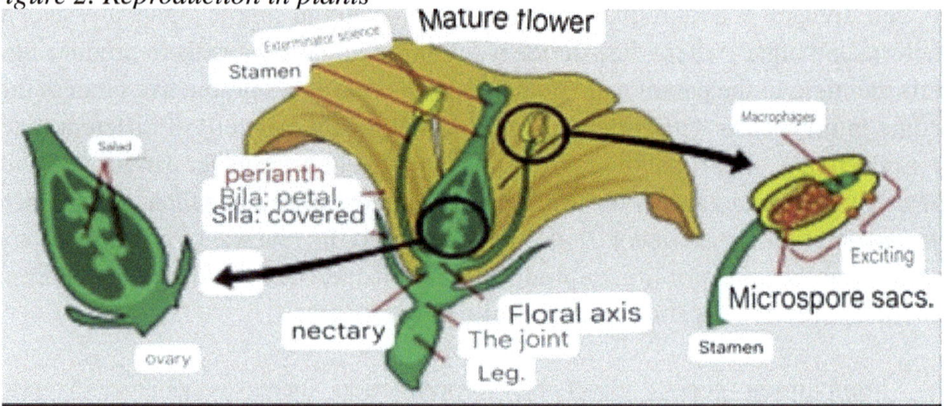

- Asexual reproduction: This type of reproduction involves the reproduction of plants without the use of sexual reproductive parts. Examples of this include propagation by bulbs and corms and roots and shoots.

- Artificial propagation: In this case, modern techniques are used to propagate plants, Such as cell amplification and plant tissue culture.

The process of plant reproduction aims to increase the number of plants and improve their genetic characteristics. Thay are contributes to the reproduction of plants in the environment and improving their environment and biodiversity. The process of reproduction in plants is considered an essential part of the natural life cycle and the biological continuity of plants and living organisms in general.

Plant reproduction is of great importance in the environment and biological system. This importance comes from the following factors:

Preserving biodiversity: Living organisms reproduce plants, which leads to increased biodiversity in the environment, which enhances the stability of the ecosystem and maintains its balance.

Environmental improvement: Plant propagation contributes to improving environmental quality and soil fertilization, as it contributes to strengthening the soil and improving its quality through the cultivation process, stabilizing nutrients and vegetative respiration.

Providing food and shelter: Plant reproduction is a major source of food and shelter for many other organisms in the ecosystem, including animals and insects.

Maintaining biological balance: Plant reproduction is an essential part of the natural life cycle, contributing to maintaining biological balance in the environment and ensuring the continuity of the biological system.

Sexual reproduction in plants is the type that greatly impacts biodiversity. Gametes from two different plants unite to produce new seeds. This means the resulting seeds will have different genetic makeup than the original two plants, increasing genetic diversity in plants.

Genetic diversity plays an important role in plants' adaptation to environmental changes and challenges such as diseases and pests. Furthermore, genetic diversity increases the chances of species surviving and reproducing in diverse environments. Sexual reproduction in plants is a key factor in increasing biodiversity and contributes to environmental sustainability and the continuity of biological systems (Michael, 2010).

physiological effects on plant reproduction

Plant physiology greatly influences the reproductive process. Physiological factors, such as light, temperature, humidity, and nutrition, can influence a plant's ability to reproduce effectively. When environmental conditions are favorable, plants can produce flowers and seeds very efficiently. While some plants reproduce by seeds, there are also plants that reproduce by growth, dispersal, and asexual reproduction. These processes are greatly influenced by environmental and physiological factors. In addition, physiological factors, such as plant hormones and growth-promoting chemicals, play an important role in the reproductive process. These factors contribute to the regulation of plant growth, flower formation, and the stimulation of reproductive processes.

Plant physiology plays a crucial role in regulating reproductive processes and their impact on plant growth and reproduction in the surrounding environment. (Cristina F., et al;2007)

ECOSYSTEM

An ecosystem is a group of organisms and environmental factors that interact with each other in an environment. A specific ecosystem includes the plants, animals, microbes, and the natural environment in which they live, as well as the biological and environmental factors that affect these organisms.

Ecosystems rely on an integrated interaction between living and non-living components, where organisms and various environmental factors exchange nutrients, energy, and resources, and are influenced by each other directly or indirectly. (Rahman;2012)

There are many diverse ecosystems that differ depending on the environmental factors and organisms present within them. The most important of these are:

- ✓ Terrestrial ecosystems: These include forests, plains, deserts, valleys, rainforests, and tropical forests. Terrestrial ecosystems vary in their terrain, plants, and animals.
- ✓ Aquatic ecosystems: These include seas, oceans, lakes, rivers, and swamps. These ecosystems are home to a wide variety of aquatic organisms, such as fish, algae, and waterfowl.
- ✓ Polar ecosystems: These include the Arctic and Antarctic, and are characterized by harsh climatic conditions and distinct biodiversity.
- ✓ Desert ecosystems: These include deserts and dry areas characterized by water scarcity and plants adapted to these dry conditions.
- ✓ Urban ecosystems: include cities and urban areas that are affected by human activity and contain diverse environmental and biological diversity.

These are some of the main types of ecosystems, and there are many other ecosystems that differ in their composition, characteristics, and impact on living organisms and the environment in general. (J.Sureesh & R. Rameeraj;2012)

The importance of the ecosystem is due to many aspects, including:

- Ecological balance: The ecosystem contributes to maintaining ecological balance and regulating the cycle of vital materials and energy in the system.
- Sustainability: An ecosystem provides essential environmental services to living organisms and humans, such as purifying air and water and providing natural resources.
- Biodiversity: An ecosystem contributes to the maintenance and increase of biodiversity through interactions between different organisms.
- Ecosystem balance: An ecosystem maintains the balance of the biological system so that the relationships between organisms and environmental factors remain stable and balanced.

The ecosystem is a fundamental factor in the sustainability of life on Earth. It must be preserved and its balances respected to ensure the continuity of the biological system. (Sekhar, et al;2024)

Figure 3. types of ecosystem

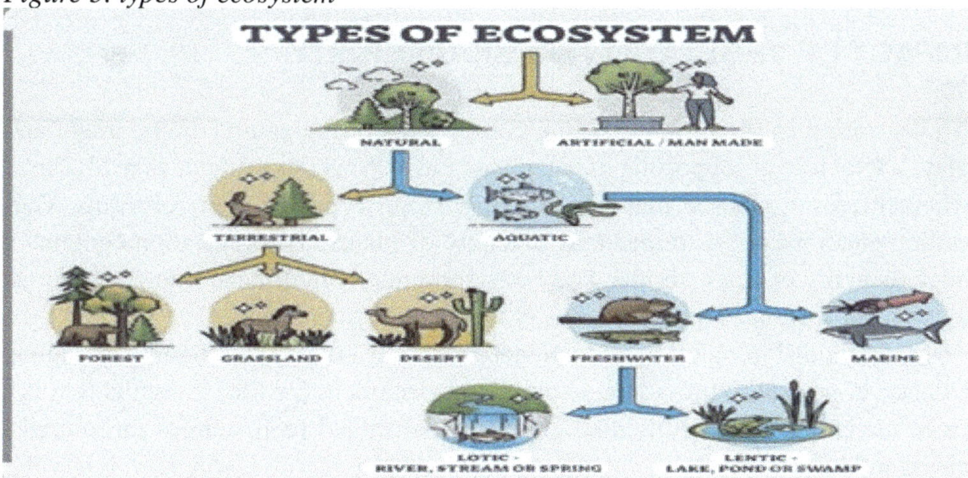

VEGETATION COVER

Vegetation refers to the grasses and trees that cover the ground in an area. Vegetation is an important part of the natural environment and significantly impacts wildlife and the environment. Vegetation can be diverse, including a variety of plants such as grass, shrubs, and trees. Vegetation can have a significant impact on the distribution of animals, insects, and birds in the environment, in addition to its role in regulating soil, water, and air.

Vegetation plays a major role and is of great importance in maintaining ecological balance and providing a suitable environment for living organisms. The importance of vegetation in providing the necessary food for living organisms through the process of photosynthesis, and preserving the soil, as plants work to stabilize the soil and prevent its erosion through their roots. Maintaining healthy soil for growing crops and supporting biodiversity, as plants regulate the climate, temperature and humidity in the region, and help mitigate the harmful effects of climate change. Plants provide shelter for animals and insects, and help protect them from outside elements such as storms and hail. It absorbs carbon dioxide and produces oxygen, improving air quality and reducing environmental pollution. Therefore, vegetation constitutes an essential part of the natural environment, contributing to the maintenance of biodiversity and ecosystem balance. (M. Getabalew, et al;2020)

IMPACT OF VEGETATION ON BIODIVERSITY

Vegetation is one of the essential components of the natural environment and plays a vital role in supporting life on planet Earth. Vegetation includes wild plants, whether trees or grasses, that cover the Earth's surface in a variety of forms. Vegetation encompasses a tremendous diversity of plants. The vegetation contains a huge diversity of trees, shrubs, and herbs, ranging from tropical rain-rich regions to dry, arid regions. The importance of the vegetation is due to:

Maintaining biological balance: Vegetation contributes to maintaining biological balance in nature by providing a suitable environment for life. It regulates water flow, maintains soil fertility, and provides shelter and food sources for animals, birds, and insects.

Reducing desertification and soil erosion: Vegetation plays an important role in preventing desertification and soil erosion, as it prevents the effects of wind and heavy rain on the soil and prevents the spread of sand and dust.

Improving air quality: Plants absorb carbon dioxide from the atmosphere and produce oxygen through photosynthesis, which contributes to improving air quality and reducing the effects of global warming.

Water purification: Plants contribute to water purification by absorbing excess nutrients and chemical pollutants from the soil, thus reducing pollution of groundwater and natural water sources.

Biodiversity: Vegetation is home to many different organisms, including plants, animals, fungi, and bacteria, contributing to biodiversity and environmental sustainability.

Despite the importance of vegetation, it is constantly threatened by excessive population growth and human activities. The destruction of vegetation can lead to a decrease in biodiversity and an increase in environmental risks and disasters, such as floods, droughts and land degradation. Therefore, it is essential to recognize the importance of vegetation in protecting the environment and maintaining biological balance on Earth's surface. This can be achieved through awareness and concern for its preservation. (E.O.Thomson;2014)

Figure 4. grassland ecosystem functions

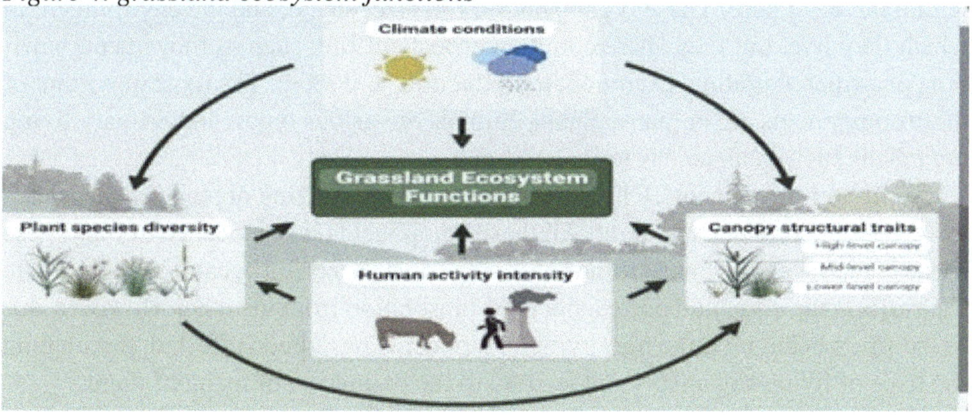

The effect of plant reproduction on the vegetation cover

Plant reproduction plays an important role in the impact of vegetation cover. There are many ways for plants to reproduce, including sexual and asexual reproduction. Sexual reproduction is where flowers pollinate and produce new seeds, which can spread and grow to form new plants.

Asexual reproduction in plants involves spreading and growing through roots, stems, leaves, or buds. These methods help plants reproduce quickly, increasing the vegetation cover in an area. These plants cover bare areas, improve soil quality, and protect it from erosion.

However, it should be noted that some plants can grow excessively and have negative effects such as competition, changing the overall composition of the vegetation, spreading diseases and pests, and changing soil properties. This leads to an imbalance in the ecological balance and negatively impacts on other animals and plants living in the same environment. This requires proper monitoring and management of plant reproduction to maintain healthy plants and ecological balance. (Pueyo et al., 2010)

BIOSPHERE

It is the space, place, or medium in which living organisms live, and it extends from the deepest depths of the seas and oceans to the greatest height above the mountains, the biosphere consists of two main parts of life:

The classification of living components includes all living organisms that live within the biosphere. These organisms vary in size, shape, and the environment in which they live, but they share common aspects of life, such as movement, nutrition, growth, respiration, excretion, reproduction, and more. The most important of these components are humans, plants, animals of various types, and primary living organisms such as algae, bacteria, and fungi.

Nonliving components: These are components that living organisms need, such as soil, water, air, heat, and light. All these components lack evidence of life. These living components are linked to the non-living components in a dynamic system where each affects the other and one cannot exist in isolation from the other. However, due to incorrect behaviors, the non-living components have been affected, threatening the lives of living organisms and leading to the extinction of many of them.

The term biosphere in ecosystems refers to the relationships and interactions between living organisms and their environment, including plants, animals, microbes, and other environmental factors such as air, water, and soil. The biosphere plays a vital role in maintaining the balance of ecosystems. The biosphere contributes to ecosystems' ecological balance and sustainability through its role in nutrient cycling, energy distribution, and regulating competitive and cooperative relationships among organisms. Understanding and supporting the biosphere in ecosystems contributes to the maintenance of the diversity of life and environmental sustainability. (Polunin & Grinevald, 1988)

Figure 5. Earths biosphere

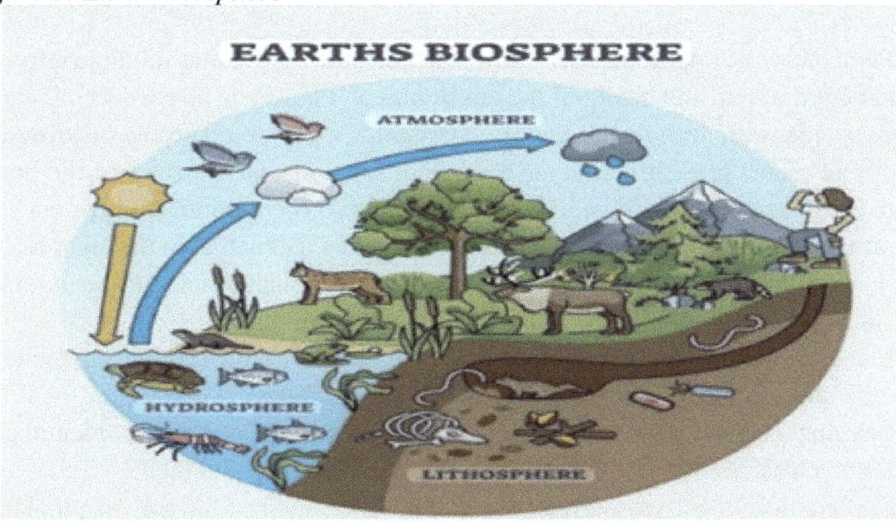

The effect of vegetation on the biosphere

Vegetation plays an important role in maintaining the biosphere and its balance. Dense, diverse vegetation helps stabilize soil and prevent erosion, purifies the air, and absorbs harmful gases. In addition, plants help regulate temperature and humidity in the environment and provide habitat for animals, insects, plants, and microorganisms, contributing to the diversity of wildlife and maintaining ecological balance. Therefore, it can be said that vegetation plays a major role in supporting and maintaining life on Earth and preserving the biosphere.

Vegetation is one of the essential components of the natural environment that plays a vital role in supporting life on planet Earth. The loss of vegetation cover affects not only soil and water, but also biodiversity, leading to the disappearance of natural habitats for plants, animals, and insects, and negatively impacting the stability of ecosystems. (Wantong Li;2024)

Plants are an important part of the ecosystem, connecting not only the lithosphere, atmosphere, and hydrosphere, but also playing a crucial role in maintaining global ecological balance. Vegetation variation is often viewed as an indicator of environmental change across spatial and temporal scales.

Vegetation as an environmental indicator

Plants serve as important environmental indicators, providing valuable information about the state and health of the environment. Plants can be used as indicators to assess many environmental factors. Vegetation cover refers to the quantity and quality of plants present in a given area. This serves as an indicator of the health of an ecosystem because it is influenced by numerous environmental factors such as climate, soil, climate change, and other activities. Human activities that lead to levels of pollution and affect soil quality, ecological balance and wildlife diversity. Reasons why vegetation is an environmental indicator:

Plant diversity: The diversity of plants in each area reflects the diversity of wildlife and its ability to adapt to environmental changes.

Seasonal change: Plants can exhibit changes in seasons and environmental conditions such as the impact of climate change and natural phenomena.

Soil quality: Plants may reflect the quality and health of the soil, as plants can be dense and diverse in healthy soil.

Impact of human activity: Vegetation can be an indicator of the impact of human activity on the environment, such as the destruction of agricultural land or environmental pollution.

Therefore, vegetation cover is considered a sensitive indicator of environmental conditions and can be used as a tool for understanding the environment, assessing its health, and working to protect and preserve it. Plants can contribute to improving environmental quality through vegetative reproduction. They also help purify the air and soil and absorb harmful gases.

VEGETATION INDICATORS

Many indicators are used to measure vegetation cover and assess its condition, such as:

Normalized Degradation Vegetation Index (NDVI): This index is used to measure the amount and strength of green vegetation.

Photochemical Reflectance Index (PRI): This index is used to evaluate the efficiency of plants in using light.

Water Stress Index (MSI): This index is used to assess the water stress status of plants.

Plants are a vital indicator of environmental health and can be used to monitor environmental changes and take measures to maintain ecological balance and biodiversity. (MN Zielaie, et al;2020)

Threats to biodiversity

Biodiversity faces numerous risks and threats to sustainability and its impact on the environment and humans. These threats include:

Human factors: -

- The use and fragmentation of habitats of animal and plant species for agricultural, residential, or commercial purposes.
- Species loss: Climate change and environmental degradation are causing rapid species loss, affecting the balance of ecosystems.
- Environmental degradation: Human activities such as environmental pollution and land use change play a major role in environmental degradation and biodiversity decline.
- Climate change: It leads to changes in the environment and natural conditions, affecting species and increasing the risk of their extinction.
- Overfishing and poaching: Many vital species are subject to overfishing and poaching, leading to a decline in their numbers and threatening their survival.
- Plant and animal diseases: Diseases spread rapidly among plants and animals, threatening the balance of ecosystems and affecting biodiversity.

These activities have already threatened 25% of all mammals, 13% of birds, and more than 21,000 other plant and animal species with extinction (Tilman et al., 2017).

Natural factors: -

Although the impact of natural factors on biodiversity is catastrophic, it is short-lived, and biodiversity often recovers to its previous state.

Examples of natural factors that destroy biodiversity:

1. Hurricanes and storms.
2. Volcanoes and earthquakes.
3. Drought.
4. Water and ocean floods (tsunami).

Small or fragmented habitats (such as small islands) or those containing few species are more affected by these factors and may destroy the entire ecological structure due to its fragility and protection. Therefore, biodiversity and its protection are essential, and the necessary measures and actions must be taken to reduce these risks and threats, and to raise awareness of the importance of preserving the environment and wildlife. (Wilson S.;2013)

CONCLUSION

Reproduction in living organisms is a vital mechanism for the continuation of life, the transmission of genetic traits between successive generations, and the preservation of these traits from extinction, especially good genetic traits. The importance of biodiversity lies in its role in enhancing the strength and efficiency of various ecosystems, and plant propagation is an important means of preserving the genetic diversity of plants.

Plants can reproduce in many ways (sexually and asexually), which helps improve their ability to adapt to change. It also contributes to the conservation of plant genetic resources and ecosystem balance.

Vegetation provides many environmental services essential for preserving the environment and biodiversity. Vegetation can contribute to resisting environmental and climate change and providing solutions for adapting to it. Plants play a vital role in improving air and water quality and mitigating the effects of climate change.

Biodiversity is important for the stability and health of ecosystems. Every living organism plays an important role in the balance of the ecosystem and contributes to the maintenance of biodiversity.

Conserving biodiversity is essential to maintaining a healthy environment and preserving natural resources for current and future generations, Which faces many pressures such as the degradation of natural resources, climate change, pollution, and unsustainable exploitation of natural resources. For these reasons, I need to protect and enhance biodiversity through sustainable and informed actions and decisions to conserve living organisms and ecosystems.

Because it is considered an essential part of the natural and cultural heritage of people and contributes to improving the quality of life and public health, everyone is required to work to protect and preserve it to ensure the sustainability of life on Earth. Reproduction contributes to increasing the diversity of living organisms and their adaptation to the surrounding environment. It also contributes to the renewal of generations and ensuring the continuity of life. This function is considered one of the most important, as it increases genetic diversity among individuals of a single species, allowing it to evolve, improve its characteristics, and adapt to the environment from generation to generation. Its importance stems from its role as the cornerstone of the life cycle of living organisms, and it is essential for the continuation of life on planet Earth.

Genetic diversity plays an important role in plants' adaptation to environmental changes and facing challenges such as diseases, pests, and unfavorable environmental conditions. In addition, it increases the chances of species surviving and reproducing in diverse environments. Sexual reproduction in plants is a key factor in increasing biodiversity and contributes to environmental sustainability and the continuity of biological systems.

Vegetation is an important part of the ecosystem and plays a crucial role in maintaining global environmental balance. Plant diversity is often viewed as an indicator of environmental change over spatial and temporal scales. They are diverse and are considered essential components of the natural environment, playing a vital role in supporting life on planet Earth. The loss of vegetation affects biodiversity, leading to the disappearance of natural habitats for plants, animals, and insects, and impacting the stability of ecosystems. The process of plant reproduction plays an important role in influencing and maintaining the vegetation cover and thus preserving biodiversity on planet Earth.

REFERENCES

Aguilar, R., Ashworth, L., Galetto, L., & Aizen, M. A. (2006). Plant reproductive susceptibility to habitat fragmentation: Review and synthesis through a meta-analysis. *Ecology Letters*, *9*(8), 968–980.

Blicharska, M., Smithers, R. J., Mikusiński, G., Rönnbäck, P., Harrison, P. A., Nilsson, M., & Sutherland, W. J. (2019). Biodiversity's contributions to sustainable development. *Nature Sustainability*, *2*(12), 1083–1093.

Clark, H., & Wu, H. (2016). The sustainable development goals: 17 goals to transform our world. Furthering the work of the United Nations, 36-54.

Getabalew, M., Alemneh, T., & Zewdie, D. (2020). Effect of rangeland vegetation cover on climate change mitigation.

Hardinsyah, H., Dwikurnia, F., & Kurniawan, A. (2007). Faktor-faktor penentu pola konsumsi pangan di Indonesia. [Factors determining food consumption patterns in Indonesia] In Thaha, R., Jalal, A., & Setiawan, B. (Eds.), *Pembangunan gizi dan pangan dalam perspektif kemandirian lokal* (pp. 45–62). IPB Press.

Johri, B. M., & Srivastava, P. S. (2012). *Reproductive Biology of Plants*. Narosa Publishing House.

Michael, G. S. (2010). *Plant reproductive biology*. Science Publishers.

Ollinger, S. V. (2003). *Forest ecosystems. Encyclopedia of Life Sciences*. Nature Publishing Group.

Polunin, N., & Grinevald, J. (1988). Vernadsky and biospheral ecology. *Environmental Conservation*, *15*(2), 117–122.

Pueyo, Y., Kéfi, S., Díaz-Sierra, R., Alados, C. L., & Rietkerk, M. (2010). The role of reproductive plant traits and biotic interactions in the dynamics of semi-arid plant communities. *Theoretical Population Biology*, *78*(4), 289–297.

Sekhar, P. R., Savithri, Y., Kavitha, N., & Kumari, D. A. (2024). Role of biodiversity and its importance in sustainable development: A comprehensive review. *Uttar Pradesh Journal of Zoology*, *45*(8), 184–188.

Tilman, D., Clark, M., Williams, D. R., Kimmel, K., Polasky, S., & Packer, C. (2017). Future threats to biodiversity and pathways to their prevention. *Nature*, *546*(7656), 73–81.

United Nations. (1992). Convention on Biological Diversity, June 5, 1992. https://www.cbd.int/doc/legal/cbd-en.pdf

Wilson, S. (2013). Biodiversity loss, natural disasters, and small island developing states [Briefing paper]. Foundation for International Environmental Law and Development. https://www.field.org.uk/wp-content/uploads/2013/07/SIDS-Vulnerability-Brief-2013.pdf

Chapter 8
Modern and Traditional Propagation Techniques for Fruit Crops

Waleed Fouad Abobatta
https://orcid.org/0000-0002-1063-3165
Horticulture Research Institute, Agricultural Research Center, Giza, Egypt

ABSTRACT

Traditional and modern propagation techniques have their place in fruit crop production. Traditional ones remain a reliable backbone of fruit cultivation, while modern ones address critical challenges, e.g., disease, sterility, and genetic conservation. Traditional methods, e.g., cuttings, layering, and grafting, are well-established and effective for many species. Modern techniques are achieving sustainable fruit production. By leveraging these methods, the fruit crop industry can address challenges related to uniformity and disease control, ultimately enhancing productivity and quality. Tissue culture and somatic embryogenesis offer advanced solutions for producing rapid and disease-free seedlings. The choice of technique depends on the specific requirements of the crop, available resources, and desired outcomes. The differences between traditional and modern techniques encompass their environmental impacts, resource efficiency, economic implications, and technological advancements. The future lies in using biotechnology to enhance traditional systems to sustain fruit production.

1. INTRODUCTION

Fruit crops are important for human health because they provide essential nutrients like vitamins, fiber, and sugars. To grow these crops, producers use both traditional

DOI: 10.4018/979-8-3373-5626-6.ch008

and modern propagation methods to produce seedlings. Fruit crops are propagated through sexual (seed-based) and asexual (vegetative) methods to maintain genetic consistency, improve yield, and enhance disease resistance. Traditional techniques, i.e. grafting, cutting, and rooting cuttings, have been used for centuries because they are simpler and cost-effective, helping to grow plants that are true to their parent traits (Prakash & Paul, 2021).

Healthy propagative material is crucial for successful fruit growing. It is essential to ensure the production of seedlings to prevent plant death and the spread of diseases, which can lead to higher costs and lower yields. Some diseases can even make growing fruit impossible in certain areas (Ventura, et al., 2019). Using budwood from resistant cultivars of fruit trees is important for producing healthy seedlings. Other recommended methods include integrated disease management, cultural and biological practices, proper substrate preparation, and balanced nutrition. Nurseries should use pathogen-free materials, maintain clean conditions, and manage water quality for irrigation. Regular cleaning and proper disposal of materials are essential, as well as keeping records of seedling production and controlling access to nurseries (Khan & Korban, 2022). The enduring importance of traditional propagation methods in fruit cultivation remains the backbone of global fruit production due to their unique simplicity, cost-effectiveness, and reliability.

Both sexual and asexual propagation techniques are used to grow fruit crops and produce seedlings. The procedure of reproducing and multiplying plants is known as Traditional techniques. It allows plants to create seeds for producing offspring. Alternatively, through an asexual propagation technique, it obtains specific traits from trees that might not be preserved through sexual reproduction (Awotedu, et al., 2021).

Traditional methods support complex agroecosystems; they are ideal for propagating the diverse mix of fruit trees, shrubs, and rootstocks that required for agroforestry systems, forests, and mixed farms, which are increasingly recognized as models for resilient and sustainable food production (Ak, et al., 2021).

Traditional propagation techniques include sexual and asexual methods to produce fruit seedlings. Sexual propagation is essential for developing new plants and rootstocks utilized in breeding programs to create genetic diversity and variations in yield and fruit quality (Rathour & Kumar, 2023). While modern propagation offers precision and scale, traditional methods provide sustainability, accessibility, and resilience (Devi, et al., 2025).

Seed propagation is the oldest and simplest method for fruit crops, mainly used for species that do not root well vegetative or for producing rootstocks (Wani, et al., 2022). Several practices are necessary for the sexual reproduction of fruit seedlings to promote healthy germination and growth of healthy seedlings. However, due to genetic variation and late bearing, most commercial fruit orchards prefer vegetative or modern propagation methods (Din, et al., 2019).

Asexual techniques include budding, grafting, and tissue culture, which help produce large numbers of plants in a short time. They are used to produce virus-free seedlings for fruits such as bananas and apples (Iqbal & Singh, 2020). A combination of these methods supports consistent fruit production, preserves selected varieties, increases disease resistance, and ensures uniformity. These methods support consistent fruit production, preserve selected varieties, increase disease resistance, and ensure uniformity.

Asexual propagation using stem cuttings is common in horticulture for multiplying fruit crops (Abobatta, 2025). Auxins such as naphthalene acetic acid (NAA) and indole-3-butyric acid (IBA) improve the rooting ability and quality of these cuttings. Dipping the base of the cutting in auxins stimulates the formation of adventitious roots that grow from stems or leaves rather than from existing roots. They promote root growth, cell division, and root patterning. Applying auxins to stem cuttings of various fruits, such as pomegranate, grape, and dragon fruit, enhances rooting success, root quality, and shoot development (Saini, 2024).

Modern propagation techniques play a crucial role in enhancing the sustainability and yield of fruit crops. Advanced biotechnologies, such as somatic embryogenesis and protoplast fusion, offer effective solutions to specific challenges. Modern advancements have revolutionized propagation, ensuring higher efficiency and scalability. Modern propagation techniques (e.g., tissue culture, somatic embryogenesis, micropropagation, and genetic engineering) offer potential environmental benefits and resource efficiency and are increasingly adopted in fruit crop production due to their precision, efficiency, and ability to overcome limitations of traditional methods. These techniques are essential for commercial nurseries, disease-free planting material, and genetic improvement (Khaitov, et al., 2024).

Modern propagation techniques offer significant advantages over traditional methods in terms of yield, disease management, and sustainability (Ahmar, et al., 2020). While they require higher initial investments and specialized skills, their long-term benefits make them a viable option for sustainable fruit crop production.

Tissue culture is more than a modern propagation technique; it is a sustainable agriculture tool that ensures, year-round, large-scale, uniform plant supply, genetic improvement for future challenges, and reduced dependence on chemical inputs. Thus, tissue culture bridges productivity, sustainability, and resilience in fruit crop production systems (Abdalla, et al., 2022).

Somatic embryogenesis (SE) is a cornerstone technology in modern plant science and biotechnology; it is considered an effective and powerful tool for plant propagation by harnessing the totipotency of plant cells. SE integrates plant propagation, conservation, and genetic improvement. Acts as a platform for synthetic seeds, genetic engineering, and biotechnology research and is considered a bridge between traditional horticulture and modern molecular breeding to support sustainable agriculture. (Aronen, et al., 2025).

Auxins are a quintessential example of appropriate technology for various reasons. It is a simple, affordable, and scalable solution that addresses a fundamental need with a disproportionately positive impact. By democratizing the ability to clone superior plants, it puts power directly into the hands of farmers and communities, fostering self-reliance, enhancing livelihoods, and promoting sustainable agricultural development (Bhatla & Lal, 2023).

For developing countries, in particular, the affordability and ease of using auxin-based rooting powders make them an accessible and transformative technology, allowing small-scale farmers to propagate improved varieties and increase their yields with minimal investment. They are a prime example of how a fundamental understanding of plant physiology can lead to profound practical benefits in agriculture (Pincelli-Souza, et al., 2024).

The environmental conditions influence the propagation and survival of seedlings. The ideal conditions for propagation are contingent upon factors, including temperature, humidity, light intensity, and soil conditions (Khezri, et al., 2024). In traditional propagation, temperature affects seeds and cuttings, which root more rapidly at higher temperatures. In modern propagation, humidity prevents desiccation in tissue culture seedlings and young plants. Light intensity influences photosynthesis and the development of seedlings, while, soil aeration and water supply are critical factors for root growth.

The environmental effects of traditional and modern propagation techniques of fruit crops differ greatly, depending on the use of resources, chemical inputs, energy usage, and ecological consequences. Traditional methods have a smaller energy footprint but run the hazards of soil erosion and chemical contamination if not handled properly (Manohar, et al., 2024). Modern methods provide accuracy and disease management, consume high energy, and produce harmful waste. The most effective way to lessen the environmental impact of fruit propagation is to strike a balance between production and ecological stewardship through site-specific solutions.

Modern propagation techniques encounter multidimensional barriers in developing countries, including high costs and weak infrastructure due to socio-political neglect. While they hold promise for addressing disease, sterility, and climate resilience, the main constraints include context-adapted protocols for local crops, affordable, decentralized production models, and strong farmer-researcher-market linkages (Singh, et al., 2021). While traditional propagation methods for fruit trees, such as cuttings and grafting, are still important for small-scale farming, modern techniques, such as tissue culture and genetic engineering, are changing the way fruit is propagated on a larger scale. Each method has its benefits; traditional techniques are reliable and accessible, while modern techniques can address issues like disease and genetic preservation.

The future of fruit cultivation lies not in replacing one with the other but in strategically combining both. Combining traditional and modern propagation methods can enhance propagation efficiency and make fruit production more efficient and sustainable (Ak, et al., 2024).

The future of fruit cultivation will not be about choosing one method over the other, but rather about integrating them effectively to ensure that fruit production is sustainable and resilient. Growers will need to consider their scale, resources, and specific goals when deciding on the right approach for their needs. Modern techniques can help improve uniformity and disease control, ultimately leading to better productivity and quality in the fruit crop industry.

This chapter discusses both traditional and modern fruit propagation methods, covering their advantages, limitations, and applications in sustaining fruit production. Furthermore, compare the environmental impacts, effectiveness, and cost-effectiveness, as well as the major differences in genetic diversity between the propagation techniques. It also explores the challenges and limitations of applying modern propagation techniques in developing countries.

2. TRADITIONAL PROPAGATION TECHNIQUES

Traditional propagation remains important because it democratizes fruit cultivation. Despite the rise of advanced biotechnological propagation techniques, traditional methods remain indispensable in fruit cultivation. These time-tested practices—such as seed propagation, cuttings, grafting, layering, and budding—continue to play a vital role in different-scale farming, biodiversity conservation, and sustainable agriculture (Yadav & Singh, 2018).

Traditional propagation is inherently aligned with the principles of agroecological, sustainable, and low-input agriculture. It empowers farmers worldwide to sustain fruit production with minimal resources, preserving cultural heritage and genetic diversity, and being simple, cost-effective, and reliable (AK, et al., 2021).

2.1. Seed Propagation

The oldest and simplest method involves sowing seeds extracted from fruits. Used for crops like mango, papaya, and citrus (though often leads to variability). Improving seed propagation methods not only aims to improve germination, but it also involves a robust and efficient system that forms the foundation for the success of fruit nurseries. Combining traditional knowledge with the use of technology is essential for enabling producers to achieve the highest crop yield, ensure high quality, and sustain productivity. The advantages of seed propagation include low cost, ease of performance, and maintenance of genetic diversity (Joshi, et al., 2023).

Advantages of seed propagation

a) Simple, economical, and easy.
b) Produces vigorous, long-lived trees with strong taproots.
c) Some species (e.g., papaya, phalsa, and loquat) only propagate by seeds.
d) Produce rootstock for asexual propagation of fruits (e.g., citrus, mango, and apple).
e) Efficient tool in breeding programs (developing hybrids, new varieties)

For fruits such as papaya (*Carica papaya* L.) and guava (*Psidium guajava* L.), seed propagation is more practical. When seeds are sown in plastic bags with well-drained soil, germination exceeds 90% within 15-20 days, with a 50-60% survival rate for seedlings (Pereira, et al., 2017).

Healthy seedling production through sexual propagation requires numerous steps, including:

a) Quality seed selection
b) Proper treatment before cultivation includes fungicide treatment to prevent seed rot, treatment with growth regulators (GA_3) to improve germination in dormant seeds, and scarification/stratification where required (e.g., apple, pear, loquat).
c) Suitable sowing media & depth
d) Good nursery management (watering, shading, weeding, and mulshing)
e) Pest/disease protection to prevent damping-off
f) Proper hardening before transplant

There are some cautions during the production of seedlings by sexual propagation, including:

a) Seed Dormancy: Many fruit seeds (e.g., apple, pear, mango) require stratification or scarification to break dormancy (Arpitha & Navya, 2025)
b) Seed Viability: Citrus and guava seeds remain viable for months, while mango seeds lose viability within a few weeks (de Carvalho, et al., 2021).

Disadvantages of Seed Propagation

a) Seedlings are not true-to-type (high genetic variability).
b) Long juvenile phase, which delays fruiting (more than 5 years for mango and guava).
c) Variable fruit quality (size, taste, color, and yield).
d) Risk of disease transmission.
e) Some seeds show dormancy or short viability (e.g., mango, citrus).

It is well documented that the limitations of seed propagation include a long juvenile phase and high genetic variability, which result in inconsistent fruit quality and negatively impact commercial production (Din, et al., 2019). Furthermore, some seedless fruits (e.g., navel orange, seedless grapes, and seedless guavas) cannot propagate sexually (Musacchi & Neri, 2019).

Advantages of traditional techniques

a) Economic Necessity: For 80% of the world's small farmers, modern methods are too expensive.
b) Ecological Wisdom: Traditional techniques co-evolved with local ecosystems.
c) Cultural Legacy: Seed-saving and grafting are living agricultural heritage.
d) Low Energy Footprint: It relies primarily on solar energy and human labor, not fossil-fuel-powered electricity, for artificial lighting and climate control.
e) Minimal Chemical Inputs: While auxins are used, traditional propagation does not require the sterile, synthetic media and chemical cocktails for tissue culture.

Therefore, traditional techniques could be incorporated into organic farming practices. The main steps to produce healthy fruit seedlings through sexual propagation include

a) Seed selection from open-pollinated (non-hybrid) fruits
b) Seed treatment before sowing seeds
c) cultivation
d) Provide optimal germination Conditions
e) Seedling care
f) Transplanting

2.2. Asexual propagation

Asexual techniques are preferred for their cost-effectiveness and simplicity. In sapling production, asexual or vegetative propagation is significant for efficiently reproducing various fruit species and rootstocks. Asexual methods ensure true-to-type seedlings, preserving desired traits, and are crucial for sustainable, high-yield, healthy fruit production (Iqbal & Singh, 2020).

Techniques such as grafting, cuttings, and tissue culture improve disease resistance, enhance fruit quality, and promote uniformity in orchards. A common method involves grafting micro-propagated plants onto seedling rootstocks, such as by T-budding of navel orange (Abobatta, 2025).

Main asexual propagation techniques

a) Cutting: Cuttings are the use of pieces of a plant (stem, leaf, or root) placed in a suitable environment and conditions to form a complete, independent new plant. The advantage of cutting propagation is the uniformity of the seedlings and their similarity to the mother plant (Yadav & Singh, 2018). Cutting include hardwood cuttings (e.g., grapes, figs) and softwood cuttings (e.g., guava, pomegranate). The chosen type of cutting depends on the plant species and the propagation season to achieve the highest success rate.

Key factors affecting the success of various types of cuttings:

1. Hormones (auxins): Using rooting hormones increase the success rate and regulation seedling growth.
2. Rooting medium: Using a mixture of sterile, well-drained, and moisture-retaining materials, such as (perlite, vermiculite, coarse sand, and peat moss), increase the success of cuttings.
3. Humidity: Providing adequate humidity, especially when using young or leafy cuttings, reduces water loss and stimulate root formation. While, over humidity increase root rot diseases.
4. Temperature: A warm medium (21-24°C) encourages callus formation and root growth.
5. Plant material: Using plant material from a vigorous mother plant with the desired characteristics and free of disease is important for successful cuttings.
   b) Budding and Budding: Joining a scion (desired variety) with a rootstock (root system), including whip grafting (e.g., apple, pear), cleft grafting (e.g., avocado, mango), and T-budding (e.g., citrus, stone fruits).

c) Layering: a technique of air layering is preferred in litchi and mango, while for raspberries and blackberries, ground layering is the proper technique for propagation (Haokip, et al., 2023).
d) Grafting: the grafting technique combines desirable traits (hardy rootstock with high-quality scion). There are numerous grafting techniques, including inarching, veneer, cleft, side, and approach. It's used to propagate many fruit varieties, e.g., citrus, mango, cashew, grapes, and pears (Shivran, et al., 2022).

Enduring importance of traditional propagation methods in fruit cultivation. They may have environmental impacts related to seed germination, agronomic management of mother plants, and rooting of cuttings (Awotedu, et al., 2021).

Limitations of traditional methods

a) Disease/Pest resistance: Limited by available resistant rootstocks
b) Sterile cultivars: Cannot propagate seedless varieties (e.g., bananas, navel orange).
c) Slow breeding: Decades to develop new rootstocks.
d) Rejuvenation: Mature trees lose vigor; hard to propagate.
e) Scale-Up: Labor-intensive for mass production

Where Traditional Methods Outperform Modern Ones?

While modern propagation techniques, e.g., tissue culture and somatic embryogenesis, offer remarkable capabilities, they are not a panacea. There are numerous critical scenarios where traditional propagation methods are not just viable, but demonstrably superior in terms of practicality, economics, and overall success (Manohar, et al., 2024).

The differences between traditional and modern propagation techniques for sustainable fruit crop production, the environmental impacts of these techniques, their resource efficiency, economic implications, and technological advancements. A balanced integration of both approaches ensures sustainable and high-quality fruit crop production (Table 1).

Table 1. Main differences between traditional and modern propagation techniques for sustainable fruit crop production.

Aspect	Traditional Techniques	Modern Techniques
Propagation Method	Seed, cutting, layering, grafting, budding	In vitro propagation, somatic embryogenesis, genome editing
Cost	Low	High
Skill Requirement	Moderate (grafting)	High (tissue culture, genetic engineering
Yield	Lower, variable	Higher, more consistent
Scalability	Limited	High (mass production)
Disease Management	Limited, higher risk of disease spread	Effective, production of disease-free plants
Time Efficiency	Slow (months/years)	Rapid (weeks/months
Genetic Uniformity	Variable (except grafting)	Highly uniform
Sustainability	Less efficient, higher chemical use	More efficient, eco-friendly practices

Traditional propagation methods remain the superior choice for developing countries, they are more democratic, more resilient, more adaptable, and more economically viable. This is because it represents a form of technology that is deeply integrated with human skill and ecological reality, ensuring its enduring importance for generations to come (Manohar, et al., 2024).

Figure (1) shows that the superiority of traditional methods versus modern techniques often boils down to a few core principles.

Figure 1. Core principles of traditional propagation methods

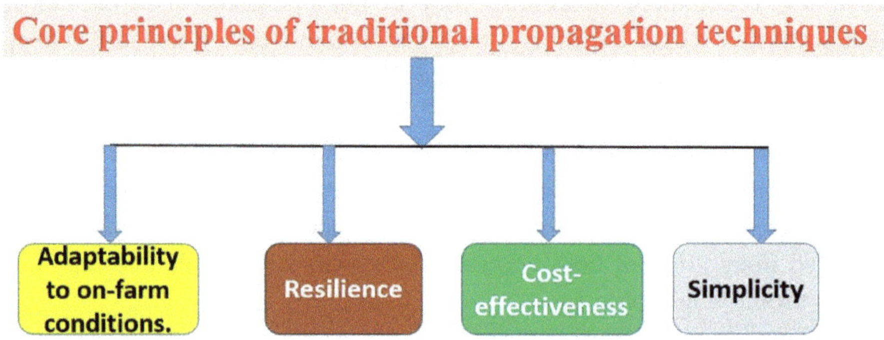

Regarding asexual propagation techniques, various factors determine the proper technique for each type or group of fruit tree. Table (2) shows the differences between asexual techniques, their advantages, and the determining factors for each technique.

Table 2. Comparing asexual propagation techniques

Technique	Benefits	Challenges
Cuttings & Micropropagation	Maintains plant characteristics, high-quality production	Low growth and survival rates
Tissue Culture	High-quality, virus-free plants, true-to-type	Requires skilled personnel, high costs
Grafting & Budding	Combines desirable traits, improves vigor and productivity	Susceptible to bacterial rot
Container Cultivation	Controlled environment, maximizes yield and quality	Substrate sustainability
CRISPR-Cas Genome Editing	Precise modifications, enhanced resilience	Ethical and regulatory considerations
Marker-Assisted Selection	Accelerates breeding, ensures high-quality traits	Requires advanced technology

Propagation of fruit trees in Egypt

Propagation techniques directly affected fruit quality, uniformity, orchard productivity, and export standards. Egypt is a leading fruit producer and exporter in the Middle East & North Africa. Main fruits include citrus (oranges, mandarins, lemon, and grapefruit), grapes, mango, guava, date palm, pomegranate, figs, and olives (Beshara, et al., 2024).

Traditional Propagation in Egypt

Seed propagation in Egypt is mainly used for papaya, loquat, guava, and producing rootstocks of citrus and mango. Vegetative propagation primarily employs cuttings for grapes, figs, pomegranates, and olives. Budding is used for producing citrus seedlings, with T-budding and patch budding being widely adopted. Other propagation methods include grafting for mango (common veneer and side grafting) and suckers or offshoots for date palms and bananas.

Egypt relies on both traditional methods (cuttings, grafting, and budding) and modern tissue culture for fruit propagation. Citrus & grapes dominate traditional propagation, while banana and date palm lead in modern lab-based micropropagation (tissue culture). The future lies in certified, disease-free, and large-scale propagation systems to support Egypt's role as a major fruit exporter.

Case Study (1): Olive propagation in Egypt "Hardwood cuttings"

Outcome: *Cuttings produce more resilient trees adapted to local drought conditions.*

Egypt is the world's largest producer of table olives, with over 100 million olive trees cultivated across New Valley, Matrouh, and North Sinai (Taha & Khalifa, 2024). Propagation methods vary from traditional cuttings to modern tissue culture, each with distinct economic and agronomic impacts.

Among the traditional propagation techniques used for the olive tree, stem cutting stands-out. It is considered the most important and widely used commercial method, providing a large number of high-quality plants from a limited source of parent material in a short time. It is also inexpensive, rapid, and simple and does not require the special techniques necessary in grafting or budding. Using auxin treatment, such as IBA, enhances cuttings' rooting (Eid, et al., 2018). Generally, previous studies have shown that June is the best time for rooting olive cuttings, with the highest success rates compared to other times of the year (Gad & Ibrahim, 2018).

Methodology Technique: Semi-hardwood cuttings treated with IBA (Indole-3-butyric acid) sown in wooden boxes or polyethylene pages directly. Variety: Picual (high oil content, drought-resistant). Location: Nursery of Horticulture Research Institute, Agricultural Research Center, Giza, Egypt.

Advantages

a) Low cost, no lab required.
b) Farmers reuse pruning waste for cuttings.
c) Adapted to conditions of arid region.
d) the highest significant rooting percentage was obtained in June collection

What Works Best in Egypt?

For Subsistence Farmers: Stick with cuttings (low risk, zero cost).
For Export-Oriented Farms: Invest in grafting (balance of yield and resilience).
For Government Projects: Limited tissue culture (niche use for disease-free stock)

Case Study (2): Citrus Propagation in Egypt

Egypt is the 6th largest citrus producer globally, with over 5 million tons in 2024 (Annual Reports 2024), primarily oranges (70%), mandarins (15%), and lemons (10%).

The country's success in citrus cultivation relies on natural climatic advantages, such as abundant sunlight and warm temperatures, fertile soils, efficient propagation techniques, effective management practices, and strategic planning (Abobatta, 2018).

Figure 2. Bud grafting technique for citrus seedlings (Photo by Dr. Waleed Abo-batta, 2023)

Dominant Method of Citrus Propagation in Egypt

Most Egyptian citrus nurseries rely on balancing traditional grafting, particularly T-budding, due to the low skill requirements and high compatibility with local rootstocks (Abobatta & Ahmed, 2023).

a) Seed Propagation to produce rootstocks (e.g., Volkameriana, Sour Orange, Rough lemon, and Cleopatra mandarin).
b) T-Budding: the most common varieties produced, about 905 of citrus seedlings (Figure2).
c) Grafting & Budding (Dominant Method) include various techniques as follows
d) Cleft grafting: use in old orchards to renovate trees
e) Side Veneer grafting: Side Veneer grafting to change the varieties or high-value varieties (e.g., 'Valencia').
f) Cutting Propagation: For produce rootstocks, particularly in Upper Egypt, for propagating Rough lemon rootstock.

Preferred Rootstocks in Egypt:

a) **Sour Orange** (*Citrus aurantium*) semi-dwarf rootstock→ Offers several advantages for citrus cultivation in Egypt and exhibits some resistance to certain citrus diseases.
b) **Volkamer Lemon** → Vigorous rootstock, preferred for sandy soils, particularly for Valencia orange production.
c) **Cleopatra Mandarin** → Tolerant of clay soil conditions, it grows well in clay soil, as in the Delta, with a delicious fruit taste.

Propagation Techniques for Fruit Crops

Modern propagation techniques are integral to achieving sustainable and high-yielding fruit crops (Figure 3). By leveraging these methods, the fruit crop industry can address challenges related to uniformity, disease control, and environmental sustainability, ultimately enhancing productivity and quality (Khaitov, et al., 2024). Modern techniques such as tissue culture and micropropagation offer benefits (e.g., disease control and management of genetic resources), potentially reducing the environmental impact.

Micro-propagation (Tissue Culture)

Micro-propagation involves the aseptic culture of plant cells, tissues, or organs in a controlled environment to produce genetically identical clones. Tissue culture is a Tool for Sustainable Agriculture, offers numerous benefits in the propagation of sustainable fruit crops, including rapid clonal propagation, high-quality planting material, and genetic improvement. It is widely used for mass multiplication of elite cultivars and conservation of rare species, where traditional methods are time-consuming and less effective in obtaining virus-free plants, such as banana, pineapple, strawberry, and date palm. Shoot tips, nodal segments, or meristems are preferred due to high mitotic activity, for instance, banana meristem culture for virus-free planting material (Jain, et al., 2023).

Figure 3. Modern propagation technology

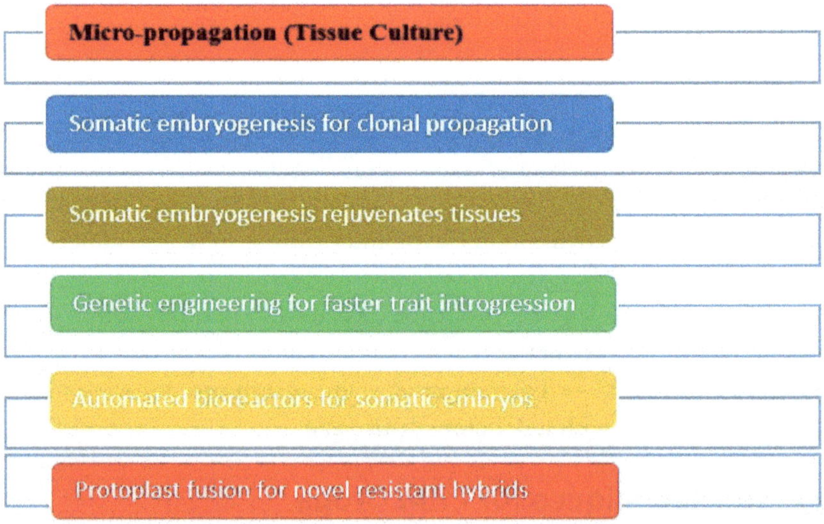

Tissue culture's potential for year-round production and genetic enhancement positions it as a valuable tool for sustainable agriculture, with ongoing research and biotechnological applications shaping its prospects (Uysal, et al., 2018).

The main advantages of the tissue culture technique involve mass production (millions/year), disease elimination (virus-free plants), and germplasm conservation and propagation of rare species.

Contribution to Sustainable Agriculture:

a) Produce disease-free plants: Tissue culture eliminates viruses, fungi, and bacteria, consequently reducing pesticide use.
b) Preserve natural resources: Produces millions of seedlings from a small quantity of plant materials.
c) Resource efficiency: Uniform plants mature together, reducing inputs for irrigation, fertilization, and harvesting.
d) Increase climate resilience through rapid production of stress-tolerant cultivars.

Numerous fruit crops that can be propagated by tissue culture, including

a) Banana: Production of disease-free Cavendish clones.
b) Pineapple: Rapid multiplication of smooth cayenne varieties.
c) Date Palm: Rescue of rare cultivars through somatic embryogenesis.

Thus, tissue culture bridges productivity, sustainability, and resilience in fruit crop production systems.

The limiting factors for tissue culture include high setup cost (laminar flow hoods, media), knowledge dissemination, risk of somaclonal variation, acclimatization, and professional labor.

Somatic Embryogenesis (SE)

Somatic embryogenesis (SE) or Cloning Plants from Somatic Cells, is a remarkable asexual process in modern plant biotechnology. Somatic cells (non-reproductive body cells, like leaf, stem, or root cells) are induced to develop into structures that resemble zygotic embryos (the embryos formed from the fertilization of an egg cell) (Aronen, et al., 2025). These "somatic embryos" can then germinate and grow into complete, fertile plants (Figure 4).

Figure 4. Process of Somatic Embryogenesis

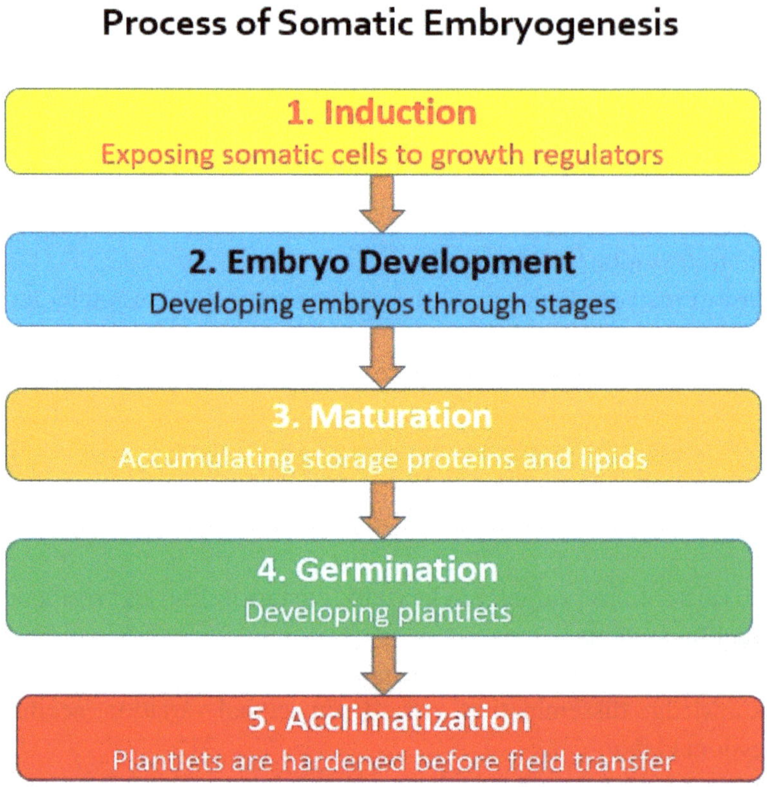

It is an effective and powerful tool for plant propagation by harnessing the totipotency of plant cells, their unique ability to dedifferentiate and regenerate into a completely new organism under the right conditions. It enables the clonal propagation of elite plants (e.g., mango, citrus), genetic engineering, conservation, and fundamental research into plant development (Henao-Ramírez, *et al.*, 2020).

Major Applications of Somatic Embryogenesis

a) **Mass Cloning** (Micro-propagation): Produces large numbers **of** true-to-type, uniform plants.
b) **Synthetic Seeds** (Artificial Seeds): Enabling easy storage, transport, and sowing.
c) **Genetic Transformation:** Embryogenic cultures serve as a platform for gene transfer and transformation (e.g., CRISPR editing, Agrobacterium-mediated transformation).

d) **Germplasm Conservation:** Embryos can be **cryopreserved** for long-term storage of elite cultivars or endangered species.
e) **Stress Tolerance & Breeding**: Somatic embryogenesis is key for producing plants with enhanced resistance to drought, salinity, or diseases.
f) **Haploid Plant Production:** Enables faster development of homozygous lines, more efficient hybrid breeding, and valuable genetic resources for fruit crops.

Advantages

a) High multiplication potential.
b) Produces virus-free and pathogen-free plants.
c) Uniformity and true-to-type propagation.
d) Suitable for both commercial fruit crops and forestry species.

Challenges and Limitations

a) Species and Genotype Dependence.
b) Low Efficiency.
c) Somaclonal Variation.
d) Abnormalities.
e) Maturation and Germination.
f) High Cost.
g) Scale-Up.

Protoplast Fusion (PF)

Protoplast Fusion (Somatic Hybridization) is a sophisticated plant breeding technique rather than a direct propagation method like grafting or cuttings. It is a powerful, albeit challenging, tool for fruit crop improvement, primarily focused on creating novel parental lines (Khalil, et al., 2023).

Developing superior rootstocks with combined disease resistance and abiotic stress tolerance from sexually incompatible wild relatives is the main goal of PF. It also offers a direct route to seedless triploid varieties. The novel rootstocks and scion varieties it generates are then propagated vegetatively (via tissue culture, cuttings, or grafting) for commercial fruit production (Ranaware, et al., 2023).

Despite significant technical hurdles and limitations, it remains a vital technique for overcoming breeding barriers and introducing valuable traits into fruit crops where conventional methods fail.

What is Protoplast Fusion technique?

a. Protoplast Isolation: Plant cells (usually from mesophyll tissue or cell suspensions) are treated with enzymes (cellulase, pectinase) to digest their rigid cell walls, leaving behind naked protoplasts (the plasma membrane and its contents) (Gopal & Sahadevan, 2025).
b. Fusion: Protoplasts from two different parents (different species, genera, or even varieties) are induced to fuse together. Common methods include, Chemical Fusion, Electrical Fusion, and Hybrid Selection (Ranaware, et al., 2023).
c. Hybrid Selection: The fused protoplasts (heterokaryons) are cultured. Selecting the true somatic hybrids from unfused parental protoplasts and homokaryons (fused protoplasts from the same parent) is critical. This is often done using complementary selection, fluorescent labeling, or manual selection (Kaur, et al., 2024).
d. Regeneration: Selected hybrid protoplasts are induced to regenerate a new cell wall, divide, and form callus. This callus is then stimulated (using specific plant growth regulators) to undergo somatic embryogenesis or organogenesis to regenerate whole plants (Klimek-Chodacka, et al., 2020).
e. Characterization: Regenerated plants are rigorously tested to confirm their hybrid nature (using morphological, cytological (chromosome counts), biochemical (isozymes), and molecular (DNA markers, GISH—Genomic in Situ Hybridization) techniques) (Ovcharenko, et al., 2023).

Advantages for fruit improvement

PF overcoming breeding barriers, It is well documented that traditional breeding techniques face significant hurdles, for instance, sexual incompatibility, long generation times, complex genetics, and sterility. Protoplast fusion bypasses sexual reproduction entirely (Reed & Bargmann, 2021).

a) Somatic hybridization creates novel combinations in a single step (though regeneration still takes time).
b) Proper for propagating many commercial seedless fruit varieties.
c) Effective in generating triploid or tetraploid.
d) Enables novel combinations of nuclear and cytoplasmic genomes.
e) Faster than multiple generations of backcrossing.

Challenges and Limitations

a) Technical Difficulty
b) Somatic Incompatibility
c) Genetic Instability
d) Unpredictability
e) Sterility
f) Linkage Drag
g) Cost and Time
h) Regulatory Hurdles

3. ROLE OF AUXINS IN IMPROVING FRUIT CROP PROPAGATION

Auxins are among the most powerful, versatile, and cost-effective tools in the arsenal of plant propagation, playing a pivotal role in both traditional and modern fruit crop cultivation. Their ability to naturally regulate plant growth makes them indispensable for improving the efficiency and success of propagation (Teiba, et al., 2023).

What are Auxins?

Auxins are a class of plant hormones (phytohormones) that act as potent chemical messengers, regulating a wide array of growth and developmental processes. Indole-3-Acetic Acid (IAA) is considered the regular auxin used to improve propagation. However, for commercial and research use, synthetic auxins are preferred because they are more stable, cost-effective to produce, and often more potent (Nemhauser, 2018).

Auxins are a cornerstone of effective and economical fruit crop propagation (Table 3). They bridge the gap between traditional wisdom and modern science, offering a simple, low-cost solution to one of the biggest challenges in horticulture: "reliably creating new plants from existing ones." (Wang, et al., 2024)

These auxins are useful for propagators in various ways that include rescuing difficult-to-root varieties, accelerating the production cycle in nurseries, standardizing and improving the quality of planting material, and enabling advanced techniques, i.e., somatic embryogenesis.

Auxin-based rooting powders are affordable and easy for fruit growers to use in traditional propagation techniques, particularly in developing countries. They help small-scale farmers grow strong plants and increase their yields at low cost (AI Makhmari, 2016).

Main roles of auxins in fruit crop propagation include:

a) Stimulate root initiation in cuttings and layers.
b) Enhance graft/bud union success.
c) Promote growth of rootstock seedlings.
d) Induce callus, embryogenesis, and rooting in tissue culture.

How auxins improve propagation?

Auxins have numerous benefits in various propagation techniques by influencing cell division, elongation, and differentiation (Table 3). For traditional propagation techniques, many desirable fruit tree and shrub varieties (e.g., apples, pears, avocados, and some citrus rootstocks) are difficult to root from cuttings naturally.

Table 3. Benefits of auxins in propagation of fruit trees

Feature	Explanation	Impact
High Efficiency	They directly target the fundamental physiological process of root formation	Dramatically increases rooting percentages
Low Cost	Synthetic auxins like IBA and IAA are inexpensive to manufacture. A small amount treats thousands of cuttings	A tiny bottle or rooting hormone powder costs a few amount of money and provides immense value
Ease of Use	Application methods are simple (dipping or ducting) and required no special equipment in traditional propagation	Accessible to everyone, from large-scale commercial nurseries to smallholder farmers and home gardens
Speed and uniformity	Auxins not only increase the number of cuttings that root but also make them root faster and more uniformly	Reduces the time and space required in propagation beds, leading to faster crop turnover.
Versatility	They are effective across a vast range of fruit species and can be used in both low-tech (field) and high-tech (laboratory) settings.	A single tool for multiple applications, making it a cornerstone of propagation science.
Reduced waste	Increasing the success rate and consuming fewer valuable plant materials that are wasted on failed cuttings.	Enhances the efficiency and sustainability of propagation programs.

Generally, IBA and IAA are the most common auxins used to enhance rooting cuttings, which is the most common and impactful use of auxins in fruit propagation (Sekhukhune & Maila, 2025).

Dipping the base of the cutting in a low concentration of synthetic auxin (IAA or IBA) for a short period stimulates the formation of adventitious roots. This could be due to triggering the dedifferentiation of cells, which reorganize to form root initials (root primordia) and develop into the root system (Nale, et al., 2024).

Harnessing the natural power of these hormones leads to numerous advantages, including

a. Rescue difficult-to-root varieties.
b. Accelerate the production cycle in nurseries.
c. Standardize and improve the quality of planting material.
d. Enable advanced techniques like somatic embryogenesis.

Here is how auxins are applied across different propagation techniques.

Methods of Application:

There are different methods to use auxins, while the common method is Talcum powder, followed by using liquid solution, and gel form is rarely used in commercial seedling production.

Various auxins are used in propagation techniques, while the most popular ones used in propagation are

a) IBA (Indole-3-Butyric Acid): The "gold standard" for rooting. It is highly effective at inducing root initiation and is less likely to cause toxic side effects than other auxins. It is also relatively stable.
b) NAA (1-Naphthaleneacetic Acid): A very effective rooting agent and often used in tissue culture. It is more stable than IBA but can be more phytotoxic (toxic to plants) at higher concentrations.
c) 2,4-D (2,4-Dichlorophenoxyacetic Acid): A very potent and inexpensive auxin. Its main use in tissue culture is to promote somatic embryogenesis and callus development. It is generally too strong and non-selective for use in rooting cuttings, as it can easily damage the plant.

Using auxins in Modern Propagation

Auxins have essential roles in various steps of modern propagation. It is a crucial component of sterile culture media used in laboratories for numerous purposes, including callus induction, somatic embryogenesis, and rooting of microshoots. In addition, auxins (especially 2, 4-D) are used in somatic embryogenesis to induce somatic cells to become embryogenic and form embryos (Babich, et al., 2020).

a) Forming callus: In modern propagation techniques such as tissue culture and somatic embryogenesis, 2,4-D stimulates the cells to divide rapidly and form an undifferentiated mass of cells called "callus," which is the first step in modern propagation (Desai, et al., 2022).
b) Induce somatic cells: in somatic embryogenesis, 2,4-D stimulates the cells to divide rapidly and form an undifferentiated mass of cells called "callus," which is the first step in modern propagation.
c) Rooting of Microshoots:
   - Microshoots were transferred after multiplication to the rooting media, which typically contains a low concentration of a milder auxin (e.g., IBA or NAA) to stimulate the microshoots to develop a functional root system, turning them into complete plantlets ready for acclimatization.
   - Organogenesis: Auxins, in combination with cytokinins, control whether a culture forms shoots or roots. A higher ratio of cytokinins to auxins promotes shoot formation, while a higher ratio of auxins to cytokinins promotes root formation.

Limitations and consecrations

While incredibly useful, the use of auxins is not without its challenges.

a) Concentration is critical: The principle of hormonal dose-response is paramount, the optimal concentration varies significantly according to species.
   - Too low: No effect
   - Optimal: Maximum rooting response.
   - Excess: can inhibit rooting and cause callus formation instead of roots, or sometimes with high doses cause phytotoxicity for the cutting and destroy it.
b) Species Specificity: Not all plants respond equally. Some root easily without auxins, while others are recalcitrant even with high doses. Finding the right protocol requires research.
c) Potential for Abnormal Growth: Excessive doses or improper use can lead to malformed, stubby, or non-functional roots. For instance, an imbalance of auxins in tissue culture can lead to hyperhydricity, where plants become waterlogged and translucent.
d) Environmental Impact: Generally, auxins are considered low-risk; improper disposal of large quantities of synthetic auxins (like 2,4-D) could potentially contaminate water sources. 2,4-D is also a widely used herbicide, so care must be taken to avoid drift or contamination of non-target plants.

For developing countries, in particular, the affordability and ease of using auxin-based rooting powders make them an accessible and transformative technology, allowing small-scale farmers to propagate improved varieties and increase their yields with minimal investment. They are a prime example of how a fundamental understanding of plant physiology can lead to profound practical benefits in agriculture (Stuepp, et al., 2017).

4. WHEN TO USE WHICH TECHNIQUE?

Choosing proper propagation techniques depend various factors include:

a) Farming scale: Generally, in small spaces, the propagation budget is very low, and there is no reliable laboratory access. It is preferable to use traditional propagation methods such as grafting or cuttings.
b) Eradicating viruses (e.g., HLP in citrus) Micropropagation (meristem culture) is the best way to produce pathogen-free plants.
c) Conserving rare varieties: In the case of rare varieties, modern techniques provide sustainable solutions by preservation grafting, long-term storage, and field revival.
d) Creating seedless triploid fruits: Creating seedless triploids: For seedless triploid varieties and sterile clone hybrids, micropropagation combines genomes efficiently. Modern propagation techniques, e.g., protoplast fusion or tissue culture, are considered efficient techniques.
e) Rapid replanting after disease or environmental disasters: As was the case with Spanish citrus cultivation in 2024, modern micropropagation is faster and can facilitate large-scale production of clean planting material.

5. ENVIRONMENTAL IMPACTS OF TRADITIONAL AND MODERN PROPAGATION TECHNIQUES

The environmental impacts of traditional and modern propagation techniques for fruit crops vary significantly, influenced by resource use, chemical inputs, energy consumption, and ecological effects (Table 4).

Table 4. Comparative Environmental and Economic impacts of Modern vs. Traditional

A. Environmental Impacts		
Impact Category	**Traditional Techniques**	**Modern Techniques**
Energy Use	Low (manual labor, sunlight-dependent)	High (lab energy, lighting, sterilization
Water Use	Moderate to high (field irrigation)	Low to moderate (closed systems)
Chemical Pollution	High (fertilizers/pesticides in runoff)	Low (controlled lab use) but hazardous waste
Waste Generation	Organic waste, plastic pollution	Plastic, e-waste, chemical waste
Biodiversity Impact	Mixed (genetic diversity vs. monoculture risks)	Negative (clonal uniformity, genetic erosion)
Carbon Footprint	Low (minimal machinery)	High (energy-intensive labs)
B. Economic Impacts		
Economic factor	Traditional techniques	Modern techniques
Initial Investment	Very Low	Extremely High
Cost per Plant (Small Scale)	Very Low	Very High
Cost per Plant (Large Scale)	Moderate	Very Low (Economies of Scale)
Labor	Low-skilled, seasonal, high volume	High-skilled, year-round, lower volume
Energy Use	Negligible (sunlight)	Very High (growth chambers, labs
Primary Economic Risk	Disease Introduction (Catastrophic long-term loss)	Technical Failure (Contamination, variation)
Production Speed	Slow (Season-bound)	Very Fast (Year-round)
Scalability	Linear (Land-limited)	Exponential (Bioreactor-limited)
Product Quality	Variable	Uniform, High (Disease-free)
Market Access	Standard Markets	Premium & Export Markets (Disease-free certification)
Ideal For	Smallholders, local nurseries, common varieties	Large corporations, export-oriented farms, new varieties, disease eradication

Traditional techniques have a lower energy footprint but risk chemical pollution and soil degradation if poorly managed. Modern techniques offer precision and disease control but demand high energy and generate hazardous waste (Ventura, et al., 2019).

The most sustainable approach integrates both, using modern methods for critical needs (e.g., disease-free plants) and traditional techniques for scaling. Ultimately, reducing the environmental footprint of fruit propagation requires context-specific solutions, balancing productivity with ecological stewardship (Tang, et al., 2024).

Below is a comparative analysis of their environmental footprints, followed by mitigation strategies and sustainable practices.

a) Environmental Impacts of Traditional Propagation Techniques (Resource Use):
-Land & Water:

Traditional propagation techniques require large areas of land, particularly for seedling beds and nurseries, and irrigation; however, water use is often localized and manageable (Martínez de Azagra Paredes, et al., 2022).

-Soil Health:

Repeated nursery cultivation can deplete soil nutrients without proper rotation/composting, moreover, in sloped terrain, it could increase erosion.

- Fertilizers/Pesticides:

Traditional nurseries often rely on synthetic fertilizers and pesticides to accelerate growth and control pests/diseases. Grafting wounds requires treatment with fungicides, which contaminate soil and groundwater.

- Organic and plastic wastes

Burned or landfilled pruned materials, rootstock trimmings, and unsuccessful grafts, releasing CO_2 or methane. Plastic pollution from seedling/cutting polythene bags contributes if not recycled.

- Biodiversity & Ecology:
- Seed propagation maintains genetic diversity, supporting adaptation to climate change.

On contrary, monoculture nurseries reduce local biodiversity and increase pest vulnerability.

Environmental Impacts of Modern Propagation Techniques

a) Energy Consumption:

Most modern propagation techniques rely on constant energy, for instance, tissue culture labs require constant energy for sterile conditions (laminar flow hoods, autoclaves, and growth chambers). Moreover, cryopreservation and liquid nitrogen production are energy-intensive (Hasnain, et al., 2022).

b) Chemical Inputs

Modern propagation techniques rely on chemical inputs, including media components, such as agar, sucrose, and growth regulators (e.g., cytokinins), which are derived from resource-intensive processes for natural resources (Tang, et al., 2024). On the other hand, disposal of hormone-laden media could contaminate water systems. While sterilization agents like ethanol, bleach, and hydrogen peroxide generate chemical waste if not neutralized properly.

c) Plastic and electronic wastes represent a challenge for protecting the environment and reducing pollution through modern techniques.

The high consumption of various plastic tools in modern propagation techniques, like single-use petri dishes, culture vessels, and cryovials, contributes to plastic pollution. On the other side, small electronic parts, bioreactors, sensors, and automated systems generate e-waste gradually.

d) Biodiversity & Genetic Risks

There are some risks associated with modern propagation techniques. They can cause genetic erosion by producing identical clones, which reduces genetic diversity and raises vulnerability to pests and diseases. Additionally, somatic hybrids (disease-resistant rootstocks) might have invasive potential and could outcompete local relatives if introduced into natural ecosystems (Shukla, et al., 2025).

6. ECONOMIC IMPACTS OF PROPAGATION TECHNIQUES

The economic impacts of propagation techniques are profound and multifaceted, influencing everything from the initial investment and operational costs to the long-term profitability and market access for fruit growers.

The choice between traditional and modern techniques is rarely just technical; it's a core business decision with significant financial consequences.

I. Economic Impacts of Traditional Propagation Techniques

Traditional methods (seedlings, cuttings, and grafting) are the bedrock of fruit cultivation. Their economic profile is characterized by low input costs but variable scalability.

The main economic advantages are cost savings & accessibility as follows:

a) Low Initial Capital Investment:
   - Infrastructure: Requires minimal specialized equipment. A nursery for grafting or cuttings needs basic tools (knives, shears), shade net, and land.
   - Labor: Labor is often semi-skilled or unskilled (e.g., how to make a graft cut can be learned through apprenticeship), and can be sourced locally at lower wages, particularly in developing countries.
   - Economic Impact: Democratizes fruit production. It allows smallholder farmers and new entrants with limited capital to start a business, fostering rural employment and entrepreneurship.

7. MITIGATION STRATEGIES FOR SUSTAINABLE PROPAGATION

The economic choice between traditional and modern propagation is not about which is "better," but which is more appropriate for a specific scale, market, and risk tolerance. While modern propagation techniques (e.g., tissue culture, somatic embryogenesis) offer transformative potential for fruit crop improvement, their application in developing countries faces significant limitations. These barriers span technical, economic, infrastructural, and socio-political dimensions, often restricting adoption to large-scale commercial operations or research institutions. Sustainable adoption requires more than technology transfer does. It also requires inclusive innovation ecosystems that prioritize equity and resilience.

1. For traditional techniques

Follow agroecological practices such as using compost and organic fertilizers to reduce synthetic inputs. Moreover, implement crop rotation in nurseries to improve soil health. Besides minimizing water consumption by using modern irrigation (e.g., drip irrigation or sprinkler irrigation) and rainwater harvesting. Furthermore, composting organic waste and recycling plastic nursery materials reduces waste.

2. For Modern Techniques

Renewable energy (solar-powered labs) and LED lighting reduce gas emissions. While recycling culture vessels and neutralizing chemical waste before disposal decreases environmental hazards. In addition to combining micropropagation with cryopreservation of wild relatives.

The most sustainable approach integrates both, using modern methods for critical needs (e.g., disease-free plants) and traditional techniques for scaling. Adopt circular economy principles (recycling, renewable energy) to mitigate impacts. Ultimately, reducing the environmental footprint of fruit propagation requires context-specific solutions, balancing productivity with ecological stewardship.

Mitigation strategies for developing countries

A. Short-Term solutions
    1. Shared facilities
    2. Mobile laboratories
    3. Farmer training
B. Long-term solutions
    1. Policy reforms
    2. Public-private partnerships
    3. Open-source technologies

Until these gaps are bridged, traditional methods will remain the primary practical choice for most smallholders, with modern techniques serving niche roles (e.g., elite cultivars, disease eradication).

8. CROSS-CUTTING SOLUTIONS

a) Using hybrid methods for plant propagation is an important means of reducing environmental impacts. Modern technologies could be used to produce disease-free plants and propagate rare species, which could then be propagated by traditional techniques to reduce environmental impacts.
b) Subsidize sustainable laboratories and ban the use of single-use plastics in nurseries. In addition, implement certification programs for nurseries to achieve a low-carbon footprint and zero waste.
c) Adopt circular economy principles (recycling, renewable energy) to mitigate impacts.

Integrating traditional & modern techniques

The integration of the two systems contributes to achieving better results, as modern techniques produce the elite, disease-free plants, rootstocks, and mother plants that provide the desired specifications. These are then propagated on a large scale using traditional methods such as grafting or cuttings in large, advanced nurseries.

a) Collaboration between farmers and scientists to improve traditional methods to achieve participatory breeding.
b) Low-Cost Modern Adaptations: Mini-grafting kits for smallholders.
c) Auxins (rooting hormones) to enhance cutting success.

9. LIMITATIONS OF APPLYING MODERN PROPAGATION TECHNIQUES IN DEVELOPING COUNTRIES

I. Technical Limitations
   1. Skill Gap: Modern techniques require trained staff in plant biotechnology, sterile protocols, and diagnostics. For instance, tissue culture laboratories in parts of Africa or Southeast Asia may rely on expatriate technicians, limiting local capacity.
   2. Protocol Development Challenges

Species-Specific Protocols: Methods optimized for temperate crops (e.g., apple, grape) often fail for tropical species (e.g., banana, mango). Therefore, laboratories may struggle to commercialize protocols for high-demand local crops.

3. Contamination and Quality Control:

Humid climates and unreliable electricity increase contamination risks in tissue culture. In addition, poorly controlled conditions lead to off-types, reducing farmer trust. For instance, banana plantlets in African labs often show high rejection rates due to fungal contamination.

II. Economic Limitations
   a) High Initial Investment:
   b) Recurring Expenses:
   c) Limited Access to Capital.
   d) High Production Costs
III. Infrastructural Limitations
   1. Unreliable energy supply and backup costs
   2. Poor Transportation and Logistics
      a) Cold Chain Failure

        b) Last-Mile Delivery
    3. Water and Sanitation Issues
        a) Water Quality
        b) Waste Disposal

IV. Socio-Political and Institutional Limitations
- a) Weak Policy Support
- b) Limited Extension Services
- c) Intellectual Property Barriers

V. **Market and Adoption Challenges**
- a) Farmer Preferences
- b) Market Fragmentation
- c) Scale-Up Issues

For reasons mentioned above, traditional propagation is more proper for developing countries, whereas the economy of accessibility and resilience is important. It has low financial barriers, making it the engine for smallholder and local nursery industries. Its weakness is the unquantified risk of disease, a risk that can bankrupt an operation and often ignored in pursuit of short-term savings. Modern propagation is the economy of scale, quality, and risk mitigation. It represents a high-stakes, high-reward industrial model. The massive upfront investment buys control, speed, and the ability to produce a premium, disease-free product that can dominate global markets. Its profitability is entirely dependent on technical excellence and achieving high-volume production.

10. FUTURE PERSPECTIVES

1. Traditional methods will remain dominant for on-farm propagation due to accessibility and cost.
2. Modern techniques will drive innovation in breeding, conservation, and disease management.
3. Hybrid systems (e.g., lab-grown rootstocks + traditional grafting) will optimize efficiency and resilience.
4. Automation (robotic grafting, AI-driven tissue culture) will reduce costs and expand modern tech adoption.

11. CONCLUSION

The economic choice between traditional and modern propagation is not about which is "better," but which is more appropriate for a specific scale, market, and risk tolerance. Traditional propagation priorities are accessibility and resilience, it has low financial barriers, making it the preferred technique for smallholder and local nursery industries. The unquantified risk of disease is the main risk that can bankrupt an operation and is often ignored in pursuit of short-term savings. Modern propagation techniques, such as tissue culture and somatic embryogenesis, offer advanced solutions for rapid and disease-free seedling production. Furthermore, it is an economy of scale, quality, and risk mitigation. It represents a high-stakes, high-reward industrial model. The massive upfront investment buys control, speed, and the ability to produce a premium, disease-free product that can dominate global markets. Its profitability is entirely dependent on technical excellence and achieving high-volume production. Modern labs produce the elite, disease-free mother plants or rootstocks, which are multiplied on a large scale using traditional methods (e.g., grafting or cuttings) in large, sophisticated nurseries. The disease-free foundation of modern biotech and the cost-effective scalability of traditional techniques create a robust and economically efficient propagation pipeline from lab to field. By leveraging these methods, the fruit crop industry can address challenges related to uniformity, disease control, and environmental sustainability, ultimately enhancing productivity and quality. The primary differences between traditional and modern techniques encompass their environmental impacts, resource efficiency, economic implications, and technological advancements. The future lies in strategic integration through the use of biotechnology to enhance traditional systems to sustain fruit production.

REFERENCES

Abdalla, N., El-Ramady, H., Seliem, M. K., El-Mahrouk, M. E., Taha, N., Bayoumi, Y., Shalaby, T. A., & Dobránszki, J. (2022). An academic and technical overview on plant micropropagation challenges. *Horticulturae, 8*(8), 677. DOI: 10.3390/horticulturae8080677

Abobatta, W. F. (2018). Development growth and productivity of orange orchards (*Citrus sinensis* L) in Egypt (delta region). *Advances in Agricultural Technology & Plant Sciences, 1*(1), 1–11. DOI: 10.63235/AATPS.180003

Abobatta, W. F. (2025). Importance of propagation techniques of fruit trees. *J Appl Biotechnol Bioeng, 12*(4), 180–182.

Abobatta, W. F., & Ahmed, F. K. (2023). Effect of Urea and Nano-nitrogen Spray Treatments on Some Citrus Rootstock Seedlings. *Horticulture Research Journal, 1*(1), 68–84. DOI: 10.21608/hrj.2023.306950

Ahmar, S., Gill, R. A., Jung, K. H., Faheem, A., Qasim, M. U., Mubeen, M., & Zhou, W. (2020). Conventional and molecular techniques from simple breeding to speed breeding in crop plants: Recent advances and future outlook. *International Journal of Molecular Sciences, 21*(7), 2590. DOI: 10.3390/ijms21072590 PMID: 32276445

Ak, B. E., Hatipoglu, I. H., & Dikmetas, B. (2021). Propagation of fruit trees. *Recent Headways in Pomology (Edited by: M. Pakyurek) Chapter, 3*, 55-92.

Ak, B. E., Hatipoğlu, İ. H., & Hatipoğlu, M. (2024). Modern Methods in Ornamental Plant Propagation: Improving Resistance and Sustainability. *Endangered Species, 1*(2), 3.

Al Makhmari, S. M. (2016). Effect of substrates, IBA and types of cuttings on rooting. *Theses*. 722. https://scholarworks.uaeu.ac.ae/all_theses/722

Aronen, T., Varis, S., & Tikkinen, M. (2025). Somatic embryogenesis: concept, principles, and applications. In *Forest Microbiology* (pp. 373–388). Academic press., DOI: 10.1016/B978-0-443-21903-0.00022-9

Arpitha, K., & Navya, V. (2025). Seed dormany and it's mechanism. *Seed Biology and Biotechnology*, 18.

Awotedu, B. F., Omolola, T. O., Akala, A. O., Awotedu, O. L., & Olaoti-Laaro, S. O. (2021). Vegetative propagation: A unique technique of improving plants growth. *World News Nat Sci., 35*, 83–101.

Awotedu, B. F., Omolola, T. O., Akala, A. O., Awotedu, O. L., & Olaoti-Laaro, S. O. (2021). Vegetative propagation: A unique technique of improving plants growth. *World News of Natural Sciences*, *35*, 83–101.

. Babich, O., Sukhikh, S., Pungin, A., Ivanova, S., Asyakina, L., & Prosekov, A. (2020). Modern trends in the in vitro production and use of callus, suspension cells and root cultures of medicinal plants. *Molecules, 25*(24), 5805. DOI: 10.3390/molecules25245805

Beshara, S., Kassem, A., Fors, H., & Harraz, N. (2024). A Comprehensive Review and Mapping Citrus Supply Chains from a Sustainability Perspective across the European Union, Middle East, and Africa. *Sustainability, 16*(19), 8582. DOI: 10.3390/su16198582

Bhatla, S. C., & Lal, M. A. (2023). Auxins. In *Plant Physiology, Development and Metabolism*. Springer., DOI: 10.1007/978-981-99-5736-1_15

de Carvalho, D. U., Boakye, D. A., Gast, T., Leite Junior, R. P., & Alferez, F. (2021). Determining seed viability during fruit maturation to improve seed production and availability of new citrus rootstocks. *Frontiers in Plant Science, 12*, 777078. DOI: 10.3389/fpls.2021.777078

Desai, P., Desai, S., Rafaliya, R., & Patil, G. (2022). Plant tissue culture: Somatic embryogenesis and organogenesis. In *Advances in plant tissue culture* (pp. 109–130). Academic Press., DOI: 10.1016/B978-0-323-90795-8.00006-0

Devi, K. V., Raj, S., Alavekar, M. S., & Bhrihavasi, G. (2025). Advancements and Future Trends in Horticultural Propagation. *Horticulture Crops*, 204.

Din, S., Wani, R. A., Ab Waheed Wani, F. N., Farwah, S., Rizvi, S., & Nisar, T. F. W. S. (2019). Enhancement in temperate fruit crops. *Journal of Pharmacognosy and Phytochemistry*, *8*(2), 1203–1216.

Eid, A. A. M., Nomier, S. A., Ibrahim, M. M., & Gad, M. M. (2018). Effect of some natural extracts, indolbutiric acid and naphthalene acetic acid on rooting of picual olive cuttings. *Zagazig Journal of Agricultural Research*, *45*(1), 119–136. DOI: 10.21608/zjar.2018.49816

Gad, M. M., & Ibrahim, M. M. (2018). Effect of IBA and some natural extracts on rooting and vegetative growth of Picual olive sucker and shoot cuttings. *Current Science International*, *7*(2), 191–203.

Gopal, G. V., & Sahadevan, N. (2025). *Plant Tissue Culture basic Laboratory Techniques and Advances in Biotechnological Standards for GM Crops*. Crown Publishing.

. Haokip, S. W., Yumkhaibam, T., Lian, H. N., & Sheikh, K. A. (2023). Propagation Techniques in Fruit Crops. *Integrated Publications TM New Delhi*, 29.

Hasnain, A., Naqvi, S. A. H., Ayesha, S. I., Khalid, F., Ellahi, M., Iqbal, S., Zeeshan, H. M.,Abbas, A., Adamski, R., Markowska, D., Baazeem, A., Mustafa, G., MOustafa, M., Hasan, M.E., & Abdelhamid, M. M. (2022). Plants in vitro propagation with its applications in food, pharmaceuticals and cosmetic industries; current scenario and future approaches. *Frontiers in plant science, 13*, 1009395. DOI: 10.3389/fpls.2022.1009395

Henao-Ramírez, A. M., & Urrea-Trujillo, A. I. (2020). Somatic Embryogenesis for Clonal Propagation and Associated Molecular Studies in Cacao (*Theobroma cacao L.*). In Chong, P., Newman, D., & Steinmacher, D. (Eds.), *Agricultural, Forestry and Bioindustry Biotechnology and Biodiscovery*. Springer., DOI: 10.1007/978-3-030-51358-0_5

Iqbal, M., & Singh, K. K. (2020). Propagation of temperate fruit crops. *Innovative agriculture and botany*, 119-135.

Jain, S., Sharma, R., Nidhi, N., Behera, S. D., Baksh, H., Sharma, R., & Kumar, V. (2023). Advancement in tissue culture techniques for fruit crops. *Int J Environ Clim Change, 13*(11), 4396–4407. DOI: 10.9734/ijecc/2023/v13i113620

Joshi, B. K., Ghimire, K. H., Neupane, S. P., Gauchan, D., & Mengistu, D. K. (2023). Approaches and advantages of increased crop genetic diversity in the fields. *Diversity, 15*(5), 603. DOI: 10.3390/d15050603

Kaur, S., Ijaz, S., & Nasir, B. (2024). Tissue Culture and Somatic Fusion in Plants. In Ijaz, S., Ul Haq, I., & Mohamed Ali, H. (Eds.), *Trends in Plant Biotechnology*. Springer., DOI: 10.1007/978-981-97-0814-7_1

Khaitov, M., Rasulov, I., & Karshiyev, J. (2024). Technology and methodology of propagation of fruit seedlings. *Science and innovation, 3*(D9), 276-279.

Khalil, M., Sarwar, M. S., Tariq, H., Farooq, M., & Ullah, M. (2023). A new method for plant breeding through the use of plant tissue culture. *Journal of Life and Social Sciences, 2023*(1), 20–20. DOI: 10.64013/bbasrjlifess.v2023i1.20

Khan, A., & Korban, S. S. (2022). Breeding and genetics of disease resistance in temperate fruit trees: challenges and new opportunities. *Theoretical and Applied Genetics, 135*(11), 3961-3985. DOI: 10.1007/s00122-022-04093-0

Khezri, M., Asghari-Zakaria, R., & Zare, N. (2024). Plant Cell and Tissue Culture: Propagation, Improvement, and Conservation of Medicinal Plants. In Kumar, N. (Ed.), *Biosynthesis of Natural Products in Plants*. Springer., DOI: 10.1007/978-981-97-2166-5_11

Klimek-Chodacka, M., Kadluczka, D., Lukasiewicz, A., Malec-Pala, A., Baranski, R., & Grzebelus, E. (2020). Effective callus induction and plant regeneration in callus and protoplast cultures of *Nigella damascena* L. *Plant Cell, Tissue and Organ Culture, 143*(3), 693–707. DOI: 10.1007/s11240-020-01953-9

Manohar, K. A., Shukla, G., Shahina, N. N., Sivasankarreddy, K., Ravuther, S. S., & Chakravarty, S. (2024). Conventional Versus Non-Conventional Methods of Propagation of Forest Tree Species: Applications and Limitations. In: Thomas T, D., Razdan, M.K., Kumar, A. (eds) Biotechnological Approaches for Sustaining Forest Trees and Their Products. Springer, Singapore. DOI: 10.1007/978-981-97-4363-6_18

Martínez de Azagra Paredes, A., Del Río San José, J., Reque Kilchenmann, J., Diez Hernández, J. M., & Sanz Ronda, F. J. (2022). Methods for watering seedlings in arid zones. *Forests, 13*(2), 351. DOI: 10.3390/f13020351

Musacchi, S., & Neri, D. (2019). Optimizing production of quality nursery plants for fruit tree cultivation. In *Achieving sustainable cultivation of temperate zone tree fruits and berries* (pp. 183–242). Burleigh Dodds Science Publishing.

. Nale, R., Sharma, G., Pal, R., Patel, R. K., & Sharma, S. (2024). Effect of IBA and NAA on the rooting and vegetative growth of hardwood cuttings in common fig (*Ficus carica* L.). *International Journal of Bio-Resource and Stress Management.*

Nemhauser, J. L. (2018). Back to basics: What is the function of an Aux/IAA in auxin response? *The New Phytologist, 218*(4), 1295–1297. DOI: 10.1111/nph.15172 PMID: 29738089

Ovcharenko, O. O., Rudas, V. A., & Kuchuk, M. V. (2023). Protoplast Fusion for Cellular Engineering of the Brassicaceae. *Cytology and Genetics, 57*(5), 432–450. DOI: 10.3103/S0095452723050043

Pereira, F. M. P., Usman, M., Mayer, N. A., Nachtigal, J. C., Maphanga, O. R. M., & Willemse, S. (2017). Advances in guava propagation. *Revista Brasileira de Fruticultura, 39*(4), e-358.

Pincelli-Souza, R. P., Tang, Q., Miller, B. M., & Cohen, J. D. (2024). Horticultural potential of chemical biology to improve adventitious rooting. *Horticulture Advances, 2*(1), 12. DOI: 10.1007/s44281-024-00034-7

Prakash, A., & Paul, P. (2021). *Recent trends in propagation of horticulture crops and application of Genomics tools in plant Breeding. Recent Trends in Propagation of Horticulture Crops*. Taran Publication.

Ranaware, A. S., Kunchge, N. S., Lele, S. S., & Ochatt, S. J. (2023). Protoplast technology and somatic hybridisation in the family Apiaceae. *Plants, 12*(5), 1060. DOI: 10.3390/plants12051060

Ranaware, A. S., Kunchge, N. S., Lele, S. S., & Ochatt, S. J. (2023). Protoplast technology and somatic hybridisation in the family Apiaceae. *Plants*, *12*(5), 1060. DOI: 10.3390/plants12051060 PMID: 36903923

Rathour, A., & Kumar, R. (2023). *Plant Propagation Techniques in Horticulture. Recent Trend in Agriculture*. Integrated Publications New Delhi.

Reed, K. M., & Bargmann, B. O. (2021). Protoplast regeneration and its use in new plant breeding technologies. *Frontiers in Genome Editing*, *3*, 734951. DOI: 10.3389/fgeed.2021.734951 PMID: 34713266

Saini, N. (2024). The potential of plant growth regulator (auxin) in enhancing fruit crop propagation via Stem Cuttings. International Journal of Advanced Biochemistry Research, 8(5), 1001-1007. DOI: 10.33545/26174693.2024.v8.i5l.1230

Sekhukhune, M. K., & Maila, Y. M. (2025). Growth Regulator Indole-3-Butyric Acid on Rooting Potential of *Actinidia deliciosa* Rootstock and Actinidia arguta Female Scion Species Stem Cuttings. *Horticulturae, 11*(1), 53. DOI: 10.3390/horticulturae11010053

. Shivran, M., Sharma, N., Dubey, A. K., Singh, S. K., Sharma, N., Sharma, R. M., Singh, M., & Singh, R. (2022). Scion–rootstock relationship: Molecular mechanism and quality fruit production. *Agriculture, 12*(12), 2036. DOI: 10.3390/agriculture12122036

Shukla, G., Pradhan, S., Debberma, S., Singh, M., Sharma, P., Roy, B., & Shukla, P. S. (2025). Revolutionizing Tree Propagation by Exploiting the Power of Synthetic Seed Technology. In *Advances in Seed Quality Evaluation and Improvement* (pp. 447–464). Springer Nature Singapore. DOI: 10.1007/978-981-96-5178-8_18

Singh, D. P., Singh, A. K., & Singh, A. (2021). *Plant breeding and cultivar development*. Academic Press.

Stuepp, C. A., Wendling, I., Trueman, S. J., Koehler, H. S., & Zuffellato-Ribas, K. C. (2017). The use of auxin quantification for understanding clonal tree propagation. *Forests, 8*(1), 27. DOI: 10.3390/f8010027

Taha, A. M., & Khalifa, H. E. H. (2024). Olive in Egypt: Cultural Practices. *Olives and Olive Related Products-Innovations in Production and Processing: Innovations in Production and Processing, 35.*

. Tang, L., Syed, A. U. A., Otho, A. R., Junejo, A. R., Tunio, M. H., Hao, L., ... & Channa, J. A. (2024). Intelligent Rapid Asexual Propagation Technology—A Novel Aeroponics Propagation Approach. *Agronomy, 14*(10), 2289. DOI: 10.3390/agronomy14102289

Teiba, I. I., El-Bilawy, E. H., Elsheery, N. I., & Rastogi, A. (2023). Microbial allies in agriculture: harnessing plant growth-promoting microorganisms as guardians against biotic and abiotic stresses. *Horticulturae, 10*(1), 12. DOI: 10.3390/horticulturae10010012

Uysal, O., Sevimli, T., Sevimli, M., Gunes, S., & Sariboyaci, A. E. (2018). Cell and tissue culture: the base of biotechnology. In *Omics technologies and bio-engineering* (pp. 391–429). Academic Press., DOI: 10.1016/B978-0-12-804659-3.00017-8

Ventura, J. A., Lima, I. D. M., Martins, M. V. V., Culik, M. P., & Costa, H. (2019). Impact and management of diseases in the propagation of fruit plants. *Revista Brasileira de Fruticultura, 41*(4), e-647. DOI: 10.1590/0100-29452019647

Wang, G., Zeng, J., Du, C., Tang, Q., Hua, Y., Chen, M., Yang, G., Tu, M., He, G., Li, Y., He, J., & Chang, J. (2024). Divergent roles of the auxin response factors in lemongrass (Cymbopogon flexuosus (Nees ex Steud.) W. Watson) during plant growth. *International Journal of Molecular Sciences, 25*(15), 8154. DOI: 10.3390/ijms25158154

Wani, A. W., Yaseen, I., Fayaz, S., Khursheed, A., Gupta, A., Hassan, G. I., Rajan, R., Qayoom, S., Gani, I., & Jayarajan, S. (2022). Advances in Propagation and Nursery Management: Methods and Techniques. In *Apples* (pp. 45-57). CRC Press.

Yadav, D., & Singh, S. P. (2018). Vegetative methods of plant propagation: I-cutting layering and budding. *Journal of Pharmacognosy and Phytochemistry, 7*(2), 3267–3273.

Chapter 9
Modern Techniques in Plant Propagation

Fahad K. Y. Al-Dulaimi
https://orcid.org/0000-0002-5958-3496
Northern Technical University, Iraq

Janan K. Al-Tarjuman
Northern Technical University, Iraq

Waad Said Faizy
Northern Technical University, Iraq

ABSTRACT

This chapter reviews the most prominent modern methods used in plant propagation, focusing on their effective role in supporting sustainable agricultural practices. Tissue culture and micropropagation are highlighted as the most accurate and efficient methods for producing homogeneous, disease-free plants, especially in light of contemporary agricultural and environmental challenges. The chapter also discusses the importance of balancing environmental factors (such as light, temperature, and humidity) and the biological materials used (such as hormones and nutritional media). It also addresses modern biotechnological applications as supportive tools for enhancing propagation efficiency and improving plant genetic traits. The chapter concludes that the use of these techniques directly contributes to increasing plant productivity, reducing dependence on natural resources, and preserving biodiversity, making them a fundamental pillar of sustainable agriculture.

DOI: 10.4018/979-8-3373-5626-6.ch009

INTRODUCTION

The Importance of Plant Propagation in Modern Agriculture

Plant propagation is one of the fundamental pillars of agricultural production, as it relies on providing sufficient numbers of plants with desirable traits, whether for commercial, research, or conservation purposes for endangered species. Thru propagation, it is possible to obtain genetically identical plants to the parent plant, or to develop new varieties with improved traits, such as disease resistance, environmental stress tolerance, and higher crop quality (Salgotra et al.,2023; Hasnain et al.,2022)

In the past, traditional propagation methods such as seed propagation, cuttings, or grafting were sufficient to meet agricultural production needs, but today they have become limited in the face of major challenges, such as the increasing demand for food, climate change, and the spread of pests and diseases (Cucu et al.,2025). These traditional methods can also take a long time and are prone to losing distinctive traits or transferring diseases. Here comes the role of modern plant propagation, which allows for the production of large numbers of plants in a short period of time, ensuring high quality and uniformity in traits. In addition, these technologies can produce plants free from seed-borne or vegetative diseases, which is of utmost importance in commercial agriculture (Akhtar et al.,2024; Paulikienė et al.,2025).

The Role of Modern Technologies in Supporting Food Security and Sustainability

The world today faces a dual challenge represented by the increase in population, which is expected to exceed 9 billion by 2050, and at the same time, the decline in arable land due to urban expansion and environmental degradation. To achieve food security, it is essential to increase agricultural production sustainably, meaning without harming the natural resources for future generations (Wang,2022; Al-Dulaimi and Al-Tarjuman,2024)

Modern techniques in plant propagation provide innovative solutions to this challenge, allowing for the production of high-yield and high-quality crops with reduced use of resources such as water, fertilizers, and pesticides. They also help in producing varieties resistant to drought, salinity, or diseases, reducing the need for chemical inputs and maintaining ecological balance (Abdul Aziz et al.,2022; Fussy and Papenbrock,2022)

Moreover, these techniques contribute to preserving biodiversity by propagating endangered plants or those of economic and cultural importance and reintroducing them into their natural environments. This, in turn, supports ecosystems and enhances their ability to adapt to climate changes (Jie et al.,2025; Higuera-Mendieta et al.,2021).

The Evolution of Propagation Methods from Traditional to Modern

Plant propagation methods have undergone significant developments throughout history. In the beginning, humans relied on natural and simple methods, such as collecting and planting seeds, or taking parts of the plant (cuttings, suckers, grafting) to propagate desired plants. With the advancement of knowledge in botany and the understanding of the principles of genetics, propagation methods began to improve in terms of efficiency and quality (Dennis et al.,2008; Kaisera ET AL.,2020).

The major transformation occurred with the discovery of the phenomenon of totipotency in the early twentieth century, which proved that every plant cell contains the genetic information necessary to produce a complete plant. This discovery opened the door to plant tissue culture techniques, which enabled scientists to propagate plants in a sterile laboratory environment and control all factors affecting growth (Su et al.,2021; Sussex et al.,2008).

In recent decades, molecular techniques such as molecular markers, genetic modification, and genome editing technology (CRISPR-Cas9) have been introduced, allowing not only the propagation of plants but also the precise and rapid improvement of their traits, and the integration of new traits that were not possible thru traditional breeding (Erdoğan et al.,2023). Thus, modern plant propagation has become a strategic tool in agriculture, not only to increase production but also to address environmental, economic, and social challenges, and to ensure the sustainability of global food production (Huang and Wang,2024).

The Scientific Basis for Modern Techniques

The success of modern techniques in plant propagation does not come from nowhere; it relies on a deep understanding of the biological and physiological principles that govern the growth of plant cells and tissues. These principles are based on fundamental scientific facts related to genetics, plant physiology, biochemistry, and controlled environments (Sussex,2008; Bidabadi and Jain,2020).

The Concept of Totipotency

Totipotency is the foundation upon which all plant tissue culture techniques are based. It refers to the ability of a single plant cell, whether from root, stem, or leaf tissue, to divide and differentiate to produce a complete plant genetically identical to the parent plant, if placed in the appropriate environmental and nutritional conditions (Fehér,2019).

The German scientist Haberlandt discovered this phenomenon in 1902 when he conducted experiments on isolated plant cells, and later proved that the genetic material in any cell contains all the information necessary to form all parts of the plant. The difference between the various cells is not due to a difference in genetic content, but rather to gene expression, meaning the activation or deactivation of certain gene groups in response to chemical and environmental signals (Kutschera and Ray,2022). The importance of the embryonic callus in propagation is:

1. It allows the production of complete plants from very small cells or tissues.
2. They allow for the preservation of desirable genetic traits.
3. Provides a basis for the propagation of rare or endangered plants from few samples (Ikeuchi et al.,2013).

Plant Growth Regulators and Their Role in Differentiation

Plant Growth Regulators are organic compounds, either natural or synthetic, that affect the physiological processes of plants at low concentrations. Controlling the ratios of these substances within the nutrient medium is key to directing the differentiation pathway of cultured tissues (Didi et al.,2022).

Auxins: The most important ones are Indole-3-acetic acid (IAA), Indole-3-butyric acid (IBA), and Naphthaleneacetic acid (NAA). And they have their functions: stimulating the formation of adventitious roots, increasing cell elongation, and promoting the formation of callus during wound healing (Khadr et al.,2020).

Cytokinins: The most important ones are Kinetin (KIN), Benzyl Amino Purine (BAP), and Zeatin (Zeatin). And they have their functions: stimulating cell division, encouraging the formation of buds and branches, delaying leaf senescence (Aremu et al.,2020).

Gibberellins: The most important of which is gibberellic acid (GA3). Functions: Stem elongation, breaking dormancy in seeds and tubers (Castro-Camba et al.,2022).

Abscisic Acid (ABA): regulating dormancy, increasing drought tolerance, controlling the opening and closing of stomata (Muhammad et al.,2022).

Table 1. The effect of the auxin to cytokinin ratio on the differentiation pathway in plant tissue culture

Auxin to cytokinin ratio:	potential outcome
Very high (10:1)	Stimulates root formation
Approximately equal (1:1)	Stimulating callus formation
Low (e.g., 1:10)	Stimulating bud formation

Components of the Nutrient Medium and Its Importance

The nutrient medium is an artificial environment that contains all the elements needed for plant growth in the laboratory. The most commonly used medium is the Murashige and Skoog Medium (MS Medium), which was developed in 1962 (Churchman and Lowe,2012). The nutrient medium consists:

1. Major mineral elements (N, P, K, Ca, Mg, S) essential for the formation of proteins, enzymes, and tissues. The major mineral elements (N, P, K, Ca, Mg, S) essential for the formation of proteins, enzymes, and tissues.
2. Minor elements (Fe, Mn, Zn, B, Cu, Mo) participate in enzyme activity and biochemical reactions. Trace elements (Fe, Mn, Zn, B, Cu, Mo) participate in enzyme activity and biochemical reactions.
3. Carbon source often sucrose at a concentration of 2–3% as an energy source. Carbon source often sucrose at a concentration of 2–3% as an energy source.
4. Vitamins such as thiamin (B1), niacin (B3), pyridoxine (B6). Vitamins such as thiamin (B1), niacin (B3), pyridoxine (B6).
5. Amino acids such as glycine to stimulate cell division. Amino acids such as glycine to stimulate cell division.
6. Growth regulators added according to the goal (rooting, bud formation, callus formation). Growth regulators added according to the goal (rooting, bud formation, callus formation).
7. Gelatin or agar to give the medium a solid consistency if necessary. Gelatin or agar to give the medium a solid consistency if necessary (Bonnet et al.,2019).

Controlled Environmental Conditions

Controlling environmental factors is important for the success of plant propagation in the laboratory (Abdalla et al.,2022), including:

Temperature: Often 24 ± 2°C for most plants.

Lighting: White fluorescent lighting with an intensity of 1000–3000 lux, with a lighting schedule of 16 hours light / 8 hours dark.

Relative humidity: between 70–90% to reduce transpiration and water loss.

Sterilization: Tools and media: in the autoclave at 121°C and 15 psi for 15–20 minutes, Plant parts: using sodium hypochlorite or ethyl alcohol at appropriate concentrations.

Genetic Response and Variation Among Plant Species

Not all plants respond equally well to modern techniques. This is due to the genetic differences between species or varieties, the presence of compounds in the tissues that inhibit cell division, and specific requirements for some species regarding growth regulators or nutrients (Garban and Ilia,2024). Example, the banana plant responds rapidly to micropropagation, while some types of woody trees require significant adjustments in the nutrient medium and growth conditions.

Modern Laboratory Techniques

Plant Tissue Culture

Plant tissue culture is a laboratory technique that relies on the cultivation of small parts of plants (such as cells, tissues, or organs) in a sterile artificial environment that provides all the necessary nutrients and hormones for their growth and development (Hasnain et al.,2022). This technique is based on the principle of cellular totipotency, which states that each plant cell contains the complete genetic information that enables it to form a complete plant under suitable conditions (Delporte et al.,2012). The basic stages:

1. Initiation Stage: Selecting a disease-free mother plant, Sterilizing the plant parts (Explants) using disinfectants such as sodium hypochlorite or ethyl alcohol, Placing the samples on a sterile nutrient medium.
2. Multiplication Stage: Stimulating cell division by adjusting the concentrations of growth regulators (such as cytokinins and auxins), Increasing the number of resulting plant organs.
3. Rooting Stage: Stimulating root formation using appropriate concentrations of auxins (IAA, IBA, NAA).
4. Acclimatization Stage: Gradually transferring small plants from sterile conditions to a natural environment, with control over humidity and temperature (Eliwa et al.,2024).

The Applications of plant tissue culture:

1. Producing virus-free plants. Rapid propagation of rare or endangered species.
2. Producing genetically identical plants for research purposes.
3. Germplasm Conservation thru Cryopreservation (Krasteva et al.,2020).

Factors Affecting the Success of Technology:

1. Composition of the growth medium: The most famous media are MS, B5, WPM.
2. Sterilization: Preventing the growth of fungi and bacteria.
3. Lighting and temperature: Adjust the light intensity (2000–3000 lux) and temperature (22–25°C) (Bekalu et al.,2023).

Micropropagation

Micropropagation is one of the applications of plant tissue culture and aims to produce large numbers of genetically similar plants in a short period, using small parts of the plants (Kulus and Tymoszuk,2024). The Advantages are producing millions of plants in a short time, the possibility of propagation throughout the year without being affected by the seasons, preserving the desired genetic traits (Abdalla et al.,2022).The Stages:

1. Induction: Placing tissues in a medium containing cytokinins to stimulate bud growth.
2. Rapid propagation: Dividing the resulting shoots and replanting them in new media.
3. Rooting: Stimulating root formation.
4. Transport and acclimatization. Transport and localization (Safana et al.,2021).

Somatic Embryogenesis

It is the formation of complete embryos from somatic (non-sexual) cells in a sterile environment, and these embryos are capable of developing into complete plants (Karami et al.,2009). The Stages are:

1. Embryonic induction using high concentrations of auxins such as 2,4-D. 2. Development of embryos: Embryos go thru cardiac stages, fish stage, and cotyledon stage.
   3. Embryonic maturation Increase in carbohydrate and hormone content.

4. Germination transforming embryos into small plants, the applications of somatic embryogenesis are producing large quantities of high-quality seedlings, production of artificial seeds that can be easily stored and transported.

Propagation thru Plant Organs (Organogenesis)

It is the formation of new plant organs (roots or buds) directly or indirectly from the cultured tissues (Long et al.,2022), Types of the propagation is direct Formation of buds or roots from the tissue without passing thru the callus stage. Indirect: Formation of callus first, then the organs. Influencing factors are type of the cultured tissue, Concentrations of growth regulators, Environmental conditions (Bidabadi et al.,2020).

Sterile Hydroponics Technique

Using sterile nutrient solutions to propagate plants without soil, in a laboratory environment with complete control over environmental factors. The advantages of hydroponics technique are Free from diseases, faster growth, the possibility of producing high-quality plants. Applications of techniques are production of seedlings prepared for cultivation in greenhouses, Propagation of medicinal plants (Rajaseger et al.,2023).

Molecular Techniques in Plant Propagation

The last few decades have witnessed significant advancements in the use of molecular techniques in plant sciences, as these tools have become an essential part of modern breeding programs, whether aimed at improving genetic traits, ensuring genetic purity, or accelerating seedling production (Kumar et al.,2024).

Molecular techniques are based on the study and analysis of genetic material (DNA and RNA) at the molecular level, allowing for a precise understanding of the genetic structure of plants, the mechanisms regulating gene expression, and how genes are linked to desirable traits such as high productivity, disease resistance, and tolerance to harsh environmental conditions (Abdulraheem et al.,2024; Mu et al.,2022).

Unlike traditional propagation methods, which primarily rely on the morphological traits and phenotypic characteristics of the plant, molecular techniques offer high precision and greater speed in evaluation and selection, reducing the time and cost required to develop new varieties or produce genetically identical plants. These techniques are also unaffected by external environmental factors such as light, temperature, and soil, making them more reliable in genetic evaluation (Ahmar et al.,2020; Kakoulidou et al.,2021).

One of the important aspects of molecular techniques in propagation is their ability to detect genetic differences early, even in the initial stages of growth. By using molecular markers, the genetic similarity between plants produced from tissue culture can be determined, thereby ensuring the stability of genetic traits and preventing any unwanted mutations or genetic changes (Espinosa-Leal et al.,2018).

These technologies also play a vital role in programs for conserving the genetic resources of rare and endangered plants, as genetic fingerprinting and molecular characterization enable the creation of precise databases that help protect and efficiently propagate them (Wang et al.,2021). From a practical standpoint, the integration of molecular techniques with tissue culture and micropropagation provides an advanced, synergistic system for producing plants with high genetic quality, contributing to sustainable agriculture, achieving food security, and improving the economic viability of plant production projects.

The rapid development in fields such as genetic modification, CRISPR-Cas technology, genomics, and proteomics has opened wide horizons for researchers and farmers to produce genetically modified plants with high precision, while reducing potential environmental risks, which will make these technologies the cornerstone of the next generation of plant propagation systems (Istiandari and Faizal,2025).

Molecular Markers

Definition of Molecular Markers

Molecular markers are specific sequences of DNA used to detect genetic differences between individuals, strains, or plant species. These markers represent precise tools thru which genetic variation can be studied at the genomic level, and they are more reliable than morphological or chemical markers because they are not affected by the environment or different growth stages (Collard and Mackill,2008; Amiteye,2021; Bidyananda et al.,2024).

These markers are sometimes referred to as the plant's genetic fingerprint because they provide a unique map thru which the genetic identity of any plant sample can be accurately determined. This is crucial in breeding programs, especially when there is a need to ensure genetic purity or avoid genetic contamination (Ben Romdhane et al.,2018).

The importance of molecular markers in plant propagation

Molecular markers play a pivotal role in micropropagation for the following reasons (Majumder et al.,2025).

1. Detection of genetic similarity enables the determination of the genetic uniformity among plants produced thru tissue culture or vegetative propagation, ensuring the stability of the desired traits.
2. Identifying genetic diversity allows for the study of genetic differences between various cultivars or strains, which helps in selecting the best genetic materials for breeding programs.
3. Early detection of mutations provides the opportunity to identify undesirable genetic changes at early stages, before any effects on phenotypic traits appear.
4. Contributing to Marker-Assisted Breeding programs can be used to select plants carrying specific genes associated with desirable traits, thereby accelerating breeding and propagation programs.

Common Types of Molecular Markers in Plant Breeding

Molecular markers can be divided into two main types (Hasan et al.,2021):

1. Indicators based on nuclear DNA (Nuclear DNA Markers)
   A. RAPD (Random Amplified Polymorphic DNA):

A simple and rapid technique that relies on the polymerase chain reaction (PCR) to amplify random parts of the genome, and is used to detect genetic variation between individuals.

B. AFLP (Amplified Fragment Length Polymorphism):

A precise technique that relies on breaking down DNA using specific cutting enzymes, then amplifying the resulting fragments, and it is more accurate than RAPD.

C. SSR or Microsatellites (Simple Sequence Repeats):

They rely on short nucleotide repeats and are characterized by their high ability to detect genetic variation.

2. Indicators based on organellar DNA markers Indicators based on plastid or mitochondrial DNA (Organellar DNA Markers) Used to detect traits inherited from the mother or father, such as traits associated with chloroplasts or mitochondria (Chung, 2025;Park et al.,2021).

Steps to Apply Molecular Markers in Plant Propagation

1. Collect plant samples by selecting young leaves or tissues to ensure high-quality DNA.
2. DNA extraction: Using specific extraction protocols that ensure the purity of the DNA and its free from contaminants that hinder molecular reactions.
3. Choosing the appropriate marker involves selecting the type of marker (RAPD, SSR, AFLP...) Based on the study's objective and the required accuracy.
4. Low-cost and easy-to-implement measures to expand their benefits in developing countries and small farms.
5. Analysis of the results involves comparing the patterns of bundles or sequences to determine the degree of similarity or difference between the samples (Collard et al.,2008; Panthee,2023).

Challenges and Limitations

Despite the high accuracy of molecular markers, their use in breeding programs may face some challenges such as: High cost compared to traditional methods, the need for advanced equipment and laboratories, the necessity of having specialized technical expertise (Kumar et al.,2024; Collard et al.,2008).

DNA Fingerprinting in Propagation

Definition of DNA Fingerprinting

DNA Fingerprinting is a technique that determines a unique pattern for each plant at the DNA level, allowing for the precise identification of each individual or variety. This technique is considered a powerful tool to ensure the stability of the genetic traits of cloned plants and to confirm that no mutations or undesirable changes occur during propagation (Hasan et al.,2021).

Scientific Principles

Genetic fingerprinting relies on the use of various molecular markers such as RAPD, SSR, AFLP, and SNP to generate specific patterns of genetic fragments. These patterns are analyzed using electrophoresis or genetic sequencing, allowing for the comparison of samples and the observation of subtle genetic differences (Debnath et al.,2023; Mammadov et al.,2012).

Applications in plant propagation

1. Verifying the genetic purity of cloned plants.
2. Monitoring genetic mutations

3. Evaluation of genetic diversity among varieties (Hasnain et al.,2022).

Some practical examples, In the propagation of bananas and rice, genetic fingerprinting techniques were used to ensure that the cloned plants matched the parent plant and were free from genetic variations, In rare medicinal plants, genetic fingerprinting was used to preserve genetic diversity and ensure the purity of samples before propagation (Feng et al.,2020; Abdullah and Al-dulaimi,2024).

Benefits and Limitations:

1. High accuracy and reliability.
2. The ability to detect subtle genetic differences.
3. Reducing the risks associated with commercial breeding.
4. The need for advanced laboratories and precise equipment.
5. The cost of sample analysis is relatively high.
6. It requires specialized technical expertise to accurately analyze the results.

Genetic Transformation

Genetic modification is the introduction or alteration of one or more genes within the plant genome with the aim of improving genetic traits, such as increasing productivity, disease resistance, or tolerance to harsh environmental conditions. It represents an advanced step in plant breeding programs, where genetically improved plants can be produced in a short time compared to traditional methods (van Esse et al.,2020; Yuan et al.,2024).

Methods of genetic modification

1. Genetic modification using *Agrobacterium tumefaciens*: This method relies on using bacteria to transfer the desired gene to the plant tissue, Used extensively in seed and vegetative plants (Hwang et al.,2017).
2. Gene gun transformation (Biolistics):, It relies on bombarding tiny particles coated with genes into plant cells, Suitable for plants that are difficult to modify using Agrobacterium (Gelvin,2003).

Practical applications

- Producing plants resistant to viruses and fungi.
- Increased tolerance to salinity and drought.
- Introducing genes responsible for productive or qualitative traits in field and vegetable crops (Parmar et al.,2017; Giordano et al.,2021).

Benefits and Limitations

- The ability to modify genetic traits with high precision.
- Integrating desirable traits without affecting other traits.
- Some plants may have difficulty accepting foreign genes.
- High cost and strict monitoring for environmental applications (Ghimire et al.,2023; Movahedi et al.,2023).

CRISPR-Cas technology in propagation

RISPR-Cas technology is a modern tool for precise gene editing, based on a natural defense system found in bacteria against viruses (Ebrahimi et al.,2023). This system has been modified to become a means of controlling the plant genome, allowing for the addition, deletion, or modification of a specific gene with utmost precision without affecting the rest of the genome (Van Eck,2020). In the context of plant propagation, CRISPR is considered a revolutionary tool because it allows for the rapid and efficient production of genetically improved plants, with the ability to enhance desirable traits such as disease resistance, increased productivity, and greater tolerance to environmental stress, without the need for lengthy traditional propagation processes (Ansori et al.,2023).

The mechanism of CRISPR-Cas

1. Identifying the target gene: Selecting the DNA sequence to be modified in a specific plant..
2. Designing a guide RNA (sgRNA): It matches the target sequence and directs the Cas9 enzyme to the correct location in the genome.
3. Cutting DNA with Cas9: The enzyme cuts both strands of DNA at the target site
4. Genome repair: Plant cells exploit natural repair mechanisms to introduce the required modifications, whether by deleting, replacing, or adding a specific sequence (Asmamaw et al.,2021; Yang et al.,2024).

Practical applications in plant propagation

1. Production of disease-resistant plants are modifying genes associated with resistance to fungi or viruses, thereby reducing the need for chemical pesticides.
2. Improving productive traits are modification of genes related to productivity, fruit size, or crop quality.
3. Increasing tolerance to harsh environmental conditions are enhancing resistance to drought, salinity, and high temperatures, making the plants suitable for propagation in various environmental conditions.

4. Acceleration of propagation programs are producing genetically improved plants in a short period compared to traditional methods (Kavhiza et al.,2022; Tyagi et al.,2021).

Challenges and Limitations

1. You need advanced laboratories and precise techniques to analyze the results
2. Some regulatory and legal restrictions on the use of genetically modified plants in certain countries.
3. The necessity of assessing long-term environmental impacts before commercial spread (Raman,2017).

Gene Cloning and Its Role in Propagation

Gene Cloning

Gene cloning is the process of producing multiple identical copies of a specific gene or a part of the plant genome. These copies are introduced into host plant cells to produce plants that carry the cloned gene, allowing for the control of gene expression for desired traits. In plant propagation, genetic cloning is an effective tool for rapidly producing genetically improved plants, while maintaining the desired genetic traits and ensuring their stability across generations (Rahbaran et al.,2021; Al-Tarjuman et al.,2024).

Stages of genetic cloning

1. Isolating the desired gene: The gene associated with a desirable trait, such as disease resistance or increased productivity.
2. Inserting the gene into a vector: A vector such as a plasmid is used to carry the gene into plant cells.
3. Transfer the gene to the host plant cells: Thru genetic modification techniques such as Agrobacterium or Biolistics.
4. Gene expression testing: Ensuring that the cloned gene functions correctly in the plant cell.
5. Producing genetically stable plants: Transforming the modified cells into complete plants using tissue culture and micropropagation (Khan et al.,2016; Zhi et al.,2022).

Practical applications in propagation

1. Multiplication of high-yield field crops: Introducing genes responsible for increasing fruit size or disease resistance.

2. Propagation of medicinal and aromatic plants: Cloning genes that control the production of active compounds, ensuring the quality of the final product.
3. Propagation of rare and endangered plants: Producing genetically identical plants to preserve genetic diversity without the need to collect seeds or plant parts from nature (Sardar,2023; Ahmad et al.,2022).

Benefits and Limitations

- Producing genetically improved plants quickly.
- Ensuring the stability of desired genetic traits.
- Integrating technology with tissue culture and micropropagation to produce high-quality genetically identical plants.
- High cost in some applications.
- The need for advanced laboratories and specialized technical expertise.
- Evaluation of environmental and legal risks before commercial use (Correia et al.,2024; Hasnain et al.,2023)

Genomic, proteomic and metabolomic technologies

Genomic Techniques in Plant Propagation

Genomics is the study of the complete genome of a plant, which is the entire set of genetic material, to understand the organization of genes, their functions, and the interactions between them. In plant breeding programs, genomics provides powerful tools to identify, Genes associated with desirable traits such as disease resistance and environmental stress tolerance.The genetic regions that change or remain stable during tissue culture or micropropagation. The genetic relationships between different varieties, which helps in selecting the appropriate genetic materials for propagation programs (Kumar et al.,2024; Kim et al.,2020; Thanoon and Khalil,2023), Genomic techniques include:

- Whole Genome Sequencing to identify all the nucleotides in plant DNA.
- Targeted Sequencing to identify specific genes associated with desired traits.
- Genetic Mapping to determine the locations of genes on chromosomes and link them to agricultural traits.

Proteomics and its role in propagation

Proteomics is the study of all the proteins expressed in the plant cell during different growth stages. In plant propagation, proteomics helps to evaluating protein expression during tissue culture and micropropagation stages, Identifying the proteins responsible for rooting, bud formation, and the growth of small plants. Comparing the plants produced from micropropagation with the mother plant to ensure the stability of genetic traits at the protein level, Proteomics techniques include:Two-dimensional polyacrylamide gel electrophoresis (2D-PAGE), Mass Spectrometry to determine protein composition and analyze changes (Chin and Tan,2018; Wendt Dos Santos et al.,2018)

Metabolomics in Plant Propagation

Metabolomics is the study of all the small chemical compounds (metabolites) produced in the plant during growth. Metabolomics allows us to understand how propagation processes, such as tissue culture or genetic modification, affect plant metabolism (Manickam et al.,2023; Hamany Djande et al.,2020; Gupta et al.,2022), including of production of important secondary metabolites in medicinal and aromatic plants, evaluation of the quality of the plants produced from propagation in terms of nutritional or medicinal content, discovering metabolic changes resulting from mutations or genetic modifications. Metabolomics techniques included Chromatography (GC-MS, LC-MS) for the analysis of chemical compounds, NMR Spectroscopy for determining chemical structures (Dehaven et al.,2010; Fiehn,2016; Saadallah et al.,2023).

The integration of genomics, proteomics and metabolomics

The integration of these three techniques allows for a comprehensive understanding of the plant at the genomic, proteomic, and metabolomic levels, providing a complete picture of How gene expression occurs during micropropagation, stability of genetic traits at different levels, the impact of the environment and laboratory practices on the quality of the resulting plants (Roychowdhury et al.,2023; Wang and Zhao,2023).

Practical Applications

1. Improving the micropropagation of medicinal plants are ensuring the production of high-quality plants containing active compounds.
2. Commercial propagation of food crops are evaluating the productivity, genetic quality, and typical characteristics of the cloned plants.
3. Conservation programs for endangered species are using genomic, proteomic, and metabolomic data to ensure genetic diversity and trait stability during propagation (Moraes et al.,2021; De Carlo et al.,2021).

Case Studies and Practical Applications

A case study on the propagation of rare plants

Molecular markers and genetic fingerprinting were applied in the propagation of rare plants such as medicinal gentian and rare aloe, where studies showed that the use of SSR and RAPD enabled researchers to verify the purity of the cloned plants and avoid genetic mutations during tissue culture. The genetic fingerprinting also helped in creating a database of genetic resources, ensuring the preservation of genetic diversity for these species (Ganie et al.,2015; Bidyananda et al.,2024).

Applications in food crops

In field crops such as rice, corn, and wheat, CRISPR-Cas technologies and genetic modification have contributed to the production of disease-resistant plants that can withstand drought and salinity conditions, while maintaining productivity and quality traits. For example, recent studies have shown that modifying fungal resistance genes in wheat has led to a reduction in yield loss by up to 30% in disease-affected environments (Sami et al.,2021; Chen et al.,2024; Al-Tarjuman et al., 2020).

Applications in medicinal and aromatic plants

Metabolomics and proteomics have been used to improve the micropropagation of medicinal plants such as lavender and mint, as these techniques helped evaluate the content of essential oils and active compounds, ensuring the production of high-quality plants genetically identical to the parent plants (Yan et al.,2022; Şimşek et al.,2024; Alobidy and Abdulsalaam,2024).

Challenges and Future Trends

The main challenges

1. The high cost of tools and technologies: Most molecular techniques require advanced laboratories, PCR machines, mass spectrometers, and data analysis software, which increases the cost for farmers and research institutions.
2. The need for advanced expertise: Complex molecular analysis requires trained personnel, which is a challenge especially in developing countries.
3. Legal and regulatory restrictions: The use of genetically modified plants or CRISPR technology is subject to strict policies in some countries, which limits immediate commercial spread.
4. Verification of long-term genetic stability: Despite the precision of the techniques, plants must be monitored over several generations to ensure trait stability and achieve breeding goals (Liu et al.,2023; Gerace et al.,2022).

Future Trends

1. Integrating molecular technologies with precision agriculture:

Using genomics, proteomics, and metabolomics to improve plant quality and increase productivity sustainably.

2. Developing low-cost and user-friendly technologies: Designing DNA and RNA analysis tools suitable for use in small laboratories or agricultural fields.
3. Expanding the use of CRISPR-Cas and modern genomic technologies: Expanding the scope of applications to include new varieties of medicinal and food plants with a focus on environmental stress resistance.
4. Conservation of genetic diversity: Establishing digital gene banks based on genetic fingerprints and molecular markers to ensure genetic diversity and sustainable breeding.
5. Artificial intelligence and advanced biological analysis:Advanced Biological Analysis:

Using artificial intelligence to analyze genomic, proteomic, and metabolomic data with greater speed and accuracy, contributing to the improvement of breeding programs (Yan et al.,2022; Zhao et al.,2025).

Practical applications of modern technologies in propagation

Applications of Tissue Culture and Micropropagation

Tissue culture is considered one of the most widely used techniques in modern propagation, as it provides the production of genetically pure plants that are identical to the parent plant (Ozyigit et al.,2023). Propagation of high-yield agricultural plants such as potatoes, bananas, and rice, where tissue culture contributes to the production of millions of disease-free seedlings in a short period. Propagation of medicinal and aromatic plants such as lavender, mint, and ginger, where the technology allows for the production of genetically identical plants while maintaining the concentration of active compounds. Propagation of rare and endangered plants such as some types of medicinal trees and conifers, where the technology helps increase the number of plants without affecting the natural environment (Abdullah and Al-dulaimi,2024; Hasnain et al.,2023; Pandey et al.,2023; Al-Nuaimy and Abed, 2024).

Applications of molecular techniques

1. Improving genetic selection using molecular markers (SSR, SNP) to identify plants carrying desirable genes before large-scale propagation.
2. Early detection of genetic mutations especially when using tissue culture and vegetative propagation, to ensure the purity of the resulting plants.
3. Developing genetically improved varieties are thru CRISPR-Cas or genetic modification to produce plants resistant to diseases or environmental stress (Amiteye et al.,2021; Hasan et al.,2021)

Integration of technologies

The integration of tissue culture with molecular markers and genetic modification creates a comprehensive system for modern propagation, characterized by:

1. Speed of producing high-quality plants.
2. Ensuring the stability of genetic traits at both the morphological and molecular levels.
3. The possibility of applying it on a wide commercial scale while preserving genetic diversity (Wikandari et al.,2021; Hesami et al.,2021)

Global Practical Examples

1. Bananas in Asia: Integrating tissue culture with molecular markers to produce fungal-resistant varieties.
2. Wheat in the United States and Europe: Using CRISPR-Cas to modify drought-resistant genes and increase productivity.
3. Aloe and medicinal plants in the Middle East: Applying proteomics and metabolomics to ensure the production of plants with a high percentage of active compounds (Tripathi et al.,2024; Tripathi et al.,2022; Sami et al.,2021).

Conclusion and Recommendations

Conclusion

Recent years have witnessed significant advancements in modern plant propagation technologies, where the integration of tissue culture, molecular markers, genetic modification, CRISPR-Cas techniques, and gene cloning has become an integral part of modern plant production programs. These techniques provide high precision, rapid production, and ensure the stability of genetic traits, while also helping to preserve the genetic diversity of rare and endangered plants.

Genomic, proteomic, and metabolomic technologies have added a new dimension to propagation, as it has become possible to study plants at the genomic level and the level of small protein and chemical compounds, enhancing the ability to produce genetically identical and high-quality plants.

Recommendations

1. Implement modern propagation practices widely in research centers and commercial farms to ensure the production of genetically pure plants. The widespread application of modern propagation practices in research centers and commercial farms to ensure the production of genetically pure plants.
2. Integrating molecular techniques with tissue culture to accelerate breeding programs and improve genetic quality. Integrating molecular techniques with tissue culture to accelerate breeding programs and improve genetic quality.
3. Training technical and research staff on the use of modern tools such as CRISPR-Cas, proteomics, and metabolomics. Training technical and research staff on the use of modern tools such as CRISPR-Cas, proteomics, and metabolomics.
4. Establish digital gene banks and plant repositories using genetic fingerprinting and molecular markers to ensure genetic diversity. Creating digital gene banks and plant stocks using genetic fingerprinting and molecular markers to ensure genetic diversity.
5. Evaluating the environmental impact of genetically modified plants before introducing them into commercial propagation to ensure sustainable agriculture. Environmental impact assessment of genetically modified plants before their introduction into commercial propagation to ensure sustainable agriculture.
6. Developing low-cost and easy-to-apply techniques to expand their use in developing countries and small farms. Developing low-cost and easy-to-apply technologies to expand their benefits in developing countries and small farms.

REFERENCES

Abdalla, N., El-Ramady, H., Seliem, M. K., El-Mahrouk, M. E., Taha, N., Bayoumi, Y., Shalaby, T. A., & Dobránszki, J. (2022). An Academic and Technical Overview on Plant Micropropagation Challenges. *Horticulturae*, *8*(8), 677. DOI: 10.3390/horticulturae8080677

Abdul Aziz, M., Brini, F., Rouached, H., & Masmoudi, K. (2022, November 8). Genetically engineered crops for sustainably enhanced food production systems. *Frontiers in Plant Science*, *13*, 1027828. DOI: 10.3389/fpls.2022.1027828 PMID: 36426158

Abdullah, Z. Y., & Al-dulaimi, F. K. Y. (2024). Evaluating the role of Ultraviolet Radiation and Ozone gas in reducing Aflatoxin in Rice. *NTU Journal of Agricultural and Veterinary Sciences*, *4*(4), 249–253.

Abdullah, Z. Y., & Al-dulaimi, F. K. Y.Fahad K. Y. Al-dulaimiZainab Yahya Abdullah. (2024). Evaluating the role of Ultraviolet Radiation and Ozone gas in reducing Aflatoxin in Rice. (2024). *NTU Journal of Agriculture and Veterinary Science*, *4*(4). Advance online publication. DOI: 10.56286/npeme074

Abdulraheem, M. I., Xiong, Y., Moshood, A. Y., Cadenas-Pliego, G., Zhang, H., & Hu, J. (2024, January 7). Mechanisms of Plant Epigenetic Regulation in Response to Plant Stress: Recent Discoveries and Implications. *Plants*, *13*(2), 163. DOI: 10.3390/plants13020163 PMID: 38256717

Ahmad, A., Ahmad, N., Anis, M., Faisal, M., Alatar, A. A., Abdel-Salam, E. M., Meena, R. P., & Sivanesan, I. (2022). Biotechnological Advances in Pharmacognosy and In Vitro Manipulation of *Pterocarpus marsupium* Roxb. *Plants*, *11*(3), 247. DOI: 10.3390/plants11030247 PMID: 35161227

Ahmar, S., Gill, R. A., Jung, K. H., Faheem, A., Qasim, M. U., Mubeen, M., & Zhou, W. (2020, April 8). Conventional and Molecular Techniques from Simple Breeding to Speed Breeding in Crop Plants: Recent Advances and Future Outlook. *International Journal of Molecular Sciences*, *21*(7), 2590. DOI: 10.3390/ijms21072590 PMID: 32276445

Akhtar, H., Usman, M., Binyamin, R., Hameed, A., Arshad, S. F., Aslam, H. M. U., Khan, I. A., Abbas, M., Zaki, H. E. M., Ondrasek, G., & Shahid, M. S. (2024). Traditional Strategies and Cutting-Edge Technologies Used for Plant Disease Management: A Comprehensive Overview. *Agronomy (Basel)*, *14*(9), 2175. DOI: 10.3390/agronomy14092175

Al-Dulaimi, F. K. Y., & Al-Tarjuman, J. K. (2024). Food security challenges. Achieving Food Security Through Sustainable Agriculture, igi-global. pp. 28–49. DOI: 10.4018/979-8-3693-4240-4.ch002

Al-Nuaimy Maha, M. T., & Abed Faten, N. A. (2024). Zearalenone toxins produced by some species of Fusarium. (2024). *NTU Journal of Agriculture and Veterinary Science*, *4*(4). Advance online publication. DOI: 10.56286/bb44fh80

Al-Tarjuman, Janan K.,Abed, Faten N. Mula and Al-Dulaimi, Fahad K.Y. (2020). Study on fungal contamination of IRAQI paper currency. *Biochemical and Cellular Archives*, *20*(1), 2201–2208.

. Alobidy A. and Abdulsalaam Sh. Riyadh.(2024).The effect of using some extraction techniques on the physical and chemical properties of essential oils of citrus peels. (2024). *NTU Journal of Agriculture and Veterinary Science, 4*(1). DOI: 10.56286/ntujavs.v4i1.770

Amiteye, S. (2021, September 30). Basic concepts and methodologies of DNA marker systems in plant molecular breeding. *Heliyon*, *7*(10), e08093. DOI: 10.1016/j.heliyon.2021.e08093 PMID: 34765757

Ansori, A. N., Antonius, Y., Susilo, R. J., Hayaza, S., Kharisma, V. D., Parikesit, A. A., Zainul, R., Jakhmola, V., Saklani, T., Rebezov, M., Ullah, M. E., Maksimiuk, N., Derkho, M., & Burkov, P. (2023, August). Application of CRISPR-Cas9 genome editing technology in various fields: A review. *Narra J*, *3*(2), e184. DOI: 10.52225/narra.v3i2.184 PMID: 38450259

Aremu, A. O., Fawole, O. A., Makunga, N. P., Masondo, N. A., Moyo, M., Buthelezi, N. M. D., Amoo, S. O., Spíchal, L., & Doležal, K. (2020, August 22). Applications of Cytokinins in Horticultural Fruit Crops: Trends and Future Prospects. *Biomolecules*, *10*(9), 1222. DOI: 10.3390/biom10091222 PMID: 32842660

Asmamaw, M., & Zawdie, B. (2021, August 21). Mechanism and Applications of CRISPR/Cas-9-Mediated Genome Editing. *Biologics*, *15*, 353–361. DOI: 10.2147/BTT.S326422 PMID: 34456559

Bekalu, Z. E., Panting, M., Bæksted Holme, I., & Brinch-Pedersen, H. (2023, July 25). Opportunities and Challenges of In Vitro Tissue Culture Systems in the Era of Crop Genome Editing. *International Journal of Molecular Sciences*, *24*(15), 11920. DOI: 10.3390/ijms241511920 PMID: 37569295

. Ben Romdhane M, Riahi L, Jardak R, Ghorbel A, Zoghlami N. Fingerprinting and genetic purity assessment of F_1 barley hybrids and their salt-tolerant parental lines using nSSR molecular markers. 3 Biotech. 2018 Jan;8(1):57. . Epub 2018 Jan 8. PMID: 29354368; PMCID: PMC5756735.DOI: 10.1007/s13205-017-1080-3

Bidabadi, S. S., & Jain, S. M. (2020). Cellular, Molecular, and Physiological Aspects of In Vitro Plant Regeneration. *Plants*, *9*(6), 702. DOI: 10.3390/plants9060702 PMID: 32492786

Bidyananda, N., Jamir, I., Nowakowska, K., Varte, V., Vendrame, W. A., Devi, R. S., & Nongdam, P. (2024). Plant Genetic Diversity Studies: Insights from DNA Marker Analyses. *International Journal of Plant Biology*, *15*(3), 607–640. DOI: 10.3390/ijpb15030046

Bonnet, M., Lagier, J. C., Raoult, D., & Khelaifia, S. (2019, November 30). Bacterial culture through selective and non-selective conditions: The evolution of culture media in clinical microbiology. *New Microbes and New Infections*, *34*, 100622. DOI: 10.1016/j.nmni.2019.100622 PMID: 31956419

Castro-Camba, R., Sánchez, C., Vidal, N., & Vielba, J. M. (2022, October 9). Plant Development and Crop Yield: The Role of Gibberellins. *Plants*, *11*(19), 2650. DOI: 10.3390/plants11192650 PMID: 36235516

Chen, F., Chen, L., Yan, Z., Xu, J., Feng, L., He, N., Guo, M., Zhao, J., Chen, Z., Chen, H., Yao, G., & Liu, C. (2024, September 23). Recent advances of CRISPR-based genome editing for enhancing staple crops. *Frontiers in Plant Science*, *15*, 1478398. DOI: 10.3389/fpls.2024.1478398 PMID: 39376239

Chin, C. F., & Tan, H. S. (2018). The Use of Proteomic Tools to Address Challenges Faced in Clonal Propagation of Tropical Crops through Somatic Embryogenesis. *Proteomes*, *6*(2), 21. DOI: 10.3390/proteomes6020021 PMID: 29734680

Chung, K. P. (2025, April 30). Cytoplasmic inheritance: The transmission of plastid and mitochondrial genomes across cells and generations. *Plant Physiology*, *198*(1), kiaf168. Advance online publication. DOI: 10.1093/plphys/kiaf168 PMID: 40304456

Churchman, J. C., & Lowe, D. J. (2012). Alteration, formation, and occurrence of minerals in soils. In Huang, P. M., Li, Y., & Sumner, M. E. (Eds.), *Handbook of Soil Sciences — Properties and Processes* (pp. 20-1–20-72). CRC Press.

Collard, B. C., & Mackill, D. J. (2008, February 12). Marker-assisted selection: An approach for precision plant breeding in the twenty-first century. *Philosophical Transactions of the Royal Society of London. Series B, Biological Sciences*, *363*(1491), 557–572. DOI: 10.1098/rstb.2007.2170 PMID: 17715053

Correia, S., Matos, M., & Leal, F. (2024). Advances in Blueberry (*Vaccinium* spp.) In Vitro Culture: A Review. *Horticulturae, 10*(6), 533. DOI: 10.3390/horticulturae10060533

Cucu, M. A., Choudhary, R., Trkulja, V., Garg, S., & Matić, S. (2025). Utilizing Environmentally Friendly Techniques for the Sustainable Control of Plant Pathogens: A Review. *Agronomy (Basel), 15*(7), 1551. DOI: 10.3390/agronomy15071551

De Carlo, A., Tarraf, W., Lambardi, M., & Benelli, C. (2021). Temporary Immersion System for Production of Biomass and Bioactive Compounds from Medicinal Plants. *Agronomy (Basel), 11*(12), 2414. DOI: 10.3390/agronomy11122414

Debnath, S. C., Bhatt, D., & Goyali, J. C. (2023). DNA-Based Molecular Markers and Antioxidant Properties to Study Genetic Diversity and Relationship Assessment in Blueberries. *Agronomy (Basel), 13*(6), 1518. DOI: 10.3390/agronomy13061518

Dehaven, C. D., Evans, A. M., Dai, H., & Lawton, K. A. (2010, October 18). Organization of GC/MS and LC/MS metabolomics data into chemical libraries. *Journal of Cheminformatics*, *2*(1), 9. DOI: 10.1186/1758-2946-2-9 PMID: 20955607

Delporte, F., Jacquemin, J. M., Masson, P., & Watillon, B. (2012, December). Insights into the regenerative property of plant cells and their receptivity to transgenesis: Wheat as a research case study. *Plant Signaling & Behavior*, *7*(12), 1608–1620. DOI: 10.4161/psb.22424 PMID: 23072995

Dennis, E. S., Ellis, J., Green, A., Llewellyn, D., Morell, M., Tabe, L., & Peacock, W. J. (2008, February 12). Genetic contributions to agricultural sustainability. *Philosophical Transactions of the Royal Society of London. Series B, Biological Sciences*, *363*(1491), 591–609. DOI: 10.1098/rstb.2007.2172 PMID: 17656342

Didi, D. A., Su, S., Sam, F. E., Tiika, R. J., & Zhang, X. (2022, September 28). Effect of Plant Growth Regulators on Osmotic Regulatory Substances and Antioxidant Enzyme Activity of *Nitraria tangutorum*. *Plants*, *11*(19), 2559. DOI: 10.3390/plants11192559 PMID: 36235426

Ebrahimi, S., Khosravi, M. A., Raz, A., Karimipoor, M., & Parvizi, P. (2023, September 1). CRISPR-Cas Technology as a Revolutionary Genome Editing tool: Mechanisms and Biomedical Applications. *Iranian Biomedical Journal*, *27*(5), 219–246. DOI: 10.61186/ibj.27.5.219 PMID: 37873636

Eliwa, G. I., El-Dengawy, E. F., Gawish, M. S., & Yamany, M. M. (2024, March 7). Comprehensive study on in vitro propagation of some imported peach rootstocks: In vitro explant surface sterilization and bud proliferation. *Scientific Reports*, *14*(1), 5586. DOI: 10.1038/s41598-024-55685-3 PMID: 38454056

Erdoğan, İ., Cevher-Keskin, B., Bilir, Ö., Hong, Y., & Tör, M. (2023, July 22). Recent Developments in CRISPR/Cas9 Genome-Editing Technology Related to Plant Disease Resistance and Abiotic Stress Tolerance. *Biology (Basel)*, *12*(7), 1037. DOI: 10.3390/biology12071037 PMID: 37508466

Espinosa-Leal, C. A., Puente-Garza, C. A., & García-Lara, S. (2018, July). In vitro plant tissue culture: Means for production of biological active compounds. *Planta*, *248*(1), 1–18. DOI: 10.1007/s00425-018-2910-1 PMID: 29736623

Fehér, A. (2019, April 26). Callus, Dedifferentiation, Totipotency, Somatic Embryogenesis: What These Terms Mean in the Era of Molecular Plant Biology? *Frontiers in Plant Science*, *10*, 536. DOI: 10.3389/fpls.2019.00536 PMID: 31134106

Feng, T., Jia, Q., Meng, X., Chen, X., Wang, F., Chai, W., & Liang, Z. (2020). Evaluation of genetic diversity and construction of DNA fingerprinting in Polygonatum Mill. based on EST-SSR and SRAP molecular markers. 3 Biotech, 10(7), 322.

Fiehn O. Metabolomics by Gas Chromatography-Mass Spectrometry: Combined Targeted and Untargeted Profiling. Curr Protoc Mol Biol. 2016 Apr 1;114:30.4.1-30.4.32. . PMID: 27038389; PMCID: PMC4829120.DOI: 10.1002/0471142727.mb3004s114

Fussy, A., & Papenbrock, J. (2022). An Overview of Soil and Soilless Cultivation Techniques—Chances, Challenges and the Neglected Question of Sustainability. *Plants*, *11*(9), 1153. DOI: 10.3390/plants11091153 PMID: 35567154

Ganie, S. H., Upadhyay, P., Das, S., & Prasad Sharma, M. (2015, December). Authentication of medicinal plants by DNA markers. *Plant Gene*, *4*, 83–99. DOI: 10.1016/j.plgene.2015.10.002 PMID: 32289060

Garban, Z., & Ilia, G. (2024). Structure-Activity of Plant Growth Bioregulators and Their Effects on Mammals. *Molecules (Basel, Switzerland)*, *29*(23), 5671. DOI: 10.3390/molecules29235671 PMID: 39683830

Gelvin SB. (2003). Agrobacterium-mediated plant transformation: the biology behind the “gene-jockeying” tool. Microbiol Mol Biol Rev. 67(1):16-37, table of contents. . PMID: 12626681; PMCID: PMC150518.DOI: 10.1128/MMBR.67.1.16-37.2003

Gerace, E., Mancuso, G., Midiri, A., Poidomani, S., Zummo, S., & Biondo, C. (2022, June 8). Recent Advances in the Use of Molecular Methods for the Diagnosis of Bacterial Infections. *Pathogens (Basel, Switzerland)*, *11*(6), 663. DOI: 10.3390/pathogens11060663 PMID: 35745518

Ghimire, B. K., Yu, C. Y., Kim, W.-R., Moon, H.-S., Lee, J., Kim, S. H., & Chung, I. M. (2023). Assessment of Benefits and Risk of Genetically Modified Plants and Products: Current Controversies and Perspective. *Sustainability (Basel), 15*(2), 1722. DOI: 10.3390/su15021722

Giordano, M., Petropoulos, S. A., & Rouphael, Y. (2021). Response and Defence Mechanisms of Vegetable Crops against Drought, Heat and Salinity Stress. *Agriculture, 11*(5), 463. DOI: 10.3390/agriculture11050463

Gupta, S., Schillaci, M., & Roessner, U. (2022, April 15). Metabolomics as an emerging tool to study plant-microbe interactions. *Emerging Topics in Life Sciences, 6*(2), 175–183. DOI: 10.1042/ETLS20210262 PMID: 35191478

Hamany Djande, C. Y., Pretorius, C., Tugizimana, F., Piater, L. A., & Dubery, I. A. (2020). Metabolomics: A Tool for Cultivar Phenotyping and Investigation of Grain Crops. *Agronomy (Basel), 10*(6), 831. DOI: 10.3390/agronomy10060831

Hasan, N., Choudhary, S., Naaz, N., Sharma, N., & Laskar, R. A. (2021, August 27). Recent advancements in molecular marker-assisted selection and applications in plant breeding programmes. *Journal of Genetic Engineering and Biotechnology, 19*(1), 128. DOI: 10.1186/s43141-021-00231-1 PMID: 34448979

Hasnain, A., Naqvi, S. A. H., Ayesha, S. I., Khalid, F., Ellahi, M., Iqbal, S., Hassan, M. Z., Abbas, A., Adamski, R., Markowska, D., Baazeem, A., Mustafa, G., Moustafa, M., Hasan, M. E., & Abdelhamid, M. M. A. (2022, October 13). Plants *in vitro* propagation with its applications in food, pharmaceuticals and cosmetic industries; current scenario and future approaches. *Frontiers in Plant Science, 13*, 1009395. DOI: 10.3389/fpls.2022.1009395 PMID: 36311115

Hesami, M., Baiton, A., Alizadeh, M., Pepe, M., Torkamaneh, D., & Jones, A. M. P. (2021). Advances and Perspectives in Tissue Culture and Genetic Engineering of Cannabis. *International Journal of Molecular Sciences, 22*(11), 5671. DOI: 10.3390/ijms22115671 PMID: 34073522

Higuera-Mendieta, D., Uriza, P. A., Cabrales, S. A., Medaglia, A. L., Guzman, L. A., & Sarmiento, O. L. (2021). Is the built-environment at origin, on route, and at destination associated with bicycle commuting? A gender-informed approach. Journal of transport geography, 94, 103120.

Huang, W., & Wang, X. (2024). The Impact of Technological Innovations on Agricultural Productivity and Environmental Sustainability in China. *Sustainability (Basel), 16*(19), 8480. DOI: 10.3390/su16198480

Hwang, H. H., Yu, M., & Lai, E. M. (2017, October 20). Agrobacterium-mediated plant transformation: Biology and applications. *The Arabidopsis Book / American Society of Plant Biologists, 15*, e0186. DOI: 10.1199/tab.0186 PMID: 31068763

Ikeuchi, M., Sugimoto, K., & Iwase, A. (2013, September). Plant callus: Mechanisms of induction and repression. *The Plant Cell, 25*(9), 3159–3173. DOI: 10.1105/tpc.113.116053 PMID: 24076977

Istiandari, P., & Faizal, A. (2025). Integrating In Vitro Cultivation and Sustainable Field Practices of Sacha Inchi (*Plukenetia volubilis* L.) for Enhanced Oil Yield and Quality: A Review. *Horticulturae, 11*(2), 194. DOI: 10.3390/horticulturae11020194

Jie, L., & Haoran, M. (2025). Deciphering popular routes in urban parks: The impact of environmental factors on the amount, intensity and diversity of physical activity. *Urban Forestry & Urban Greening, 105*, 128684. DOI: 10.1016/j.ufug.2025.128684

Kaisera, N., Douchesa, D., Dhingrab, A., & Kevin, C. (2020). The role of conventional plant breeding in ensuring safe levels of naturally occurring toxins in food crops. *Trends in Food Science & Technology, 100*, 51–66. DOI: 10.1016/j.tifs.2020.03.042

Kakoulidou, I., Avramidou, E. V., Baránek, M., Brunel-Muguet, S., Farrona, S., Johannes, F., Kaiserli, E., Lieberman-Lazarovich, M., Martinelli, F., Mladenov, V., Testillano, P. S., Vassileva, V., & Maury, S. (2021, August 11). Epigenetics for Crop Improvement in Times of Global Change. *Biology (Basel), 10*(8), 766. DOI: 10.3390/biology10080766 PMID: 34439998

Karami, O., Aghavaisi, B., & Mahmoudi Pour, A. (2009, November). Molecular aspects of somatic-to-embryogenic transition in plants. *Journal of Chemical Biology, 2*(4), 177–190. DOI: 10.1007/s12154-009-0028-4 PMID: 19763658

Kavhiza, N. J., Zargar, M., Prikhodko, S. I., Pakina, E. N., Murtazova, K. M.-S., & Nakhaev, M. R. (2022). Improving Crop Productivity and Ensuring Food Security through the Adoption of Genetically Modified Crops in Sub-Saharan Africa. *Agronomy (Basel), 12*(2), 439. DOI: 10.3390/agronomy12020439

Khadr, A., Wang, G. L., Wang, Y. H., Zhang, R. R., Wang, X. R., Xu, Z. S., Tian, Y. S., & Xiong, A. S. (2020, December 8). Effects of auxin (indole-3-butyric acid) on growth characteristics, lignification, and expression profiles of genes involved in lignin biosynthesis in carrot taproot. *PeerJ, 8*, e10492. DOI: 10.7717/peerj.10492 PMID: 33354430

Khan, S., Ullah, M. W., Siddique, R., Nabi, G., Manan, S., Yousaf, M., & Hou, H. (2016). Role of Recombinant DNA Technology to Improve Life. *International Journal of Genomics, 2016*, 2405954. DOI: 10.1155/2016/2405954 PMID: 28053975

Kim, K. D., Kang, Y., & Kim, C. (2020, October 28). Application of Genomic Big Data in Plant Breeding:Past, Present, and Future. *Plants*, *9*(11), 1454. DOI: 10.3390/plants9111454 PMID: 33126607

Krasteva, G., Georgiev, V., & Pavlov, A. (2020, December 18). Recent applications of plant cell culture technology in cosmetics and foods. *Engineering in Life Sciences*, *21*(3-4), 68–76. DOI: 10.1002/elsc.202000078 PMID: 33716606

Kulus, D., & Tymoszuk, A. (2024). Advancements in In Vitro Technology: A Comprehensive Exploration of Micropropagated Plants. *Horticulturae*, *10*(1), 88. DOI: 10.3390/horticulturae10010088

Kumar, R., Das, S. P., Choudhury, B. U., Kumar, A., Prakash, N. R., Verma, R., Chakraborti, M., Devi, A. G., Bhattacharjee, B., Das, R., Das, B., Devi, H. L., Das, B., Rawat, S., & Mishra, V. K. (2024, November 7). Advances in genomic tools for plant breeding: Harnessing DNA molecular markers, genomic selection, and genome editing. *Biological Research*, *57*(1), 80. DOI: 10.1186/s40659-024-00562-6 PMID: 39506826

Kutschera, U., & Ray, P. M. (2022, January). Forever young: Stem cell and plant regeneration one century after Haberlandt 1921. *Protoplasma*, *259*(1), 3–18. DOI: 10.1007/s00709-021-01683-5 PMID: 34292403

Liu, Q., Jin, X., Cheng, J., Zhou, H., Zhang, Y., & Dai, Y. (2023, May). Advances in the application of molecular diagnostic techniques for the detection of infectious disease pathogens [Review]. *Molecular Medicine Reports*, *27*(5), 104. DOI: 10.3892/mmr.2023.12991 PMID: 37026505

Long, Y., Yang, Y., Pan, G., & Shen, Y. (2022, June 30). New Insights Into Tissue Culture Plant-Regeneration Mechanisms. *Frontiers in Plant Science*, *13*, 926752. DOI: 10.3389/fpls.2022.926752 PMID: 35845646

Majumder, S., Igamberdiev, A. U., & Debnath, S. C. (2025). Somaclonal Variation and Clonal Fidelity in Commercial Micropropagation: Challenges and Perspectives. *Agronomy (Basel)*, *15*(6), 1489. DOI: 10.3390/agronomy15061489

Mammadov, J., Aggarwal, R., Buyyarapu, R., & Kumpatla, S. (2012). SNP markers and their impact on plant breeding. *International Journal of Plant Genomics*, *2012*, 728398. DOI: 10.1155/2012/728398 PMID: 23316221

Manickam, S., Rajagopalan, V. R., Kambale, R., Rajasekaran, R., Kanagarajan, S., & Muthurajan, R. (2023, November 8). Plant Metabolomics: Current Initiatives and Future Prospects. *Current Issues in Molecular Biology*, *45*(11), 8894–8906. DOI: 10.3390/cimb45110558 PMID: 37998735

Moraes, R. M., Cerdeira, A. L., & Lourenço, M. V. (2021, March 21). Using Micropropagation to Develop Medicinal Plants into Crops. *Molecules (Basel, Switzerland)*, *26*(6), 1752. DOI: 10.3390/molecules26061752 PMID: 33800970

Movahedi, A., Aghaei-Dargiri, S., Li, H., Zhuge, Q., & Sun, W. (2023, November 13). CRISPR Variants for Gene Editing in Plants: Biosafety Risks and Future Directions. *International Journal of Molecular Sciences*, *24*(22), 16241. DOI: 10.3390/ijms242216241 PMID: 38003431

Mu, H., Wang, B., & Yuan, F. (2022, November 15). Bioinformatics in Plant Breeding and Research on Disease Resistance. *Plants*, *11*(22), 3118. DOI: 10.3390/plants11223118 PMID: 36432847

Muhammad Aslam, M., Waseem, M., Jakada, B. H., Okal, E. J., Lei, Z., Saqib, H. S. A., Yuan, W., Xu, W., & Zhang, Q. (2022, January 19). Mechanisms of Abscisic Acid-Mediated Drought Stress Responses in Plants. *International Journal of Molecular Sciences*, *23*(3), 1084. DOI: 10.3390/ijms23031084 PMID: 35163008

Ozyigit, I. I., Dogan, I., Hocaoglu-Ozyigit, A., Yalcin, B., Erdogan, A., Yalcin, I. E., Cabi, E., & Kaya, Y. (2023, June 29). Production of secondary metabolites using tissue culture-based biotechnological applications. *Frontiers in Plant Science*, *14*, 1132555. DOI: 10.3389/fpls.2023.1132555 PMID: 37457343

Pandey, V. K., Tripathi, A., Srivastava, S., Dar, A. H., Singh, R., Farooqui, A., & Pandey, S. (2023, March 24). Exploiting the bioactive properties of essential oils and their potential applications in food industry. *Food Science and Biotechnology*, *32*(7), 885–902. DOI: 10.1007/s10068-023-01287-0 PMID: 37123062

Panthee, D. R. (2023). Application of Molecular Markers in Crop Improvement and Beyond. *Agronomy (Basel)*, *13*(8), 2041. DOI: 10.3390/agronomy13082041

Park, H. S., Lee, W. K., Lee, S. C., Lee, H. O., Joh, H. J., Park, J. Y., Kim, S., Song, K., & Yang, T. J. (2021, January 28). Inheritance of chloroplast and mitochondrial genomes in cucumber revealed by four reciprocal F_1 hybrid combinations. *Scientific Reports*, *11*(1), 2506. DOI: 10.1038/s41598-021-81988-w PMID: 33510273

. Parmar N, Singh KH, Sharma D, Singh L, Kumar P, Nanjundan J, Khan YJ, Chauhan DK, Thakur AK. Genetic engineering strategies for biotic and abiotic stress tolerance and quality enhancement in horticultural crops: a comprehensive review. 3 Biotech. 2017 Aug;7(4):239. . Epub 2017 Jul 12. PMID: 28702937; PMCID: PMC5507805.DOI: 10.1007/s13205-017-0870-y

Paulikienė, S., Benesevičius, D., Benesevičienė, K., & Ūksas, T. (2025). Review—Seed Treatment: Importance, Application, Impact, and Opportunities for Increasing Sustainability. *Agronomy (Basel)*, *15*(7), 1689. DOI: 10.3390/agronomy15071689

Rahbaran, M., Razeghian, E., Maashi, M. S., Jalil, A. T., Widjaja, G., Thangavelu, L., Kuznetsova, M. Y., Nasirmoghadas, P., Heidari, F., Marofi, F., & Jarahian, M. (2021, November 30). Cloning and Embryo Splitting in Mammalians: Brief History, Methods, and Achievements. [Erratum in: Stem Cells Int. 2024 Jul 17;2024:9837676. doi: 10.1155/2024/9837676. PMID: 34887927; PMCID: PMC8651392]. *Stem Cells International*, *2347506*, 1–11. Advance online publication. DOI: 10.1155/2021/2347506 PMID: 34887927

Rajaseger, G., Chan, K. L., Yee Tan, K., Ramasamy, S., Khin, M. C., Amaladoss, A., & Kadamb Haribhai, P. (2023, September 30). Hydroponics: Current trends in sustainable crop production. *Bioinformation*, *19*(9), 925–938. DOI: 10.6026/97320630019925 PMID: 37928497

Raman, R. (2017, October 2). The impact of Genetically Modified (GM) crops in modern agriculture: A review. *GM Crops and Food: Biotechnology in Agriculture and the Food Chain*, *8*(4), 195–208. DOI: 10.1080/21645698.2017.1413522 PMID: 29235937

Roychowdhury, R., Das, S. P., Gupta, A., Parihar, P., Chandrasekhar, K., Sarker, U., Kumar, A., Ramrao, D. P., & Sudhakar, C. (2023, June 16). Multi-Omics Pipeline and Omics-Integration Approach to Decipher Plant's Abiotic Stress Tolerance Responses. *Genes*, *14*(6), 1281. DOI: 10.3390/genes14061281 PMID: 37372461

. Saadallah H, AL-Tarjuman, Janan, K. & Sultan Fatimah I..(2023). Evaluation of Phytochemical Constituents by GC-MS and Biological Activity of Eugenia caryophyllus extract. (2023). *NTU Journal of Agriculture and Veterinary Science, 3*(4).

Safana, H. S., Abd, A. M., & Ibrahim, M. A. (2021). Study of genetic variations of In vitro micropropagation stages of Kumquat (Citrus japonica Thunb.). Plant Cell Biotechnology And Molecular Biology, 292-299.

Salgotra, R. K., & Chauhan, B. S. (2023, January 9). Genetic Diversity, Conservation, and Utilization of Plant Genetic Resources. *Genes*, *14*(1), 174. DOI: 10.3390/genes14010174 PMID: 36672915

Sami, A., Xue, Z., Tazein, S., Arshad, A., He Zhu, Z., Ping Chen, Y., Hong, Y., Tian Zhu, X., & Jin Zhou, K. (2021, December). CRISPR-Cas9-based genetic engineering for crop improvement under drought stress. *Bioengineered*, *12*(1), 5814–5829. DOI: 10.1080/21655979.2021.1969831 PMID: 34506262

Sardar, A. (2023, September 29). Genetic amelioration of fruit and vegetable crops to increase biotic and abiotic stress resistance through CRISPR Genome Editing. [Erratum in: Front Plant Sci. 2024 Apr 25;15:1418620. doi: 10.3389/fpls.2024.1418620. PMID: 37841604; PMCID: PMC10570431]. *Frontiers in Plant Science*, *14*, 1260102. DOI: 10.3389/fpls.2023.1260102 PMID: 37841604

Şimşek, Ö., Dalda Şekerci, A., Isak, M. A., Bulut, F., İzgü, T., Tütüncü, M., & Dönmez, D. (2024). Optimizing Micropropagation and Rooting Protocols for Diverse Lavender Genotypes: A Synergistic Approach Integrating Machine Learning Techniques. *Horticulturae*, *10*(1), 52. DOI: 10.3390/horticulturae10010052

Su, Y. H., Tang, L. P., Zhao, X. Y., & Zhang, X. S. (2021, January). Plant cell totipotency: Insights into cellular reprogramming. *Journal of Integrative Plant Biology*, *63*(1), 228–243. DOI: 10.1111/jipb.12972 PMID: 32437079

Sussex, I. M. (2008, May). The scientific roots of modern plant biotechnology. *The Plant Cell*, *20*(5), 1189–1198. DOI: 10.1105/tpc.108.058735 PMID: 18515500

Thanoon, A., & Khalil Younes, A. (2023). Fusarium oxysporum f. sp. tuberosi causing wilt disease on potato plants in Iraq. (2023). *NTU Journal of Agriculture and Veterinary Science*, *3*(4). Advance online publication. DOI: 10.56286/ntujavs.v3i4.706

Tripathi, L., Ntui, V. O., & Tripathi, J. N. (2022). Control of Bacterial Diseases of Banana Using CRISPR/Cas-Based Gene Editing. *International Journal of Molecular Sciences*, *23*(7), 3619. DOI: 10.3390/ijms23073619 PMID: 35408979

Tripathi, L., Ntui, V. O., & Tripathi, J. N. (2024, August 16). Application of CRISPR/Cas-based gene-editing for developing better banana. *Frontiers in Bioengineering and Biotechnology*, *12*, 1395772. DOI: 10.3389/fbioe.2024.1395772 PMID: 39219618

Tyagi, S., Kumar, R., Kumar, V., Won, S. Y., & Shukla, P. (2021, January 2). Engineering disease resistant plants through CRISPR-Cas9 technology. *GM Crops and Food: Biotechnology in Agriculture and the Food Chain*, *12*(1), 125–144. DOI: 10.1080/21645698.2020.1831729 PMID: 33079628

Van Eck, J. (2020, September 18). Applying gene editing to tailor precise genetic modifications in plants. *The Journal of Biological Chemistry*, *295*(38), 13267–13276. DOI: 10.1074/jbc.REV120.010850 PMID: 32723863

van Esse, H. P., Reuber, T. L., & van der Does, D. (2020, January). Genetic modification to improve disease resistance in crops. *The New Phytologist*, *225*(1), 70–86. DOI: 10.1111/nph.15967 PMID: 31135961

Wang, M., Li, R., & Zhao, Q. (2023). Multi-Omics Techniques in Genetic Studies and Breeding of Forest Plants. *Forests, 14*(6), 1196. DOI: 10.3390/f14061196

Wang, M.-R., Bi, W., Shukla, M. R., Ren, L., Hamborg, Z., Blystad, D.-R., Saxena, P. K., & Wang, Q.-C. (2021). Epigenetic and Genetic Integrity, Metabolic Stability, and Field Performance of Cryopreserved Plants. *Plants, 10*(9), 1889. DOI: 10.3390/plants10091889 PMID: 34579422

Wang, X. (2022). Managing Land Carrying Capacity: Key to Achieving Sustainable Production Systems for Food Security. *Land (Basel), 11*(4), 484. DOI: 10.3390/land11040484

Wendt Dos Santos, A. L., Souza Reis, R., Schuabb Heringer, A., Segal Floh, E. I., Santa-Catarina, C., & Silveira, V. (2018). Proteomics as a Tool to Study Molecular Changes During Plant Morphogenesis In Vitro. *Methods in Molecular Biology (Clifton, N.J.), 1815*, 339–349. DOI: 10.1007/978-1-4939-8594-4_24 PMID: 29981134

Wikandari, R., Manikharda, , Baldermann, S., Ningrum, A., & Taherzadeh, M. J. (2021, December). Manikharda, Baldermann S, Ningrum A, Taherzadeh MJ. Application of cell culture technology and genetic engineering for production of future foods and crop improvement to strengthen food security. *Bioengineered, 12*(2), 11305–11330. DOI: 10.1080/21655979.2021.2003665 PMID: 34779353

Yan, S., Bhawal, R., Yin, Z., Thannhauser, T. W., & Zhang, S. (2022, July 23). Recent advances in proteomics and metabolomics in plants. *Molecular Horticulture, 2*(1), 17. DOI: 10.1186/s43897-022-00038-9 PMID: 37789425

Yang, P., Zhang, S., Hu, D., Li, X., Guo, Y., Guo, H., Zhang, L., & Ding, X. (2024). Research Progress on the Mechanism and Application of the Type I CRISPR-Cas System. *International Journal of Molecular Sciences, 25*(23), 12544. DOI: 10.3390/ijms252312544 PMID: 39684256

Yuan, P., Usman, M., Liu, W., Adhikari, A., Zhang, C., Njiti, V., & Xia, Y. (2024, December). Advancements in Plant Gene Editing Technology: From Construct Design to Enhanced Transformation Efficiency. *Biotechnology Journal, 19*(12), e202400457. DOI: 10.1002/biot.202400457 PMID: 39692063

Zhao, J., Yang, W., Cai, H., Cao, G., & Li, Z. (2025). Current Progress and Future Trends of Genomics-Based Techniques for Food Adulteration Identification. *Foods, 14*(7), 1116. DOI: 10.3390/foods14071116 PMID: 40238250

Zhi, H., Zhou, S., Pan, W., Shang, Y., Zeng, Z., & Zhang, H. (2022). The Promising Nanovectors for Gene Delivery in Plant Genome Engineering. *International Journal of Molecular Sciences, 23*(15), 8501. DOI: 10.3390/ijms23158501 PMID: 35955636

Chapter 10
Plant Propagation and Sustainable Agriculture

Wasan S. Salih Hussain
https://orcid.org/0000-0002-1537-9979
Mosul University, Iraq

Muheddin Mahmoud Rteba
https://orcid.org/0000-0003-1804-600X
Libyan Authority for Scientific Research, Libya

Abdullah K. Mohammad
Independent Researcher, Iraq

ABSTRACT

Plant propagation is a fundamental pillar of improving agricultural production and developing crops., plant propagation plays a pivotal role in achieving a balance between increased production and environmental conservation. It enables the dissemination of pest- and drought-resistant varieties, provides high-quality planting material that reduces the need for chemical inputs, and limits the depletion of natural resources. It also supports the preservation of plant genetic diversity, a critical component of ensuring the resilience of agricultural systems to climate change. Propagation techniques complement the Sustainable Development Goals by enhancing food security, increasing income opportunities for farmers, and reducing the environmental impact of agriculture. Therefore, developing the capabilities of workers in this field and expanding the adoption of these technologies in agricultural production represents a strategic step towards building a more efficient, equitable, and sustainable agricultural future.

DOI: 10.4018/979-8-3373-5626-6.ch010

INTRODUCTION

In an era where the world faces increasing challenges in food security and climate change, the importance of adopting sustainable agricultural technologies emerges as a strategic solution to ensure a sustainable and productive agricultural future. In this context, we highlight sustainable agriculture, which is the optimal solution to address the environmental and economic challenges related to the agricultural sector. By raising awareness, providing government support, and providing appropriate technology, more sustainable agriculture can be achieved that meets present needs without compromising the future of the planet. This is a means to develop the agricultural sector and improve crop productivity. There are many unconventional methods of agriculture, such as aeroponics, which involves growing plants suspended in the air without a growth medium, and obtaining food and moisture through spray distributed by sprinklers or nozzles (Hussain and Abbas, 2025).

There are also other methods of agriculture, such as hydroponics and Sandponic farming (Hussain and Mohammad, 2024).

Sustainable breeding techniques ensure crop availability by producing high-quality seeds and planting materials, transferring the distinctive genetic traits of the parent plant through cuttings, and diversifying genetic resources to address environmental challenges and climate change, thereby enhancing food security and reducing dependence on external resources.

Reproduction is defined as one of the applied branches of botany, which is concerned with increasing the number of economic plants to maintain them to meet human needs and requirements for food, clothing, and shelter. Although most field crops reproduce sexually, primarily through seeds, there are some crops that reproduce asexually. Sexual reproduction is a common method of reproduction among higher plants, and in particular, most field crops reproduce through seeds. Although some crops reproduce by seeds only a few times, sexual reproduction is the most common method. Some plants can reproduce sexually without undergoing meiosis, gamete formation, or fertilization. Common methods of asexual reproduction include corms, bulbs, rhizomes, stolons, tubers, and other vegetative parts. For example, potatoes are propagated through tubers, and sugarcane is propagated through cuttings.

Horticultural Plants Propagation

Propagation is defined as increasing the number of individuals and plant species by sexual (seeds) or asexual (vegetative) means in order to preserve and spread them.

Horticultural plants are propagated in two ways:

First - Sexual Propagation:

This refers to the use of seeds containing sexual embryos to propagate plants. These seeds are the result of pollination and fertilization in plants.

Some plant seeds can be planted for propagation after being harvested directly from the fruit, as in the case of various vegetable crop seeds, provided that the environmental conditions are favorable for germination.

Another category of plants has seeds that cannot germinate after being harvested from the fruit, even if grown under suitable environmental conditions, as in the case of some types of fruit, such as stone fruit. Apple seeds require special treatment before planting, such as exposure to low temperatures. Another category requires seed treatment, i.e., scarification, to facilitate germination.

Advantages of Sexual Propagation(Advantages and Disadvantages of Reproductive Methods).

Humans Unlike plants that plant cannot move around their environment, which means they need to use other strategies to transport pollen to ovules to form grains. For cross-pollinated plants, male plants do not have to expend energy on grain production and can therefore direct more energy toward producing and dispersing high-quality pollen. Similarly, since female plants do not have to produce pollen, they can direct more of their energy toward producing high-quality ovules, meaning these plants can pass on more abilities to their offspring to improve their chances of survival. Furthermore, because dioeciously plants require two different plants to reproduce, their offspring will have more genetic diversity from their two parents. This is crucial when the environment changes, as offspring with this greater genetic diversity will be better able to adapt to the new environment thanks to their genetic diversity. On the other hand, self-pollinating plants, which obtain their genes from only one parent, will suffer from low genetic diversity and may therefore be susceptible to many problems related to their adaptation to environmental changes. Therefore, this (abundance of genetic diversity) is considered an advantage for plants in which cross-pollination occurs! I previously learned that only 5% of plant species reproduce this way, which means that these plants are not numerous.

1. The ability to propagate large numbers of plants from a single plant or produce plants on a large scale (mass propagation), as in forest trees.
2. Plants produced from seeds, i.e., sexually propagated, are stronger than their asexually propagated counterparts (vegetative), especially in fruit trees, which are characterized by their greater resistance to environmental conditions and climatic factors due to the spread and size of the root system.

3. Sexual propagation is a means of obtaining disease-free plants, especially viral diseases, as is the case with fruit trees, particularly citrus fruits.
4. It is a means of obtaining varieties and types of seeds in fruit and vegetable trees.
5. It is a means of obtaining new varieties and types of plants through hybridization between different varieties or species.

Disadvantages of sexual propagation:

1- Trees produced from seed differ in characteristics from the parent plant from which the seeds were taken, as in palm trees, citrus fruits, and other fruit species.
2- Trees propagated by seed are often delayed in reaching fruiting. For example, palm trees propagated by seeds yield fruit after 10-20 years, or sometimes more, while those propagated by offshoots require 6-12 years, or sometimes less.
3- Trees propagated by sex exhibit undesirable traits, such as the appearance of thorns, as these impede various maintenance processes.

Second: Asexual Propagation (Vegetative Propagation):

This refers to the use of any part of the plant other than the sexual embryo of the seed. The part used for propagation may be the stem, leaf, bud, a portion of leaf tissue, the growing tip, etc.

Vegetative Propagation Methods:

1- Cuttings: These are taken from branches or stems one year or older and divided into stem cuttings, leaf cuttings, or rootstock cuttings (apples and olives).
2- Budding: A bud is taken from the plant, called the scion, and placed on another plant, called the rootstock (stone fruit, citrus fruits, roses, gardenias).
3- Grafting: This refers to the fusion of a plant stem with a branch or stem of another plant of the same variety.
4- Propagation by specialized parts such as corms (gladiolus), bulbs and tubers (potatoes), or tuberous roots (sweet potatoes and dahlias).
5- Propagation by offshoots (palm trees and bananas).
6- Propagation by plant tissue culture: This method uses very small plants (bud, part of the growing tip, pollen grain, anther, or part of the ovary) grown under completely sterile conditions in media containing all growth requirements.
7- Propagation by layering: This method is used for plants that have difficulty forming roots on their cuttings.

Advantages of asexual propagation:

1. Ensuring the production of plants with similar characteristics to the parent plant (palm propagated by offshoots).
2. Vegetative propagated plants begin fruiting earlier than sexually propagated plants. Grafted oranges produce fruit 4-6 years after grafting, while olives produce fruit 6-8 years after grafting.
3. Propagating plants that do not produce seeds under normal conditions, such as bananas, currants, garlic, and navel oranges.
4. Sexual propagation is a means of overcoming certain environmental conditions by combining two plants through the grafting process. One is resistant to environmental conditions, called the stock, and the other is resistant, called the scion.
5. The size of the resulting plants can be controlled to facilitate harvesting and processing.

Disadvantages of asexual propagation:

1- Asexually propagated plants are less resistant to environmental conditions than their sexually propagated counterparts.
2- Asexual propagation is a means of transmitting various diseases, especially viral diseases.
3- It requires more experience and skill on the part of the person performing the propagation process, as well as more specialized supplies than sexual propagation.

The Importance of Vegetative Propagation in Plants

- Vegetative propagation is used in plants for many purposes.
- To propagate seedless plant species.
- To propagate hybrid plants without alteration, as not all of their seeds produce plants identical to the parents.
- To propagate plants with low seed germination rates.
- To increase the speed of plant propagation and accelerate fruiting.
- To adapt and acclimatize plants to new and different environments. For example, pear tree roots grow well in sandy soil, but they can be successfully transplanted to such soil by grafting onto peach trees, which thrive in such soil.
- Vegetative propagation of various types prevents infection by certain parasites that attack the roots of some plants but not others. For example, European

grape roots are usually susceptible to a type of parasite that does not affect American grapes.

THE ROLE OF PROPAGATION IN SUPPORTING NUTRIENT CYCLING SYSTEMS AND MAINTAINING SOIL FERTILITY

Propagation supports nutrient cycling systems and maintains soil fertility by increasing organic matter through cover crops and green manure, fixing nitrogen with legumes, improving soil structure and structure through the deep roots of trees and native crops, and enhancing the biodiversity of microorganisms that decompose organic matter. Practices such as crop rotation and reduced tillage also contribute to reducing nutrient depletion and maintaining soil structure.

1. Soil Cover

Keeping the soil covered is a fundamental principle of conservation agriculture. Crop residues are left on the soil surface, but cover crops may be needed if the gap between harvesting one crop and establishing the next is too long. Cover crops improve the stability of a conservation agriculture system, not only by improving soil properties, but also by promoting increased biodiversity in the agro-ecosystem. (FAO, 2022) While commercial crops have market value, cover crops are primarily grown for their impact on soil fertility or as livestock feed. In areas where small amounts of biomass are produced, such as semi-arid regions or areas of eroded and degraded soils, cover crops are beneficial because they: protect the soil during fallow periods; mobilize and recycle nutrients; improve soil structure and break up compacted layers and hard beds; and allow for rotation in monocultures.

Different plants, with diverse rooting systems, explore different depths in the soil. They may also have the ability to absorb different amounts of nutrients and produce distinct root exudates (organic acids), resulting in benefits for both the soil and living organisms.

Vegetation is important in conservation agriculture to protect the soil from the effects of raindrops, to keep the soil shaded, and to maintain the highest possible moisture content. We have seen its importance for nutrient recycling; however, it also has a physical and possibly allelopathic effect on weeds, reducing their incidence and leading to a reduction in the use of agrochemicals and, consequently, production costs.

Straw residues act as a cushion that reduces pressure on the soil under wheels and hooves and therefore plays an important role in reducing soil compaction.

2. Seed Selection, Treatment, and Propagation in Organic Systems

The crop performance can be impacted by seed quality directly, which affects germination, seedling growth, resistance to disease, and yield. In organic agriculture, where not used of pesticides and synthetic fertilizers are, the essential starting with seeds to have high quality to ensure plant health and productivity. Studies we shown that the quality of poor seeds can reduce germination by 30% to 50%, which can result in lower plant populations and increased susceptibility to pests and diseases.

The selection of Organic seed depends on purity and adaptability to ecological conditions. High quality organic seeds exhibit superior viability and resistance to stresses factors, that may be reducing dependency on external inputs.

The selection of organic seeds follows principles of ecological that depend on biodiversity, stability of genetic, and ecological adaptability. In contrast to conventional systems that focus on high yield hybrids or seeds that are genetically modified, farming of organic relies on naturally resilient varieties that grow without chemical interventions. are principles of selection of organic seeds guiding, include: genetic diversity, purity of seed, and local conditions adaptation.

For that, Organic seed selection must be free from treatments of chemicals, organisms that genetically modified, and any synthetic that are in line with the ecological principles of sustainability. Seed choice ensures improved to resilience crop, better efficiency of nutrients, and higher environmental adaptation to conditions. Seeds selection in organic farming follows principles, including purity of genetics, capacity of germination, and local climatic suitability for conditions. Studies have shown that farms using organically adapted seed varieties foster long-term sustainability and biodiversity.

BIOTECHNOLOGY APPLICATIONS IN PLANT PROPAGATION FOR SUSTAINABILITY

Agricultural biotechnology is a set of technologies that utilize genetic modification and engineering in crops, livestock, forestry, fisheries, aquaculture, and agricultural industries. These technologies aim to improve plants or animals or develop microorganisms that are more easily and cost-effectively resistant to diseases, insects, and pesticides. These technologies also help adapt to climate change and conserve natural resources, such as biodiversity, land, and water, which are essential for food and agricultural production (Khatik 2023).

Environmental biotechnology plays a crucial role in addressing global environmental challenges by using biological processes to achieve sustainable solutions. This is achieved through three approaches:

1. Molecular Biology Techniques:

Plant molecular genetics is a branch of genetics that studies the molecular structure and function of genetic material (DNA) in plants, including how genetic information is stored and transmitted, and how genes express plant traits. Research in this field focuses on genetic engineering to produce crops resistant to diseases and pests, improve their characteristics, and develop techniques for diagnosing plant diseases. Using molecular biology techniques, selection and the search for new and improved genetic combinations are among the most important breeding methods. Therefore, this task becomes difficult for plant breeders if the sources of genetic variation are limited. However, if these sources are available, their task is also somewhat difficult, which is to conduct screening and selection, especially if the number of plants and genetic combinations is very large. The genetic factors (genes) responsible for traits are located in the genetic material (DNA). Therefore, by understanding the molecular structure of the gene and studying the systems that control various genetic traits, we can identify the factors that control the appearance or absence of certain traits despite the presence of the gene or genes for these traits. This is known as gene expression. This facilitates the identification of varieties that are tolerant and resistant to environmental conditions in a short time, accelerating breeding programs to produce new varieties of major grain crops that are tolerant to environmental stresses such as drought, soil salinity, and high temperatures, and that are high-yielding and high-quality. Extensive studies of viruses and bacteria that infect plants have provided a significant opportunity to advance molecular studies of these organisms.

Molecular and Plant Pathology

Numerous viral genes have been isolated and transferred either to plant hosts to attempt to confer a degree of resistance to the virus, or injected into the bacterial genetic material to enable the bacteria to produce other proteins suitable for study. The first attempts in this direction were in 1956, demonstrating that viral RNA in TMV is responsible for infecting plants and also for producing complete virus particles. In the 1970s, extensive studies were conducted on the crown gall bacteria Agrobacterium tumefaciens in dicotyledons. It was discovered that tumors are caused by a small segment of DNA called tumor DNA (TiDNA), located on the plasmid carrying the disease genes. This segment was injected into the tumor-inducing plasmid (Ti-plasmid), also known as the chromosomes of plant cells. It contains two genes responsible for overgrowth. Scientists were subsequently able to replace these two genes with others, whether from bacteria, viruses, or even animals. This enabled them to multiply foreign genes in plant cells. This discovery confirmed the possibility of introducing a foreign gene into a plant and then, through tissue culture, producing an entire plant containing these genes. This marked the beginning of genetic engineering and its applications in agriculture. This led to the development of plants with new genetic traits, including resistance to diseases and insects, and improved quality and quantity of production (Al wakel,2010). As for how DNA is introduced into host cells, this is done in various ways, including using viruses as a vector, or bombarding the plant with DNA. This involves using a special bombardment device filled with DNA and forcibly inserting it into the host cells. Some DNA molecules may successfully bind to the plant's chromosomes and replicate with them, thus producing a plant with new genetic traits.

It has been possible to produce plants resistant to the action of cell wall-degrading bacteria and fungi by introducing the genes of these bacteria and fungi responsible for producing cell wall-degrading enzymes into plant cells. These genes replicated in the host cells, resulting in plants resistant to the degradation effect of these organisms.

Molecular biology applications have had a significant impact on the design of modern methods for diagnosing plant diseases in very short periods of time, even when these organisms are present in very limited numbers or mixed with other, closely related pathogens. These methods include:

- 1Monoclonal antibodies
- 2Fatty acid profiles of pathogens
- 3 Analysis of fragments of nucleic acids produced by specific enzymes
- 4Determination of nucleotide sequences of the pathogen

In the 1980s, it was possible to label a portion of DNA in microorganisms with radioactive isotopes or colorimetric compounds, becoming a means of identifying pathogens through the design of special kits.

In general, the use of molecular biology in diagnosing plant diseases is still in its infancy, despite important previous discoveries, and its high cost currently stands as an obstacle to implementation at the field level.

2. Tissue Culture:

Plant tissue culture is a scientific technique for growing plant cells, tissues, or organs in a laboratory under sterile conditions using a specially formulated medium that provides the nutrients and hormones necessary for their growth. The technique aims to rapidly propagate plants, produce genetically identical copies, obtain disease- and virus-free plants, produce plants with desirable traits, create entire plants from genetically modified cells, and use them to preserve rare species. Plant tissue culture relies on the ability of plant tissue to form a complete new plant when provided with a suitable growth medium and environment. This ability of plant cells or tissues is called pluripotency. In the propagation of plants that have problems with seed germination (non-traditional plants), or that have short seed viability. by seed these orchids can be propagated, it takes approximately (2-3) years to produce mature orchids. Therefore, plant tissue culture has become an ideal method for protecting this species from extinction and achieving commercial viability.

Microscopic propagation can be performed in three different ways:

- Using an apical or axillary bud.
- Using an adventitious bud.
- By forming a somatic embryo using somatic plant cells.

The success of each technique depends on the genetic background, culture media, and incubation conditions.

Using tissue culture technology, which involves culturing any part of a plant (leaf, embryo, cell, root, stem, or protoplast, which is a cell without a cell wall) and growing it in an artificial medium containing all the necessary auxins, hormones, vitamins, and nutrients. Tissue culture is used as a rapid means of propagation (Altpeter et al., 2016).

Traditional vegetative propagation is not fast enough to meet the growing demand for plants, especially ornamental and vegetable plants, date palms, and bananas. As a result, the prices of these plants are constantly increasing in most countries around the world, prompting many to use tissue culture technology to provide these numbers of plants at an affordable price within a limited space, in addition to obtaining disease-free strains (Efferth, 2019).

Genetic Engineering:

Genetic engineering has been used for hundreds of years to improve plant varieties and achieve desirable traits through breeding and selection programs. These programs can take 10-12 years to implement, and sometimes extensive crossbreeding is performed between genetically distant and diverse individuals, such as different species and genera. Such crossbreeding is sometimes difficult or impossible due to the numerous genetic barriers within the cells of these strains. This requires significant effort and financial costs.

It also requires that the crops to be improved reproduce sexually so that hybridization can be done between them. Therefore, genetic engineering came with the hope of improving genetic traits in a short period of time and at lower costs by transferring the desired genes, trying to find new genetic combinations (genotypes) that carry in their cells the best genetic factors that control their production and resistance to the adverse environmental conditions surrounding them (Shetty et al., 2018).

It is now possible to transfer genes not only from one plant species or genus to another, but also to transfer genes from microorganisms to higher plants or animals. Also, through genetic engineering, we can resist viral diseases that are difficult to control in any other way.

Biotechnology has several goals in various fields, including the following: Providing consumers with nutrient-rich, long-lasting foods, or foods that contain lower levels of certain naturally occurring toxic substances found in some food plants. Developers are using biotechnology to try to reduce saturated fats in cooking oils, reduce allergens in foods, and increase disease-fighting nutrients in foods. Using genetically modified crops to produce new medicines, which could lead to improved plant-based pharmaceutical industries that could reduce production costs by using a sustainable resource. Developing genetically modified plants for a purpose known as phytoremediation, where plants detoxify contaminants in the soil or absorb and accumulate contaminants from the soil so that the plants can be harvested and disposed of safely, is intended to improve soil quality at a contaminated site.

Producing more resilient crops that thrive even in the harshest environments and require less fuel, labor, fertilizer, and water, helping to reduce stress on land and soil. Protecting crops from deadly viral diseases, as biotechnology has genetically engineered plants to resist viral diseases that are otherwise difficult to control. Increasing crop quality, leading to increased production and financial returns. Providing improved quality traits such as increased beta-carotene levels in rice to help reduce vitamin A deficiency and improving oil compositions in canola, soybeans, and corn.

Plant propagation and its role in the sustainable harvest of medicinal and aromatic crops

Plant propagation plays a pivotal role in the sustainable harvest of medicinal and aromatic plants by providing effective methods for increasing production, preserving desirable species traits, and ensuring resource sustainability through sexual and vegetative propagation. This supports the pharmaceutical and aromatic industries, reduces pressure on wild plants, and improves the quality of the produced crops.

Plants play a significant role in human life, making life possible since the beginning of time. They also provide shelter, food, clothing, and medicine. Therefore, it was not surprising that man turned to the plant kingdom to treat his ailments. Undoubtedly, he began experimenting with plants and found in them what was a cure for his ailments, so he kept them and prescribed them to his family and loved ones. As for what was harmful, he avoided it. In this way, man was the doctor, the pharmacist, and the patient all at the same time.

Since ancient times, humans used medicinal plants in their natural state, but with the increase in population and their gathering in cities, these plants were traded in concentrated form, and extracts, pills, and tablets containing active plant substances appeared. With the development of civilization and scientific progress, a new era of science was born at the beginning of the ninth century, which is "Modern Pharmacognosy." People all over the world knew the therapeutic benefits of plant extracts and herbs and preferred to use them since ancient times. People's attention did not shift from the real, proven benefits of this ancient natural treatment until the last century, when medical sciences began to search for drug treatment among industrial chemical preparations, supporting their discoveries with large-scale advertising campaigns. Due to the quick, temporary results of chemical treatments, this trend succeeded in deceiving people for some time - even after a period that may be short or long. In many cases, disease conditions appeared in another part of the body that were worse than the original disease condition, and thus the beginnings of the science of treatment using herbs appeared recently (Qandil and Ibrahim 2007).

Using agricultural residues to improve propagation processes and production systems.

Propagation processes and production systems can be improved by using agricultural residues by converting it into organic fertilizer to improve soil quality and nutrition, or converting it into animal feed to provide a source of protein and nutrients for animals. It can also be used to produce energy through biogas, or as raw materials in other industries such as paper and wood panels. These practices contribute to sustainable agriculture, reduce environmental pollution, and decrease dependence on natural resources.

Agricultural residues is a vital product of food production processes in various plants, remaining in fields after the main product or crop is harvested. Some tend to call it agricultural residues, but it has significant economic value and sometimes constitutes significant wealth that is often lost due to lack of exploitation. There is a long list of these materials, the availability and availability of which varies from country to country. Perhaps the most important include wheat straw, barley straw, rice field residues, rice husks, corn field residues, corn cobs, sunflower stalks, vegetable crop residues, and others. These residues, if annual plants, some bushes, and pruning products in orchards, forests, and trees are added together, will take on a more general title, which is what is called "lignocellulosic materials," which is an expression for materials that contain cellulose and lignin in their composition, other than tree wood.

Agricultural residues recycling systems and methods vary depending on the type of residues, the available methods and technologies, and the purpose of recycling. One of the most important uses is converting residues into organic fertilizers or animal feed, and using it to produce clean energy. This contributes to achieving clean agriculture, protecting the environment from pollution, improving agricultural products, and providing rural employment opportunities, thereby improving economic and environmental conditions and raising health and social standards.

In Iraq, there are no accurate statistics on the quantities of residues in agricultural fields because we still do not consider it a significant issue. Iraq, with its environmental diversity and vast agricultural land, produces millions of tons of these materials annually. Although some of them are used in limited areas, the vast majority is not only left unused, thus wasting resources, but is also burned on farms, thus becoming a factor in environmental pollution, leaving a negative impact on the soil (Kirkby and Fattore, 2006), the air, and living organisms.

It can be used for growing on agricultural residues. This method is used in lands suffering from soil problems such as severe salinity or soil diseases, whether in greenhouses or open fields. The residues is stacked on the soil, similar to rice husks, and vegetables are grown on them. Rice husks, cotton stalks, or sawdust can also be used to grow and produce mushrooms. Agricultural residues is used to control plant diseases, whether directly by mixing it with the soil or after converting it into organic fertilizers or compost. Some beneficial microorganisms are loaded onto these residues and added to the soil to combat pathogens present in the soil, as occurs in inhibiting the growth of the fungus *Phytophthosa Pythium* by compost, and using green fertilizers to inhibit the fungus *Rhizoctonia solani*, which causes root rot. Converting agricultural residues into organic fertilizers improves soil properties and increases growth rates. Because it makes the soil environment more alkaline, it becomes unsuitable for the growth of nematodes. In addition, the microorganisms that grow on it play a role in combating these harmful organisms (Youngquist et al., 1993).

Therefore, it is necessary to focus on raising rural awareness of how to deal with agricultural residues by developing their knowledge in this vital field. Among the most prominent areas of benefiting from agricultural residues are the production of unconventional feed, the production of organic fertilizers, and energy production. The benefit from agricultural residues can be maximized by using it as raw materials for many important industries, which increases its economic value and opens up job opportunities for young graduates. As a result of the rising prices of animal and poultry feed due to the increasing demand for it, interest has increased in trying to exploit agricultural residues in many different ways as an attempt to fill the deficit in animal feed and its high prices (Zaka et al., 2018). Agricultural residues is considered a source of income for the farmer in addition to the value of the main crop output (Sadh et al., 2018).

The Role of Plant Breeding in Enhancing Food Security

Many challenge Food security was faces for (nine – ten) billion people at 2050, this problem may be solve for that there are more options, including: new technologies use to increase production of crops and investment in research to reduce residues(Rteba and Hussain,2024), It increases agricultural production costs and reduces the farmer's economic profit margin, which negatively affects the standard of living of farmers and threatens the survival of small farms. Food security is the ability of people, at all times, to access sufficient, safe, and nutritious food to lead healthy and active lives. This concept is based on four key dimensions: physical availability of food, economic and physical access to it, proper utilization of food, and its stability over time. Food security is affected by factors such as conflict, poverty, climate change, and lack of infrastructure, and also includes combating food waste and ensuring proper nutrition for sustainable development(Murd,2010).

More than 50 million people in the Arab region or one in ten suffer from malnutrition. This number is expected to increase unless appropriate national policy measures are taken to achieve food security for all. Tracking food security is a complex task, but it is essential to help communities in the Arab region become more resilient to climate change, curb the depletion of natural resources, shift towards preserving terrestrial ecosystems, and adopt more sustainable consumption and production patterns. Food security is directly linked to plant reproduction, as the availability of sufficient and healthy food depends on the ability of plants to reproduce and produce abundant and healthy crops. Plant reproduction is influenced by multiple factors, such as soil health and nutrient availability, while pests and diseases can reduce crop production and create food insecurity. Therefore, improving plant health and reproduction through efficient agricultural practices and the use of appropriate genetic resources is essential to achieving global food security. Seed production or plant sustainability means increasing food supply, ensuring its availability for the population.

Plant breeding plays a pivotal role in enhancing food security by increasing crop productivity and ensuring a sustainable and sufficient food supply to meet the growing needs of the population. It also improves crop quality, reduces crop loss, and supports the local economy by providing job opportunities and increasing farmers' income. Strategies adopted in this area include adopting advanced agricultural technologies, innovation, focusing on plant genetic resources, sustainable pest management, and enhancing soil and seed health to ensure sustainable food systems.

Given that climate change poses a major threat to the agricultural and environmental sectors, including water shortages, rising temperatures, and increased soil salinity, which in turn threatens food security for society, biotechnology plays a significant role as one of the modern solutions to address these challenges. This is achieved by developing crops that are resistant to environmental conditions and improving production using advanced technologies such as genetic engineering and tissue culture. Sustainable development is one of the most important topics formulated from a human or peoples' rights perspective. The relationship between development and the environment emphasizes that the realization of the human or peoples' right to development cannot be fully realized unless it is coupled with the provision of a healthy environment. The human right to a healthy environment requires rational and targeted programs to achieve it. International efforts to reconcile environmental protection, sustainable development, and biodiversity conservation have contributed to this. Environmental conservation issues have become a top priority for the international community, especially after technological advances and the emergence of new biotechnology techniques to improve food and medicine, which have impacted human life, health, nutrition, and the environment.

The Industrial Revolution led to the emergence of many modern innovations, which led to scientific and technological progress. This resulted in significant results in many fields, enabling humans to utilize the Earth and its ecosystems, utilize its renewable and non-renewable resources, and develop new energy sources. Preserving biodiversity means preserving biological resources, investing in them sustainably, and reclaiming and improving these resources within a healthy environment. Therefore, cooperation among countries is essential to ensure the conservation of biodiversity, the sustainable use of biological resources, and the sharing of benefits resulting from the use of genetic resources.

The Role of Breeding in the Conservation of Genetic Resources

Crop cultivation has a long history in human civilization, with records indicating that the cultivation of cereal plants began around 8000 BC. This early plant breeding was primarily developed to provide a steady supply of food.

As human understanding of genetics advances, improved crop varieties can be achieved more rapidly. Artificial selection can be more targeted, and crop varieties can be optimized for favorable traits more quickly, producing better, more vigorous, or more palatable plants.

However, traditional plant breeding methods are slow and do not always produce the desired crop varieties. Later, biotechnology tools have made it easier to engineer desirable traits into plants that are difficult to breed using traditional methods. For example, improving nutritional deficiencies in plants is difficult through artificial selection, and vitamin A and iron deficiency pose a particular challenge. Rice, for example, lacks the genes for beta-carotene, a precursor to vitamin A. However, it does contain genes for geranyl pyrophosphate, which can be sequentially converted to beta-carotene using four enzymes. Rice was engineered using genes for two enzymes derived from daffodils and the remaining two from the bacterium *Erwinia uredovora*. The resulting crop is known as golden rice. Given that rice is the staple food for more than half the world's population, such bioengineered crops could potentially play a role in preventing blindness among children caused by vitamin A deficiency, or improving the health of rice-dependent countries.

Many commonly grown crop plants now have some degree of genetic modification introduced using biotechnology. For example, corn, papaya, and many potato varieties have been modified to resist herbicides, diseases, or pests. Genetic modification can be applied to reduce the production of allergens, as is the case with soybeans.Breeding plays a pivotal role in the conservation of genetic resources by increasing genetic diversity, enabling species to adapt to changing environmental conditions and cope with diseases and threats, and reducing the risk of inbreeding that is detrimental to survival. This can be achieved through in situ conservation (in its natural habitat) and ex situ conservation (in dedicated facilities).

Genetic resources are genetic material of actual or potential value, meaning they are any material of plant, animal, microbial, or other origin that contains functional units of heredity. Genetic resources therefore refer to that part of genetic diversity that has or could have a practical use, such as in plant breeding.

Genetic resources are one of the three levels of biological diversity identified in Article 2 of the Convention on Biological Diversity (CBD) in Rio de Janeiro, 1992.

The Importance of Reproduction in Biodiversity

Biodiversity, or biodiversity, encompasses all life on Earth, from genes to entire ecosystems. It is essential for climate stability, the provision of food, water, and medicine, and the prevention of climate crises. Humans depend heavily on biodiversity for their basic needs, with more than a billion people deriving their livelihoods from forests alone. However, biodiversity is under global threat due to human activities and climate change, leading to species loss and ecosystem degradation.

Reproduction is essential for biodiversity because it generates genetic diversity that allows organisms to adapt to changing environments, preventing species extinction and maintaining the continuity of organisms and ecosystem services such as the provision of food, water, and clean air, and supporting the evolution of species and their ability to survive in the face of pests, diseases, and climate change. Plants differ in their methods of reproduction. They either reproduce by seeds or by vegetative parts, as mentioned previously. Vegetative reproduction in plants results from the lack of seed formation, or because seeds only form under special conditions, or because of their short lifespan, as well as the wide range of genetic variations within the plant. All commercial varieties of sugarcane, bananas, and potatoes are generally sterile due to their failure to flower due to genetic factors and unsuitable climatic conditions. The nature of agricultural processes also significantly influences the failure of some plants, such as garlic and ornamental plants, to flower. All these obstacles call for the use of other methods for propagating these plants, including vegetative parts.

Sexual reproduction, by its nature, provides plant breeders with genetic variations, while vegetative propagation involves no genetic changes at all unless a genetic mutation occurs in the plant population. Grafting, in its various forms, usually occurs between varieties of the same plant species. However, grafting can sometimes be performed between varieties of different plant species. Hybrids obtained through this method of grafting are called chimeras. Chimeras are composed of tissues from two parent plants, which are naturally different in their genetic makeup. As a result of the grafting process, their different tissues overlap or accumulate. When these tissues multiply, the accumulated tissues either separate or overlap. Chimeric hybrids are propagated only vegetatively. An example of chimeric hybrids is what is observed in grapevines, where clusters with black fruits and others with yellow fruits are formed as a result of vegetative hybridization between varieties with different genetic makeup. When wind or animals transfer pollen from one plant to another, the advantage of cross-pollination is that it helps maintain the genetic diversity of species. Genetic diversity becomes beneficial when conditions change and organisms need to adapt to this change. Some plants have characteristics that prevent self-pollination, such as pollen and ovules that develop at different times.

Pollinators are animals that transfer pollen between plants. Many pollinators are insects, such as bees, butterflies, moths, and beetles. Some birds, including hummingbirds, also play a role. Similarly, some mammals, such as bats and rodents, transfer pollen between plants. The colors and scents of flowers often attract pollinators. Pollen sticks to the pollinator's body as it feeds on the flower's leaves. The seeds produced by fertilization contain genetic material from both parents. As a result, the offspring are not genetically identical to either parent plant. This genetic diversity enables them to survive in the event of a change in the environment. Offspring inherit one set of chromosomes from each parent. Therefore, each offspring has a combination of chromosomes (and alleles) from both parents. Sexual reproduction also results in genetic variation among siblings . During reproduction, genetic recombination occurs through processes such as crossing over during meiosis, resulting in the formation of new gene combinations. Genetic diversity is increased through sexual reproduction, as offspring inherit a unique set of genes from both parents.

REFERENCES

Altpeter, F., Springer, N. M., Bartley, L. E., Blechl, A. E., Brutnell, T. P., Citovsky, V., Conrad, L., Gelvin, S. B., Jackson, D., Kausch, A. P., Lemaux, P. G., Medford, J. I., Orozo-Cardenas, M., Tricoli, D., VanEck, J., Voytas, D. F., Walbot, V., Wang, K., Zhang, Z. J., & Stewart, C. N.Jr. (2016). Advancing crop transformation in the era of genome editing. *The Plant Cell*, *28*(7), 1510–1520. DOI: 10.1105/tpc.16.00196 PMID: 27335450

Efferth, T. (2019). Biotechnology applications of plant callus cultures. *Engineering (Beijing)*, *5*(1), 50–59. DOI: 10.1016/j.eng.2018.11.006

Hussain, W. S., & Abbas, M. M. (2025). Importance of Aeroponics in Vegetable Crops. In *Utilizing Aeroponics Techniques for Improved Farming* (pp. 57–70). IGI Global Scientific Publishing. DOI: 10.4018/979-8-3693-2320-5.ch003

Hussain, W. S., & Mohammad, A. K. (2024). Sustainable Agriculture Is a Way to Increase Wheat Production. In *Achieving Food Security Through Sustainable Agriculture* (pp. 252–267). IGI Global. DOI: 10.4018/979-8-3693-4240-4.ch010

Khatik, N. (2023). Role of Plant Biotechnology in Agricultural Sustainability and Improvement of Crops. *Journal of Agricultural Research*, *11*(6), 929–936.

Kirkby C A & Fattore, A. (2006). Effect of Rice Stubble Burning on Soil Health . A report for the Rural Industries Research and Development Corporation. RIRDC Publication

Qandil, A. M. A., & Ibrahim, A. K. (2007). *Production of medicinal, aromatic and ornamental plants*. Open Education Center, Faculty of Agriculture, Ain Shams University.

Rteba, M. M., & Hussain, W. S. (2024). Sustainable Agriculture: Way to Achieve Food Security. In Achieving Food Security Through Sustainable Agriculture (pp. 182-202). IGI Global.

Sadh, P. K., Duhan, S., & Duhan, J. S. (2018). Agroindustrial wastes and their utilization using solid state fermentation. *Bioresources and Bioprocessing*, *5*(1), 1–15. DOI: 10.1186/s40643-017-0187-z

Shetty, M. J., Chandan, K., Krishna, H. C., & Aparna, G. S. (2018). Genetically modified crops: An overview. *Journal of Pharmacognosy and Phytochemistry*, *7*(1), 2405–2410.

Youngquist, J., Krzysik, A., English, B., & Spelter, H. P, Chow (1993): Forest Agricultural fibers for use in building components. Proc7268 Madison, WI: product Society: 123 – 134.

Zaka, M., Morsi, Y. El-Din Mohamed, B. and Amin, I. F. (2018) The Economic Impacts of Recycling Agricultural Waste on Agricultural Production in Egypt, Journal of the Association of Arab Universities for Agricultural Sciences, AinShams University, 26(2B) (Special Issue): 1289-1302.

Chapter 11
Application of Plant Growth-Promoting Bacteria PGPB in Propagation StagesTowards a More Sustainable Agricultural System

Hiyam Adil Altaii
https://orcid.org/0000-0002-1537-9979
Mosul University, Iraq

Salim Jasim Mohammed
Mosul University, Iraq

Wasan Salih Hussain
https://orcid.org/0000-0002-1537-9979
Mosul University, Iraq

ABSTRACT

Plant growth-promoting bacteria (PGPB) play a growing role in sustainable agriculture promoting, due to their ability to support growth of plant in a natural and environmentally safe manner. During plant propagation, whether by seed, vegetative methods, or tissue culture, PGPB contribute to improved germination, enhanced root growth, and increased plant resistance to environmental stresses and diseases,

DOI: 10.4018/979-8-3373-5626-6.ch011

reducing used of chemical pesticides and fertilizers. These bacteria operate through multiple mechanisms, including atmospheric nitrogen fixation, decomposition of insoluble phosphates, and the hormones plant production ex: auxins, Growing scientific evidence suggests that applying these microorganisms in the first plant production stages can enhance propagation efficiency and improve produced of plants, while also supporting the environmental and economic objectives of sustainable agriculture. Therefore, exploring and using PGPB in propagation systems is a strategic step towards developing more resilient and sustainable agricultural systems

INTRODUCTION

Globally, we are heading towards reducing the use of artificial fertilizers because of their harmful impacts on ecology and human health. This has led to using microorganisms as biofertilizers (some of which fix nitrogen) while others dissolve important elements to plants (Hanapi et al., 2013). Bacteria of nitrogen-fixing that found in root nodules. These are used as bio-fertilizers that fix nitrogen and supply the nitrogen for plants, and obtain their organic carbon requirements through symbiosis with plants. these bacteria are characterized by a legume host-specific specialization. Therefore, in last year, bio stimulants have been used to improve soil properties and provide nutrients to plants, which accelerate growth and increase production. Vitamins are among the plant growth stimulants in the litel last years to improve the crops' productivity and increase their resistance without harming the environment. They have an important role in increasing plant disease resistance and improving yields.

Bio-organic fertilizers are one of the modern technologies used in most countries as an attempt to reduce the use of artificial fertilizers. These fertilizers are characterized by being environmentally friendly, as they work to supply the nutrients necessary for the growth of plants in an available form to the plant, especially major plant nutrients (nitrogen, phosphorus, and potassium). Many studies have proven that biological methods can be used to increase the possibility of increasing of fixation of nitrogen, and phosphorus dissolution in the soil (Osip, 2000). This highlights the urgent need to pay attention to bio-fertilizer (Ben Mahmoud, 2016). Several studies have demonstrated the significant effect of bio-fertilization with Azotobacter bacteria on potato productivity (Abdel and Al-Asafi, 2018) and bio-fertilization with Bacillus bacteria (Adnan et al., 2018). These nutrients are converted from their unavailable forms to easily absorbed forms.

PGPR includes several types of microorganisms that inhabit the rhizosphere. Bio fertilizers are microbial products that include bacteria, fungi, or both, especially PGPR that produce plant hormones (Akbarinia et al.,2003). Phytohormones Bio fertilizers are produced by (isolation, purifyied, and Dignosis) selected of beneficial soil strains microorganisms and in suitable cultures propagating them until used them, may be by mixing with seeds before sowing, contaminating the seedlings roots, or adding them to soil to provide importnat nutrients to the plant by activating them in rhizosphere and gradually making them ready to take by plant (Hari et al.,2010). Dual organic and biological fertilizers application reduces the environmental pollution that results from irrational use of artificial fertilizers, in addition to increasing the nitrogen and phosphorus supply, thus improving crop quality (Salman and Al-Whaab,2016). The activity of these bacteria increases when they are present on the surfaces of plant roots because they benefit from root secretions as a source of carbon. Carbon availability is main factors that determines activity and nitrogen fixation rate by microorganisms in heterotrophic soil. Therefore, the addition of organic waste activates the fixation process under aerobic and anaerobic conditions (Ben Mahmoud, 2019). They depend on providing the carbon and energy they need from their organic compounds oxidation in the soil. Organic residues addition stimulates the bacteria to grow, thus increasing plant growth and improving its productivity(Rteba and Hussain,2024). For bacterial inoculation to be beneficial in increasing the crop quantity and quality, inoculation must be combined with the addition of an organic or nitrogen source.

It has been observed that biofertilizer could be used to enhance soil fertility and crop yield in sustainable agriculture. Applications of helpful microorganisms as biofertilizers have since acquired primary significance in the agricultural industry because of their possible contribution to food safety, sustainable crop productions, and integrated nutrient controls.

The fixation of nitrogen biologically (BNF) is one possible solution as it forms an important part of the production of legumes and even non-legumes that grow well on a small quantity of fertilizer, and form a substantial part of the nitrogen supplied to farmers.

The continued utilization of the same piece of land annually with the increase in population has led to soil being rendered sterile such that even when chemical fertilizers are applied there is little gain realized.

Conventional agriculture plays a major role in nourishing an ever-growing population; this has had increased reliance on the use of chemical fertilizer and pesticides to boost the quantity.

Chemical fertilizers are synthetic substances composed of known quantities of nitrogen, phosphorus, and potassium. The use of chemical fertilizers causes air and groundwater pollution by enriching water bodies with nutrients. It also weakens plant roots, making them susceptible to unwanted diseases.

BIO FERTILIZERS

Securing food for the entire population of world is one of most complex problems. The global trend is towards clean agricultural or sustainable agriculture, while reducing pollution by using natural materials to increase production. This trend toward using natural resources in the fertilization process is considered a modern agricultural technology that can reduce pollution by using these materials to increase plant yields (EL-Akabawy, 2000). Such technologies include bio fertilizers, which are microbial additives containing bacteria or fungi cells, or both. Bio fertilizers are produced by isolation, purifying, and diagnosing beneficial soil microorganisms. They occur in advance of planting in mixed form, are inoculated into root seedlings, or are placed onto soil. Of much significance is the role of bio fertilizers in relation to the fertility of agricultural land, because of the fact that, they enhance the capacity of the plant to absorb nutrients and water through soil, and strive to furnish vital components of the growth of plants, which may alleviate the usage of mineral fertilizers. The agricultural method that is based on the use of beneficial microorganisms to improve the fertility of the soil and the growth of plants is known as bio fertilization. Did that through fixation of nitrogen using bacteria that then transform it into compounds that are absorbable by plants, hence eliminating the use of chemical fertilizers. Biofertilizers are materials with live organisms that are applied to plants as a source of nutrients. They serve as substitute to fertilizers made of chemicals; which pollute the environment and boost agricultural production. These organisms also operate by repairing atmospheric nitrogen or dissolving phosphate in the ground which is then transformed into a form that is more easily absorbed by plants. Biofertilizers are classified as bacterial fungal and algal. They are Rhizobacter bacteria and blue-green algae.

Examples of bacteria used in bio fertilization include

:• **Rhizobium bacteria**, which live symbiotically with the roots of legume plants, forming root nodules that aid in nitrogen fixation.

- **Azotobacter bacteria**, which fix nitrogen in the soil without requiring a specific plant association, increasing the nitrogen content available to plants.
- **Phosphate Solubilizing Bacteria**, which convert insoluble phosphorus in the soil into a form that plants can absorb, enhancing their growth.

Benefits of Bacterial Bio fertilization:

Important role bio-fertilization plays in improving fertility of soil by amassed the nitrogen and phosphorus content available to plants, that may be reduce the chemical fertilize used, which contributes to lowering costs and reducing negative environmental impacts. It also combats plant diseases. Some beneficial bacteria act as biological control of pathogens, reduce the chemical pesticides used.

Fertilization is a farming method which involves use of good microorganisms to increase the soil fertility, and growth of plant. Rhizobium bio fertilization is one of them. These bacteria fix nitrogen in the atmosphere, which is converted to compounds absorbed and used by plants, thereby decreasing fertilizers applied to plants.

There are many different types of agricultural fertilizers, some of which come from natural sources through natural methods, such as compost. Farmers often resort to agricultural fertilizers to improve soil quality and provide a suitable environment for plants to perform their various vital functions and grow healthily. There are specific types of agricultural fertilizers used to provide plants with a set of specific elements that they lack. Fertilizers vary in their role in plant growth and supplying it with the necessary nutrients. They also vary in their absorption rate and other factors, making each fertilizer of paramount importance to plant life. Among the most popular fertilizers currently in vogue are: Biogen fertilizer, a bio-fertilizer containing beneficial bacteria that fix nitrogen in the plant. It also secretes a group of substances that stimulate roots, which significantly aids in the absorption of various nutrients(Abobatta and .Hussain, 2024).

Biofertilizers and Their Role in Plant Nutrition

Over the past two decades, advanced science and major countries have focused on organic and biodynamic agriculture, two terms that encompass numerous technologies that have been developed, tested, and applied in numerous research centers. This is because the state aims to increase agricultural land and increase productivity per unit area to bridge the food gap. Studies using biodynamic and organic agriculture have achieved the highest productivity. These two systems maintain a low consumption of agricultural chemicals, which are used as fertilizers or pesticides in new agricultural lands, to avoid pollution, reduce costs, and rationalize irrigation water consumption. Fertilizers and biofertilizers are also inexpensive sources of plant nutrition, an alternative to mineral fertilizers, which pollute the environment, both soil and water, when overused. Fertilizers produce microorganisms and are used as inoculants, where they are added to agricultural soil by scattering, mixing with the soil, or mixing with plant seeds during planting. The bio-vaccine combines microorganisms that fix atmospheric nitrogen, which guides the consumption of mineral nitrogen fertilizers, as well as dissolves phosphates and facilitates microelements present in the soil, especially in calcareous soils, as well as produces amino acids, sugars, vitamins, and also produces antibiotics to protect plant roots from infection with root rot and wilting, and secretes growth regulators and stimulants, which play a role in regulating biochemical processes in the plant, which helps in the depth and spread of roots, increasing the rate of photosynthesis. Its shelf life is three months from the date of production and is stored in the refrigerator until use. It is safe and does not contain artificial hormones. The concept of bio-fertilizers and their role in plants: - Bio-fertilizer (microbial vaccine) is a microbial vaccine from diverse, highly active microbial species. The bio-fertilizer works to encourage the biological role in the soil, as it represents the natural environment for the living of these living organisms, as they play a nutritional role for the plant. Bio-fertilizers are divided into two types:

1- Fertilizers that fix atmospheric nitrogen, whether symbiotic or non-symbiotic, provide 25% of nitrogen fertilizers. Examples include cerialin, biogen, rhizobacterin, and azolla.
2- Fertilizers that dissolve and mineralize organic phosphates, converting them from an unusable form to a readily available form that is readily absorbable by plants, such as phosphorus. These fertilizers provide 50% of phosphate fertilizers. Plants can use them to replace part or all of their mineral nutrient needs. The modern trend involves gradually eliminating chemical fertilizers, either partially or completely, by relying on highly efficient biological sources that supply the soil and plants with the nutrients they need to increase productivity and improve yields. This, in turn, results in a reduction in the use of conventional chemical fertilizers.

Application of organic agriculture in the field of bio control:

Chemical pesticides have long played a significant role in combating pests and diseases that affect plants. However, they are expensive, and they are also one of the most important causes of environmental pollution in which humans, animals, plants, insects, and soil microbes live. Hence, the need to use alternative methods to chemical pesticides, namely biopesticides. Among these microbes used are Azotobacter, Bacillus subtilis, and Choococcus fleuresense, especially for wilt and root rot diseases. The harmful effects of chemical pesticides are well known, as the products of nitrate reduction to nitrite combine with pesticide decomposition products, forming nitrosamine compounds, which are mutagenic and carcinogenic. The use of chemical pesticides also affects some important biological processes in the soil, such as nitrification, a process in which ammonia is converted to nitrite and nitrate by microbes. Chemical pesticides also affect symbiotic atmospheric nitrogen fixation, as well as non-biological decomposition. Pesticides lead to the accumulation of decomposition products. Types of biofertilizers:

1. Microbin:

Abiofertilizer composed of a large group of microorganisms that increases soil fertility, fixes atmospheric nitrogen, and increases germination.

2. Nitrobin:

A nitrogen biofertilizer used for field crops, fruits, and vegetables. It contains nitrogen-fixing bacteria, saving 35% of the amount of nitrogen fertilizers used.

3. Rhizobacterin:

A biofertilizer used for field crops, vegetables, and fruits. Its effectiveness is due to its high content of symbiotic and non-symbiotic nitrogen-fixing bacteria carried on peat moss, which reside in plant roots and the surrounding soil with high efficiency throughout the plant's lifespan.

4. Bluegin: A biofertilizer containing blue-green algae capable of fixing atmospheric nitrogen in their bodies and converting it into nitrogen compounds that the plant can utilize.

5. Nodule:

A nitrogenous biofertilizer for summer legumes such as soybeans, peanuts, cowpeas, beans, and winter crops such as broad beans, clover, lentils, and fenugreek.

6. Contagin:

Alkaline soils have a detrimental effect, leading to a decrease in the rate of utilization of micronutrients. Recently, it has been used in plant nutrition to address micronutrient deficiencies by coating seeds with Contagin, a chelated compound containing zinc, iron, and manganese in balanced proportions.

7. Phosphorin:

Phosphorus is one of the main elements in plant nutrition. Plants obtain their phosphorus requirements through phosphate fertilizers added to the soil or through the decomposition of various organic materials. Phosphorin fertilizer contains highly active bacteria that convert unavailable tricalcium phosphate, which is present in Egyptian soils in high concentrations due to the concentrated use of phosphate fertilizers, into monophosphate available to the plant. These bacteria quickly multiply and spread to the plant's root zone, providing the plant with the useful phosphorus it needs during the various stages of plant growth.

Mycorrhizae:

Fungi that live symbiotically within the roots of some leguminous plants and increase the uptake of soil phosphates, which the host plant benefits from.

9- Azolla:

One of the primary fern plants that coexists with blue-green algae, Azolla fixes atmospheric nitrogen and contains the major and minor elements needed by plants. It grows on water surfaces in rice fields.

General benefits of biofertilizers

Less polluting the environment, rationalizing the utilization of the chemical fertilizer, restoring a biological balance that guarantees re-sharing the land between human beings, animals and plants, making more suitable nutrients and promoting their absorption, and secreting growth-promoting substances. They increase also the physical, chemical, and biological characteristics of the soil directly affecting plants. They contribute to the growth in the quantity of microorganisms in the soil, which makes the soil more fertile. This adds organic matter to the soil, replaces soil that has lost its fertility, and releases some antibiotics that fight some plant diseases. They also amend atmospheric nitrogen, thus saving mineral nitrogen fertilizer consumption (rationalizing), and solve the phosphates of calcareous soils and transform them to plant-nutritional form. They play a role in strengthening root and vegetative growth, and increasing the suitability of nutrients by activating the specialized microbes used (Hussain et al,2023).

Bio-fertilizers effect on growth, root and Shoot of plant traits.

The success of a bio-fertilizer used in studies and research as inoculum added to a medium of plant growth effected by efficiency in added of organism, its coexist ability with host plant, and its compete ability with organisms already present in the soil in the rhizosphere (Al-Shabini, 2006). Microorganisms also vary in nature, behavior, activity, and nutrient types they provide to plants. Lordanis et al. (2013) found that the use of rhizobia inoculum as a bio-fertilizer for legumes improves the vegetative growth parameters of plants, especially beans, which are considered a legume crop that is weak in fixing atmospheric nitrogen. Beans suffer from root nodule formation due to the limited presence and spread of specialized rhizobia in Iraqi soils.

PLANT GROWTH PROMOTING RHIZOBACTERIA (PGPR)

These bacteria are beneficial microorganisms for plants, found freely in the soil, promoting growth and protecting plants against stress factors. They do this by fixing nitrogen, dissolving phosphorus, and making it available to plants to facilitate their uptake and production of volatile organic compounds and plant growth regulators. They also reduce the effect of certain enzymes and compounds produced by plant pathogens, such as glucanases and lipases. They stimulate induced resistance, leading to increased plant resistance to diseases. that belonging to Streptomyces, Rhizobium, Xanthomonas, Bacillus, and Pseudomonas have been shown to stimulate induced resistance (Sharma et al., 2023). Studies have shown that Rhizobacteria spp. strains are able to increase nutrient use efficiency and mitigate water stress effects in high water-demanding crops and have potential applications for maintaining productivity under stress of water (Pereira et al., 2020).

Plant roots communicate with microorganisms in a sophisticated manner through chemical contact within the root zone. This stimulates defenses or induced resistance in the plant host. Chemical contacts stimulate the induction of defenses in the plant (Bashir et al., 2021). Rhizobacteria, which promote plant growth, produce volatile organic compounds (VOCs), that improve growth of plant and inhibit of pathogens. They are also capable of dispersing in the soil, facilitating their spread over longer distances(Dobrzyński et al. 2023; Ayuso-Calles et al. 2020).

Mechanisms of PGPB

Scientific data shown soil contains many of bacteria. These bacteria are influenced by soil abiotic factors, such as "temperature, salt, humidity, and the plant species present. It is noted that bacteria are highly concentrated around plant roots, a narrow area of soil called the rhizosphere. PGPB belong to a beneficial microorganisms and heterogeneous group that found in the rhizosphere, capable of stimulating growth of plant. Furthermore, they protect plant from diseases and abiotic stresses inhibiting pathogens activity, and improvement soil structure. Effect of bacteria on growth of plant and its mechanisms vary depending on strains. These mechanisms are generally classified as "direct and indirect". Indirect mechanism: involves activating the disease defense mechanism, known as ISR "induced systematic resistance ".

Direct Mechanisms of Action of PGPR Bacteria

The direct mechanism involves influencing growth of plant by increasing the supply of nutrients and plant hormones through:

- **Facilitating access to resources**
- The best studied mechanisms for promoting growth of plants included: bacteria that supply nutrients lacking for plants, such as" nitrogen, iron, and phosphorus". One or more compounds are lacking in many soils, rendering plant growth suboptimal. To avoid this problem and increase productivity, farmers worked to increasingly relied on chemical sources of phosphorus and nitrogen. In addition to their high cost, the production of fertilizer chemicals depletes resources of non-renewable resources, including oil and natural gas, used in their production, and poses human and environmental risks. Therefore, it is beneficial to use effective biological methods to provide nitrogen and phosphorus to plants, replacing some use of chemical nitrogen and phosphorus currently.
- **Bio fertilizer**

  consists microorganisms added to seeds, soil, or plant surfaces to increase their supply of nutrients and aid growth. The various microorganisms in bio-fertilizer enhance the fertility of soil, the cycling of natural nutrients, and the availability of organic material. PGPB bacteria are among the most widely used bacteria.
- Fixation of Nitrogen

  Is process of that atmospheric nitrogen converting into a usable N, which is converted to ammonia. It is essential for plants. Biological fixation of nitrogen occurs in nature by microorganisms. PGPB variety can promote the growth of the plant and productivity by increasing of nitrogen uptake. In addition to Rhizobia species, many bacteria, for example, (Azospirillum spp), are capable of nitrogen fixation and provide to plants (Salman, 2000). Generally believed that a small amount is required to be provided by the bacteria to the host plant associated with the bacteria.
- Solubilization of Phosphate

  Phosphors(P) is essential nutrient for optimal growth of plant. However, P is often found in insoluble forms. PGPB helps solubilize inorganic forms of phosphorus and mineralize organic phosphorus. Since fertilizers of phosphate are more expensive for farmers, using PGPB as an inoculant to take up phosphate is beneficial. Fe is nutrient that is rare in soil. Bacteria PGPB produce compounds that known as "siderophores", contain Fe^{+3}. Roots absorb these compounds, and use them for growth of plant (Ahmed,2015).

- **Plant hormones**

  Plant hormones play a key role in normal growth of plant, regulating and determining many processes, included leaf formation, flower development, and fruit Maturation

  . exposed Plants to stress due to many environmental factors. PGPB can alter plant hormone levels, helping them respond to stress.

Indirect Mechanisms of PGPR Action

Consists of systemic resistance is a plant's defense mechanism, that is activated when plants are infected with a pathogen or environmental stressors are present. This helps in combating diseases or resisting stress. The inducible bacteria trigger a reaction in roots, which serves as a signal that reaches all plant parts to trigger systemic resistance. Today's agriculture relies on chemical compounds to promote the growth of plants and protect crops. However, excessive chemical use can lead to unforeseen ecological problems. for that, we need to utilize the growth of plant-promoting solutions to achieve agriculture sustainable free of ecological problems.

Figure 1. Mechanisms of action of bacteria promoting plant growth (Jakubowska et al.,2025)

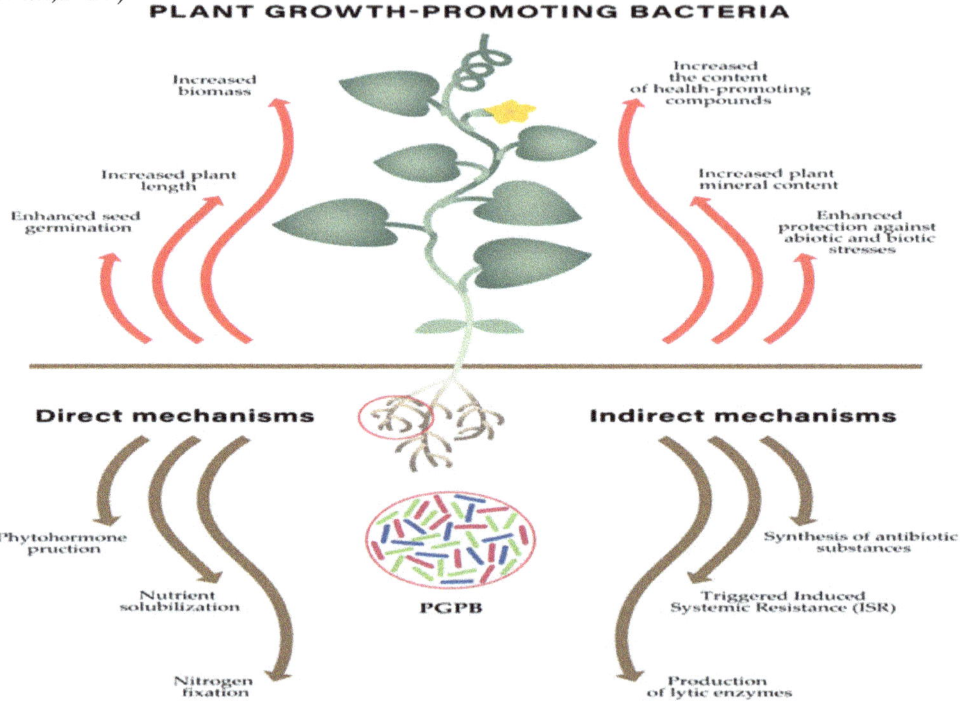

Using bacteria of soil as phytoremediation

Phytoremediation, the plants use to remove pollutants from ecology, can be significantly enhanced by using soil bacteria. These bacteria, particularly those in the rhizosphere, have vital role to supporting plants growth and increasing pollutant removal efficiency.

Figure 2. The function of phosphate-dissolving bacteria in the detoxification of heavy metal remediation (Ahmed, 2015).

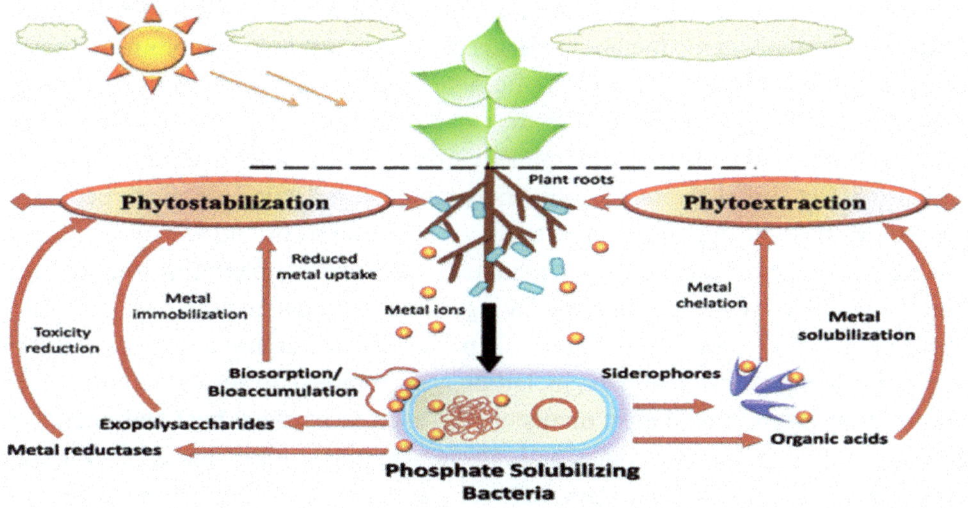

PGPR capable of enhancing the growth of plants by colonizing their roots, these bacteria have been used primarily to help plants absorb material from soil or prevent diseases. Phytoremediation is a promising approach for removing pollutants from environment. For that, using plants for remediation alone has limitations several. Application of PGPR expanded to include the remediation of plant-associated contaminated soil. The vast impact of the organic pollutants and the heavy metals has been the most discussed pollutant among all the present pollutants. PGPR in the rehabilitation of the soils contaminated by these two sources. In the last 2 decades or so, scientists have been trying to use plants to help them eliminate organic and inorganic anic pollutants in the environment, especially in soil. These phytoremediation methods have achieved a lot in a very short period of time.

The bulk of this has, however, been carried out in more controlled laboratory conditions, as opposed to in the field. Besides other types of phytoremediation techniques, and as scientists attempt to further enhance the performance of this technology, some have considered the concept of having several of these types of soil bacteria associating with plants. These are biodegradable bacteria, plant growth-promoting bacteria, and bacteria which promote phytoremediation in other ways.

The efficiency of phytoremediation of metal-contaminated soils depends primarily on bioavailability of the metals in soils and improves uptake of the metals by plants. Therefore, PSB bacteria that promote the solubilization of phosphates are viable alternatives to the use of other plant growth-promoting bacteria to enhance this process. These living organisms produce organic acids and bio surfactants- dissolve poorly soluble metal complexes- and improve bioavailability of metals and nutrient delivery to soil. Thus, PSB bacteria, with their multifunctional activity (including the synthesis of siderophores, para-aminobenzoic acid, ACC deaminize, organic acids and anti-pathogenic metabolites), are the best to promote phytoremediation of metal-contaminated soils. There are a number of physical, chemical and biological processes that are currently used to treat metal-contaminated soil. Among these, physical and chemical traits-based remediation processes are extremely expensive and also affect soil properties, biodiversity, and soil fertility. Various remediation procedures have been compared in terms of cost. Based on the above-mentioned remediation plans, it is evident that plant extract (a form of phytoremediation) is a highly cost effective method of remediating metal contaminated soil (Padmavathiamma and Li 2007).. Phytoremediation is performed at a minimal expense, due to harvesting and field management, including weed removal. In addition, the biomass produced by phytoremediation plants can be used to produce heat and energy in special plants. Unlike the physical and chemical remediation methods, phytoremediation is a green and holistic approach because these methods have no impact on the soil structure and health (Ahmed,2015).

Applications of PGPB in Various Propagation Methods

- Seed propagation: Improves germination rate and speed, and increases seedling uniformity by increasing nutrient uptake and stimulating root growth.
- Cuttings propagation: Accelerates adventitious root formation and increases rooting success, with the potential to reduce reliance on chemical growth regulators.
- Tissue culture propagation: Improves plant viability from tissue cultures, reduces transport shock, and lowers microbial contamination rates, especially in high-value crops.

The efficacy of PGPB bacteria in propagation has been shown by a number of studies, including:

- Azospirillum brasilense inoculation of cereal seeds enhanced the field germination and growth of seeds.
- Treating ornamental plant cuttings with Bacillus subtilis, which is highly efficient at converting insoluble soil phosphates due to its secretion of organic acids that dissolve insoluble phosphate. These fertilizers thus help reduce chemical phosphate fertilization by 50%, thus saving costs. They also reduce soil, environmental, and product pollution, while increasing production in

quantity and quality, thus raising prices. This has helped accelerate rooting and increase survival rates.

- Incorporating *Pseudomonas fluorescens* into banana tissue culture has led to increased growth vigor and a 30% reduction in fungal contamination. The Role of Plant Growth-Promoting Bacteria (PGPB) and Their Importance in Abiotic Stress Tolerance in Plants

PGPB bio fortification of crops is a modern idea in intensive agriculture. Increasing plant growth-promoting bacteria, such as Pseudomonas aeruginosa, Bacillus, and Azospirillum, stimulate plant growth and development through the production of phytohormones, enhanced nutrient availability, and enhanced biosynthesis of secondary metabolites by the activation of an induced systemic resistance ISR.

The phenyl propanoid route is the primary biosynthetic pathway of polyphenolic compounds such as phenolic acid and flavonoids that is triggered by ISR of PGPB in plants. It is proposed that PGPB may increase phenolic content (up to more than 200) and antioxidant activity simultaneously. PGPB can also reduce abiotic stresses such as aridity, saltiness and heavy metal contamination by releasing phenolic compounds. Phenolic compounds that PGPB can increase the synthesis of in other parts of the plant include flavonoids like quercetin, procyanidin B1, EGCG, catechin, and phenolic acids, including caffeic acid, ferulic acid, and chlorogenic acid.

The excess production of reactive oxygen species (ROS)-induced oxidative stress in plants is caused by non-living stresses such as drought, salinity, and environmental pollution. The impacts of the latter can, however, be alleviated by the stimulation of the biosynthesis of phenolic compounds as anti-oxidants to counter the effects of ROS by using PGPB. Flavonoids are among the most effective antioxidants which can be secreted by PGPB... Drought is one of the main problems associated with interfering with the growth and development of plants. Plant growth-promoting bacteria or PGPs are applied in order to mitigate the impact of drought on vegetation (Ahluwalia et al., 2021; Lucas et al., 2023). One major way of alleviating drought stress is by enhancing the synthesis of secondary metabolites, including flavonoids and polyphenols. In one study, there was a 21.5 percent increase in phenolic content of tomato (Solanum lycopersicum L.) leaves under drought conditions when subjected to Pseudomonas sp. N 5.12. Pseudomonas fulminans and Paenibacillus lentimorphus CHM12 have been observed to, individually and in combination, elevate zea mays L. (maize) phenolic content. The same was improved in faba bean (Vicia faba L.) roots by Cellulomonas pakistanis (NCCP11) and Sphingobacterium pakistanis (NCCP246).

Notably, Lucas et. al (2023) showed that adding Pseudomonas glutamate with salicylic acid notably amplified the presence of flavonoids (39%) and phenolic substances (40%) in rapeseed plants. This boost correlated with enhanced drought tolerance and increased action of antioxidant enzymes, encompassing superoxide dismutase (SOD), ascorbate peroxidase (APX), and catalase (CAT). These enzymes efficiently lessened oxidative stress, augmenting plant endurance under difficult environmental circumstances. Most studies have additionally concentrated on the impact of PGPB on the overall quantity of phenolic compounds. Parihar and colleagues (2015) strengthened the plant's capability to adjust to saline situations by inoculating maize with Bacillus sp. PM31, which augmented the creation of phenolic compounds. The elevated biosynthesis of these metabolites improved the plant's capacity to withstand antioxidants, directly contributing to plant survival. Incorporating PGPB into tissue culture methods.

Incorporating plant growth-promoting bacteria (PGPB) into tissue culture methods involves utilizing these bacteria to enhance the growth of plants cultivated in the lab. This approach seeks to boost the plants' capacity to develop and flourish in a tissue culture environment, potentially boosting propagation efficacy and upgrading the quality of the resultant plants.

Plant tissue culture involves generating a complete plant, in aseptic environments, from an individual plant section or cell. This is founded on the occurrence of totipotency (the capability of any single cell to generate a set of specialized cells distinctive of organs) and reforming into a full plant (Trigiano and Gray, 2016). This attribute is utilized for swift and widespread propagation of genotypes at a large magnitude during a relatively brief timeframe. Agriculture, horticulture and industry are also concerned with micropropagation since it produces healthy seeds all year round and does not require as many plant cycles (Suman, 2017). It is also an important technology in protecting the plant genetic material, in improving crops and production of new varieties by genetic enhancement, but it is a costly exercise. PGPB is a potential solution to use this technology to reduce the cost of the procedure (Verma et al., 2019). Some of these microbes have the ability to generate plant hormones and organic materials that may increase plant growth and production. Plant growth-promoting microorganisms (bacteria and fungi) inhabiting the rhizosphere and feeding on fallen plant cells and on the proteins and sugars released by roots (Al-Ani, 2019) have been utilized in agriculture. The majority of studies done on micropropagation and microbes have been aimed at identifying and remedying contaminants.

Nevertheless, numerous plant-based organic matter (PGPB) can assist root development and shoot lengthening, and are also key in the successful acclimatization stage. They even safeguard against the biotic and abiotic pressures that arise during in vitro propagation. Plant-based organic matter (PGPB) is a crucial element for sustainable agriculture, and thus, boosting its application in tissue culture is a difficulty. The first in vitro bacterial incubation was done by Degat et al (1987) who demonstrated a positive effect of Pseudomonas fluorescens and Pseudomonas fluorescens type on root development and bud development of primroses. Rhizobacteria are inclined toward the production of auxins and cytokinins (Amara et al., 2015). In the rhizosphere, more than 80 percent of the soil bacteria can synthesize auxins, namely IAA (Souza et al., 2015). Genetic material of the microorganism plants has an auxiliary role, particularly in the case of the propagation of plants in an unproductive environment with insufficient internal synthesis. These plant genetic indicators are measured, defined and evaluated in plant tissue culture to propagate a variety of plants such as peanuts, sorghum, carrots, potatoes, maize (Zea mays), wheat (Triticum austerum L.), and rice (Oryza sativa). Today, it is estimated that it provides more than 50 per cent of the global population with food(Respedia, 2020). As an expansion strategy, introduction of beneficial microorganisms (bacteria and fungi) has been proposed (Srivastava et al., 2002).

BIO-FERTILIZERS AND THEIR IMPORTANCE IN SUSTAINABLE AGRICULTURAL DEVELOPMENT

Bio-fertilizers are one of the most important tools in sustainable agriculture and achieving renewable productivity (Hussain and Mohammed,2024). These fertilizers are a natural source of nutrients essential for plant growth, in addition to their ability to improve soil properties and maintain ecological balance. Bio-fertilizers are a fundamental pillar of sustainable agriculture, contributing to food security without harming natural resources and helping restore the biological balance of the soil, making agriculture more environmentally friendly. The role of biofertilizers: Improving soil quality: Biofertilizers help increase the organic composition of the soil, improving its structure and its ability to absorb and retain water and nutrients. They provide nutrients by analyzing and converting available nutrients in the soil into forms that are absorbable by plants, which contributes to improving the availability of nutrients to plants. They also improve plant resistance, as bio-fertilizers stimulate root growth and enhance plant immunity, making plants more resistant to diseases and pests. Reducing dependence on chemical pesticides: By using bio-fertilizers, the need for chemical pesticides can be reduced, reducing environmental pollution and maintaining soil and plant health.

In short, bio-fertilizers are an essential tool for achieving sustainable agriculture and improving crop productivity in a healthy and environmentally friendly manner. They form an essential part of modern agricultural strategies aimed at improving the environmental and economic performance of agriculture.

What is the economic feasibility of bio-fertilizers?

- Research done on employing bio-fertilizers for harvests suggests they boost output and yield superior, residue-free crops, as their costs are quite modest when contrasted with chemical fertilizers. They also make sensible the use of mineral fertilizers and decrease environmental contamination, since they furnish a substantial amount of the essential plant nutrients (25% atmospheric N + 50% stabilized dissolved phosphate).They also improve growth traits(root and fruit systems), and increase the final yield in both quantity and quality in grain crops. They also stimulate the secretion of certain growth regulators (GA, IAA) and antibiotics, promote disease resistance, and reduce toxicity in products due to reduced chemical residues and higher yields "clean agriculture". Increasing organic matter in the soil, thus improving its properties and improving water absorption by roots, improves the properties of sandy soil by secreting sugars that work to collect them, and increases the root surface area and thus the absorption rate. It also restores the microbial balance of the soil, activates its vital processes, reduces environmental pollution, and preserves human and animal health. It furthermore decreases manufacturing expenses owing to its inexpensive price and output costs.

How do biofertilizers contribute to environmental sustainability?

Biofertilizers support environmental sustainability by enhancing soil fertility through beneficial microorganisms, reducing reliance on chemical fertilizers and soil, water, and air pollution. They also improve plant health, increase productivity, and reduce the use of pesticides, providing safe food and contributing to food security through sustainable agriculture and reducing the harmful environmental impacts of intensive agriculture.

One of the most valuable tools of sustainable agriculture and attaining renewable productivity is biofertilizers. These fertilizers can be naturally obtained and contain the nutrients needed by plants plus they can enhance the properties of soil and balance the ecology as well (Hussain et al., 2023).

A plant requires 16 nutrients throughout its life cycle, and biofertilizers provide these nutrients continuously. Biofertilizers provide the essential nutrients for plant growth in a balanced manner. These fertilizers contain nitrogen, phosphorus, and potassium, in addition to many other important elements such as calcium, magnesium, and iron. These elements contribute to promoting leaf growth, fruit formation, and enhancing plant immunity. Biofertilizers are an environmentally friendly alternative to traditional chemical fertilizers. Instead of using synthetic chemicals that can pollute soil and groundwater, biofertilizers improve soil health and maintain biological balance in the agricultural system. Furthermore, the use of biofertilizers reduces farmers' dependence on external sources of chemical fertilizers, saving costs and increasing agricultural sustainability(Abbas and Hussain,2020).

Biofertilizers and Soil Conditioners

Biofertilizers are products based on living microorganisms that enhance soil fertility and promote plant growth by naturally providing nutrients and improving the physical and chemical structure of the soil. Soil conditioners generally improve the physical, chemical, and biological properties of the soil. Biofertilizers include nitrogen-fixing microbes, phosphate-solubilizing microbes, mycorrhizal fungi, and beneficial algae, while soil conditioners may be organic materials such as vermi-composting.

Biofertilizers rely on microorganisms to improve soil, including:

- Beneficial bacteria: These fix nitrogens from the air into the soil.
  -Mycorrhizae: these increase the absorption of soil nutrients by roots.
- Algae: Algae are applied to enhance the fertility of soil and also enhance growth in plants. Biofertilizers improve the ecological balance in the soil and reduce the need for conventional fertilizers

REFERENCES

Abbas, M. M., & Hussain, W. S. (2020). Biostimulants of pepper and eggplant by using plants aqueous extract. *Plant Cell Biotechnology and Molecular Biology, 21*(65&66), 78–82.

Abobatta, W. F., & Hussain, W. S. (2024). *Achieving food security through sustainable agriculture Achieving Food Security Through Sustainable Agriculture*. IGI Global. DOI: 10.4018/979-8-3693-4240-4

Ahluwalia, O., Singh, P. C., & Bhatia, R. (2021). A review on drought stress in plants: Implications, mitigation, and the role of plant growth-promoting rhizobacteria. *Resour Environ Sustain, 5*, 100032. DOI: 10.1016/j.resenv.2021.100032

Akbarinia, G., Sefidkonf, R., & Sharifia, M. B. (2003). Study on the effect of different rates of chemical fertilizer, manure and mixture of them on seed yield and main - composition of essential oil of Ajowan (*Trachyspermum copticum*). *Iranian Research and Development, 61*, 32–41.

Al-Shabini, J. M. (2006). The phosphorus in the soil and plants. Egyptian Library for printing and publishing MSc. Thesis University of Muthanna.

Amara, U., Khalid, R., & Hayat, R. (2015). In Maheshwari, D. K. (Ed.), *"Soil bacteria and phytohormones for sustainable crop production" in Bacterial metabolites in sustainable agroecosystem. Sustainable development and biodiversity. Vol. 12* (pp. 87–103). Springer.

Ayuso-Calles M, García-Estévez I, Jiménez-Gómez A, Flores-Félix JD, Escribano-Bailón MT, Rivas R (2020) *Rhizobium laguerreae* improves productivity and phenolic compound content of lettuce (*Lactuca sativa* L.) under saline stress conditions. Foods 9:1166. 10.3390 .

Bashir, M. A., Rehim, A., Qurat-Ul-Ain Raza, H. M., Raza, A., Zhai, L., Liu, H., and Wang, H. (2021). Biostimulants as Plant Growth Stimulators in Modernized Agriculture and Environmental Sustainability. Journal of Technology in Agriculture, 311-322.

Ben Mahmoud, M. E.-T. (2016). The effect of inoculation with Burkholderia Burkholdriaspp. on tomato plant productivity. *Syrian Journal of Agricultural Research, 3*(2), 235–242.

Ben Mahmoud, M. E.-T. (2019). Study of the interaction effect between organic fertilizer (cow manure) and biofertilizer (Burkholderia spp.) on wheat plant productivity. *Syrian Journal of Agricultural Research, 6*(1), 314–320.

Dobrzyński, J., Wróbel, B., & Górska, E. B. (2023). Taxonomy, ecology, and cellulolytic properties of the genus *Bacillus* and related genera. *Agriculture, 13*(10), 1979. DOI: 10.3390/agriculture13101979

El-Akabawy, M. A. (2000). Effect of some bio fertilizers and farmyard manure on yield and nutrient uptake of Egyptian clover grown on lomy sand soil. *Egypt. J. Agric. Res., 78*(5).

Hanapi, S. Z., Awad, H. M., Ali, S. I. S., Sarip, S. H. M., Sarmidi, M. R., & Aziz, R. (2013). Agriculture wastes conversion for biofertilizer production using beneficial microorganisms for sustainable agriculture applications. Malaysian Journal of Microbiology, 60-67.

Hari, M., Seshadri, S., & Perumal, K. (2010). Booklet on Bio-fertilizer (Phospho-Bacteria).Shri AMM Murugappa Chettiar Research Center. *Taramani, Chennai, 600*, 113.

Hussain, W. S., Ahmed, H. F., Al-Saffar, R. S., & Taher, N. A. H. A. (2023). Effect of bio fertilization using banana and apple peels on flax plant's vegetative characteristics. *Al-Azhar Journal of Agricultural Research, 48*(3), 395–400.

Hussain, W. S., & Mohammad, A. K. (2024). Sustainable Agriculture Is a Way to Increase Wheat Production. In *Achieving Food Security Through Sustainable Agriculture* (pp. 252–267). IGI Global. DOI: 10.4018/979-8-3693-4240-4.ch010

Jakubowska, Z., Gradowski, M., & Dobrzyński, J. (2025). Role of plant growth-promoting bacteria (PGPB) in enhancing phenolic compounds biosynthesis and its relevance to abiotic stress tolerance in plants: A review. *Antonie van Leeuwenhoek, 118*(9), 1–17. DOI: 10.1007/s10482-025-02130-8 PMID: 40707699

Lordanis, C., Io, K., Anastassia, V., & Wilhelm, H. (2013) Commissionnon genetic resources for food and agriculture, Background study paper. FAO, 64, 2-23.

Lucas, J. A., Garcia-Villaraco, A., Montero-Palmero, M. B., Montalban, B., Ramos Solano, B., & Gutierrez-Mañero, F. J. (2023). Physiological and genetic modifications induced by plant-growth-promoting rhizobacteria (PGPR) in tomato plants under moderate water stress. *Biology (Basel), 12*(7), 901. DOI: 10.3390/biology12070901 PMID: 37508334

Osip, C., Ballescas, A., Osip, L. P., Besarino, N. L., Bagayna, A. D., & Jumalon, C. B. (2000). Philippine council for Agr. Forestry and Natural Resources. *Resources and Technology, 134*, 17–18.

Padmavathiamma, P. K., & Li, L. Y. (2007). Phytoremediation technology: Hyper-accumulation metals in plants. *Water, Air, and Soil Pollution, 184*(1-4), 105–126. DOI: 10.1007/s11270-007-9401-5

Parihar, P., Singh, S., Singh, R., Singh, V. P., & Prasad, S. M. (2015). Effect of salinity stress on plants and its tolerance strategies: A review. *Environmental Science and Pollution Research International, 22*(6), 4056–4075. DOI: 10.1007/s11356-014-3739-1 PMID: 25398215

Rteba, M. M., & Hussain, W. S. (2024). Sustainable Agriculture: Way to Achieve Food Security. In Achieving Food Security Through Sustainable Agriculture (pp. 182-202). IGI Global.

Salman, A. H. (2000).The Effect of Interaction Between Saline Water Irrigation and Organic Manures on Some Soil Properties and Onion Yield (*Allium cepa L.*).MSc. Thesis. Agriculture College–Baghdad University.Pp.107.

Salman, N. D. and Abd Al-Whaab N. R.(2016) Effect of bacterial biofertilizer and compost on the growth of Tomato plant The Iraqi Journal of Agricultural Sciences – 74(2):0261-0264.

Sharma, P., Gupta, A. K., Meena, H. M., Dubey, A., Singh, M., & Mallick, S. (2023). Inoculation of plant-growth-promoting rhizobacteria and earthworms in the rhizosphere reinstates photosynthetic attributes and secondary metabolites in *Brassicajuncea* L. under chromium toxicity. *Plant and Soil, 483*(1), 573–587. DOI: 10.1007/s11104-022-05765-y

Souza, R., Ambrosini, A., & Passaglia, L. M. (2015). Plant growth-promoting bacteria as inoculants in agricultural soils. *Genetics and Molecular Biology, 38*(4), 401–419. DOI: 10.1590/S1415-475738420150053 PMID: 26537605

Srivastava, P. S., Bharti, N., Pande, D., & Srivastava, S. (2002). In Mukerji, K. G., Manoharachary, C., & Chamola, B. P. (Eds.), *"Role of mycorrhiza in in vitro micropropagation of plants" in Techniques in mycorrhizal studies* (pp. 443–468). Springer.

Suman, S. (2017). Plant tissue culture: A promising tool of quality material production with special reference to micropropagation of banana. *Biochemical and Cellular Archives, 17*, 1–26.

Trigiano, R. N., & Gray, D. J. (2016). *Plant tissue culture, development, and biotechnology.* CRC Press. DOI: 10.1201/9781439896143

Verma, M., Mishra, J., & Arora, N. K. (2019). In Sobti, R. C., Arora, N. K., & Kothari, R. (Eds.), *"Plant growth-promoting rhizobacteria: diversity and applications" in Environmental biotechnology: For sustainable future* (pp. 129–173). Springer.

Chapter 12
Tissue Culture and Micropropagation

Amjad Abdul-Hadi Mohammed
https://orcid.org/0000-0002-5821-7523
Department of Biology, College of Science, University of Mosul, Iraq

Rana Tariq Yahay
Department of Medical Physics, College of Science, University of Mosul, Iraq

ABSTRACT

Tissue culture involves the aseptic cultivation of plant cells, tissues, or organs in controlled environments, leveraging the totipotency of plant cells to regenerate whole plants. Micropropagation, a key application, uses small explants (like axillary buds) to produce many identical plants rapidly. The field traces its roots to early 20th-century experiments, with major advances in the 1930s–1940s establishing protocols for callus and organ cultures. Tissue culture and micropropagation are essential for sustainable, large-scale, and high-quality plant production. Ongoing innovations are addressing cost, efficiency, and scalability, ensuring these technologies remain central to agriculture, conservation, and biotechnology (Khan et al., 2024).

INTRODUCTION

Principles and Historical Context

Tissue culture involves the aseptic cultivation of plant cells, tissues, or organs in controlled environments, leveraging the totipotency of plant cells to regenerate whole plants. Micropropagation, a key application, uses small explants (like axillary buds) to produce many identical plants rapidly. The field traces its roots to early

DOI: 10.4018/979-8-3373-5626-6.ch012

20th-century experiments, with major advances in the 1930s–1940s establishing protocols for callus and organ cultures. Tissue culture and micropropagation are essential for sustainable, large-scale, and high-quality plant production. Ongoing innovations are addressing cost, efficiency, and scalability, ensuring these technologies remain central to agriculture, conservation, and biotechnology (Khan et al., 2024).

SECTION 1

Define the concept of Tissue Culture and Micropropagation

Plant tissue culture is a set of techniques where plant cells, tissues, or organs are grown under aseptic and controlled conditions in a defined nutrient medium. Micropropagation, a specialized form of tissue culture, allows the rapid multiplication of genetically uniform and disease-free plants. These techniques are vital for commercial agriculture, conservation of endangered species, and plant biotechnology applications (Rahmawati et al., 2023).

Techniques and Innovations

Micropropagation typically follows stages: explant selection, proliferation, shoot/root induction, and acclimatization. Recent innovations include photoautotrophic micropropagation, which eliminates the need for sugar in the medium by increasing air exchange and CO2 supply, resulting in higher shoot and root biomass(Yahya et al.,2024)

Applications and Benefits

Micropropagation is vital for producing disease-free, high-quality planting material, conserving endangered species, and supporting genetic engineering and breeding programs. It is widely used for crops, ornamentals, forestry species, and medicinal plants, offering rapid multiplication, year-round production, and easier international exchange due to reduced quarantine risks (Patel & Sahay, 2025)

Basic Concepts

Tissue culture involves placing explants (small plant parts) into a medium containing macro- and micronutrients, vitamins, and plant growth regulators (PGRs). Micropropagation typically follows four main stages: 1) Initiation of culture from an explant, 2) Shoot proliferation using cytokinins, 3) Root induction with auxins, and 4) Acclimatization of regenerated plants into soil under greenhouse conditions (Al-Tai and Mohammed,2024).

Techniques and Approaches(Khan et al., 2024)

- Meristem Culture: Produces virus-free plants by culturing shoot tips.
- Callus Culture: Involves dedifferentiation of plant cells into callus, which can later regenerate.
- Somatic Embryogenesis: Formation of embryos from somatic cells, useful for synthetic seed production.
  -Bioreactor Systems: Modern temporary immersion systems (TIS) such as Plantform™ and SETIS™ enable large-scale, automated micropropagation with minimal somaclonal variation.
- Photoautotrophic Culture: Utilizes light and CO2 without added sugars, enhancing plant vigor.

Factors Affecting Success(Tamimi and Othman, 2023 ;Mahdi and Yahya, 2024)

Several parameters influence, and genetic instability is essential for reliable outcomes.

Applications(Khuder and Mohammed, 2025 ; Sultan et al., 2024).

- Commercial Propagation: Large-scale production of ornamentals, bananas, and medicinal plants.
- Germplasm Conservation: Long-term storage of elite or endangered genotypes.
- Crop Improvement: Development of stress-tolerant and disease-resistant varieties.
- Secondary Metabolite Production: In vitro production of valuable phytochemicals (e.g., curcuminoids in turmeric).
- Food Security: Mass propagation of staple crops like banana and potato.
- Integration with Nanotechnology: Improving efficiency of plant tissue culture through nanoparticle supplementation.

SECTION 2

Tissue Culture, Micropropagation, and Sustainable Development

Plant tissue culture and micropropagation play a significant role in achieving the goals of sustainable development. By providing rapid, large-scale propagation of disease-free and genetically uniform plants, these techniques contribute to global food security, biodiversity conservation, and sustainable agriculture. They minimize the reliance on chemical pesticides by producing pathogen-free plants, thus reducing environmental impact. Additionally, tissue culture enables the preservation of endangered species and rare genotypes through *in vitro* conservation and cryopreservation, supporting ecosystem balance. Integration of nanotechnology and renewable energy-based bioreactor systems further enhances resource efficiency, reducing costs, waste, and carbon footprint. Sustainable agriculture requires innovative tools, and tissue culture provides a pathway to produce climate-resilient crops capable of withstanding abiotic stresses such as drought, salinity, and temperature fluctuations. By reducing dependency on land, water, and external inputs, micropropagation contributes directly to several United Nations Sustainable Development Goals (SDGs) (Zhang et al., 2023 ; Yahya et al., 2024)

Role of tissue culture in sustainability

Tissue culture plays a crucial role in advancing sustainability by enabling rapid, disease-free plant propagation, conserving endangered species, and reducing environmental impact in agriculture and biotechnology (Khan et al., 2024).

Sustainable Agriculture and Food Security

Rapid, large-scale propagation of high-yield, disease-free crops through tissue culture supports food security. Consistent, high-quality planting material reduces the need for chemical inputs and land expansion, promoting efficient resource use and minimizing environmental impact.

Tissue culture enables the development of improved crop varieties with enhanced resistance to diseases and environmental stresses, supporting sustainable agricultural practices (Patel & Sahay, 2025)

Conservation and Biodiversity

Tissue culture is vital for the conservation of endangered and rare plant species, allowing for mass propagation and reintroduction into natural habitats, thus preserving biodiversity.It supports the sustainable use of medicinal plants by enabling year-round, controlled production, reducing pressure on wild populations(Ramadan et al., 2022)

Environmental and Industrial Sustainability(Burak, and Meltem, 2021)

In tissue engineering, sustainable biomaterials (e.g., silk, recycled plastics) are used to create long-term, eco-friendly scaffolds, aligning with circular economy principles and reducing laboratory waste.

Tissue culture techniques are season-independent and can be integrated with green technologies, further reducing the environmental footprint of plant and medical biotechnology.

Tissue culture is a powerful tool for sustainability, enabling efficient crop production, conservation of biodiversity, and the development of eco-friendly biotechnological solutions. Continued innovation and responsible practices are essential to maximize its sustainable impact(Cuthill,2010).

Role of tissue culture in sustainability(Fithian and Powell, 2009 ; El-Sherif, 2018)

Tissue culture plays a crucial role in advancing sustainability by enabling rapid, disease-free plant propagation, conserving endangered species, and reducing environmental impact in agriculture and biotechnology.

Sustainable Agriculture and Food Security(Grierson, 2009).

Rapid, large-scale propagation of high-yield, disease-free crops through tissue culture supports food security, especially in regions with limited arable land.

Consistent, high-quality planting material reduces the need for chemical inputs and land expansion, promoting efficient resource use and minimizing environmental impact.

Tissue culture enables the development of improved crop varieties with enhanced resistance to diseases and environmental stresses, supporting sustainable agricultural practices.

Conservation and Biodiversity(Parodi, 2011),

Tissue culture is vital for the conservation of endangered and rare plant species, allowing for mass propagation and reintroduction into natural habitats, thus preserving biodiversity. It supports the sustainable use of medicinal plants by enabling year-round, controlled production, reducing pressure on wild populations.

SECTION 3

Environmental and Industrial Sustainability (Iqbal et al., 2024)

In tissue engineering, sustainable biomaterials (e.g., silk, recycled plastics) are used to create long-term, eco-friendly scaffolds, aligning with circular economy principles and reducing laboratory waste.Tissue culture techniques are season-independent and can be integrated with green technologies, further reducing the environmental footprint of plant and medical biotechnology.

Challenges and Future Directions (Pithiya *et al.*, 2022)

Despite its benefits, tissue culture faces challenges such as contamination, genetic instability, and high production costs, which must be addressed for broader adoption.

Ongoing research focuses on improving protocols, reducing costs, and integrating sustainable materials and practices.

Tissue culture is a powerful tool for sustainability, enabling efficient crop production, conservation of biodiversity, and the development of eco-friendly biotechnological solutions. Continued innovation and responsible practices are essential to maximize its sustainable impact(Kasoju and Sunilkumar, 2024).

Impact of Plant Tissue Culture on Agricultural Sustainability

Plant tissue culture is an important agricultural biotechnological tool that contributes in the production of crops with improved food, fiber, fuel, and feed. It is one way toward commercialization to face the food availability challenge in developing countries and allow them to cope with their fast-growing population in a restricted area of land. In addition, plant tissue culture enables some rare and nearly extinct plant species to be rescued and propagated. Conventional methods of propagation thus need to be supplemented with modern breeding techniques. In this way, higher levels of agriculture, afforestation, plant improvement as well as *in vitro* production of metabolites and plant secondary products can be reached and fulfilled on a year-round basis and under disease-free conditions. The main applications of plant tissue culture in the agricultural field, plant micropropagation, inducing new varieties and constrains of plant tissue culture and challenges this technique is facing as an industry helping the agricultural field (Kasoju and Sunilkumar, 2024).

Culture comprises of behaviors and attitudes. For change in attitude or behavior to be achieved, change in culture will be required. It is difficult to change cultural values. For this to happen, people are expected to re-evaluate their beliefs, attitudes, and values. When this has been achieved, that is when change in behavior can become a reality. Most the time, it is almost impossible to unlearn certain behaviors. Also, performing new behaviors is not an easy task. This doesn't imply that cultural practices can't be changed or modified. When people in the society are determined to make such change happen, it will be very possible. It is all about the willingness to change behaviors, attitude, values, and beliefs (Berri et al., 2025)

Environments are recognized based on instrument value (and not intrinsic value). Therefore, addressing environmental problems goes beyond creating of rules and laws which prohibit such environments/surroundings from being destroyed. Because such problems are traced to cultural practices, there is need for mindsets to be changed. Changing culture implies changing some aspects of the people's lives like value, and views. When this has been done, that is when their behavior or attitude towards nature or environment can change. New views regarding such environment will have to be introduced for people to adopt. This is how old views and beliefs are replaced. According to psychologists, maladaptive human behavior is the cause of most environmental problems(Erol et al., 2023 ; Thakur et al., 2024)

Culture and Sustainable Development

Just like environmental, social, and economic dimensions of sustainable development, economic sustainability is also very crucial. Therefore, it should be grouped as one of such dimensions which supports sustainability. There is a connection between sustainable development and culture. This is because culture explains how natural resources are appreciated by the society. Based on this, it plays a crucial role towards the economic advancement of the society. Every sustainable society focuses on 3 dimensions – economic viability, environmental responsibility, social equity, and justice. Although these are important, they are inadequate when it comes to supporting mankind's future and well-being according to Hawkes (2001). Packalén (2010) believes that sustainability should be regarded as a compass that leads societies in the right direction to achieve development instead of a to do list with boxes to be ticked off. He also states that communication and action between environmental, economic(Zuzarte et al., 2024) .

Cultural heritage tends to play various roles like source of energy, social capital, employment, and cultural diversity. These are to pursue economic, environmental, and social problems of sustainability. It is possible for continuous cultural heritage to be achieved when it is used as tool or channel for economic development. This is because sustainable development is the result of the development which has been acquired (Kumar, et al, 2024).

Cultural Sustainability

Human needs as well as interest in matters related to sustainable development are defined by culture. According to Cortese (2003), for a society that is more sustainable to be achieved, there is need for a cultural shift. In every society, culture is central. It doesn't only define the attitudes of people. Also, it defines their goals. It is the basis of our general behavior in the society. Culture tends to place culture at the center of a sustainable development process (Bosch, 2009).

Soini and Birkeland (2014) state that, the term "cultural sustainability" is used in different ways in various contexts, however, there is a lack of attempts in combining culture and sustainability systematically and analytically. Even today, culture is not analyzed as a concept by itself in context of sustainability, but it is rather analyzed as a part or sub concept of social sustainability (Cuthill, 2010). Soini and Dessein (2016) point at the importance of integration of culture into sustainability discourse, as they state that human's actions, behaviors, and accounts, which are culturally embedded, are what plays a crucial role in reaching sustainability objectives. According to Clammer (2013), culture is not only an essential component of sustainable development, but also a direct result of the development itself. Through cultural sustainability, various ways on how human lives can be improved are examined. The ways of setting an example for younger generations are also examined. If the society wants future generations to be self-dependent and meet needs of their own, it is important that resources exploitation is stopped. The way of life of any society is sharpened by its cultural value. Therefore, it can bring about change in attitudes which is required for sustainable development to be achieved. Culture plays a crucial in building of communities and lively cities where people live and work. It is also important when it comes to supporting economic and social wellbeing. According to Hawkes (2001), for a sustainable society to become a reality, there is need for a sustainable culture to exist. In other words, members of the community or society need to understand their cultural values. Cultural sustainability is all about ensuring that efforts are made for the preservation of cultural elements. These could be both tangible as well as intangible elements. This is the only way social, economic, and environmental sustainability can be achieved (Duxbury 2012). Now, there has been a shift with regards to how sustainability is being viewed in the world. These views don't have any effect on our environment, social wellbeing, and ecosystem. According to Hawkes (2001), the 3 dimensions of environmental balance, social inclusion, and economic growth do not reflect our global societies' dimensions. The world is presently facing environmental, social, and economic challenges. For peace to be achieved, there is need for a combination of elements or features like diversity, creativity, and knowledge.

Two major components of every society are culture and values. However, the values of culture are dynamic since they evolve as the years are going by. First of all, there needs be a shift in society's definition of sustainability. After that, there will be a shift in how the world defines sustainability (Fithian and Powell, 2009). In 2012, UNESCO director, Irina Bokova pointed out that culture is what defines us as humans. It provides solutions and answers to most of the problems we are encountering today. Due to this, there is need for culture to be at the center of global sustainability (UNESCO, 2012a). Culture should not just be introduced but also integrated into economic, environmental, and social sustainability dimensions.

According to Parodi (2011), culture is a condition and aim of sustainable development. Also, Hauser and Banse (2011), there is a strong relationship between sustainability and culture. Furthermore, Opoku and Fortune (2011), are of the opinion that for sustainability to be achieved, organizational leadership needs to be effective. It is important for a society to be developed where the people's culture reflects sustainable behavior. According to most experts, the environment and culture function as one within a system. Therefore, trying to understand them separately is not possible. Culture is a driving force which brings about sustainability. It is a major factor that determines whether sustainability will be achieved or not (Fithian & Powell, 2009). When processes are culturally embedded, such can play a major role towards the achievement of sustainable development. When cultural issues or problems are looked at in sustainability, such can be grouped under social sustainability. Hawkes (2001), however, pointed out that there is need for a sustainability model which integrates or interlinks 4 dimensions of social equity, cultural vitality, economic health, and environmental sustainability(Figure.1)

Figure 1. Proposed four dimensions model of sustainability (Opoku, 2015)

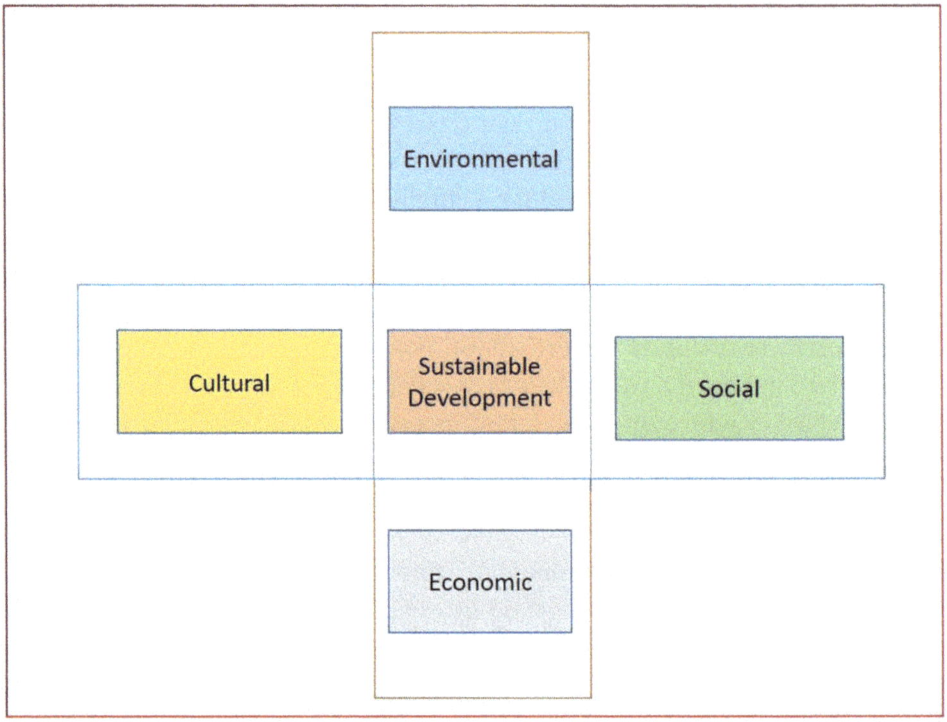

Tissue culture and micropropagation are cornerstones of modern plant biotechnology. They enable rapid multiplication, disease elimination, and genetic conservation of plant resources. Advances such as nanoparticle integration and bioreactor systems continue to expand their potential, making them indispensable for sustainable agriculture and global food security.

REFERENCES

Al-Tai, A.A. and Mohammed, A.A.-H. (2024). Germination of lavender (Lavandula angustifolia) plant seeds and production of its seedlings callus cultures. Malaysian Journal of Microbiology., 20(Special issue 6): 36–39.

Berri, N., Moise, S., Keirouz, A., Jennings, A., Castro-Dominguez, B., & Leese, H. (2025). Repurposing Laboratory Plastic into Functional Fibrous Scaffolds via Green Electrospinning for Cell Culture and Tissue Engineering Applications. *ACS Biomaterials Science & Engineering*, *11*(6), 3573–3585. DOI: 10.1021/acsbiomaterials.5c00146 PMID: 40207865

Burak, Ç., & Meltem, I. Y. (2021). The Role of Culture in Environmental Sustainability. *Present Environment and Sustainable Development.*, *15*(2), 259–272. Advance online publication. DOI: 10.15551/pesd2021152021

Cortese, A. D. (2003). The critical role of higher education in creating a sustainable future. *Planning for Higher Education*, *31*(3), 15–22.

Cuthill, M. (2010). Strengthening the 'social' in sustainable development: Developing aconceptual framework for social sustainability in a rapid urban growth region in Australia. *Sustainable Development (Bradford)*, *18*(6), 362–373. DOI: 10.1002/sd.397

El-Sherif, N. (2018). Impact of Plant Tissue Culture on Agricultural Sustainability. 93-107.DOI: 10.1007/698_2017_160

Erol, M. H., Dönmez, D., Biçen, B., Şimşek, Ö., & Aka Kaçar, Y. (2023). Modern approaches to In Vitro clonal banana production: Next-generation tissue culture systems. *Horticulturae*, *9*(10), 1154. DOI: 10.3390/horticulturae9101154

Fithian, C., & Powell, A. (2009). *Cultural aspects of sustainable development. Student paperfor sustainable design seminar/studio, fall 2009, seminar in sustainable architecture*. University of Texas School of Architecture.

Grierson, D. (2009). Towards a sustainable built environment. *CIC Start Online Innovation Rev*, *1*, 70–77.

Hauser, R., & Banse, G. (2011). Culture and culturality: approaching a multi-faceted concept. In *Parodi, O., Ayestaran, I. Banse, G.B., Sustainable Development- Relationships to Culture, Knowledge and Ethics*. KIT Scientific Publishing. DOI: 10.5771/9783845271156-31

Hawkes, J. (2001). *The fourth pillar of sustainability: culture's essential role in public planning*. Common Ground Press.

Iqbal, M., Aftab, Z.-H., Anjum, T., Rizwana, H., Akram, W., Aftab, A., Sajid, Z. A., & Li, G. (2024). Nano-Integrated plant tissue culture to increase the rate of callus induction, growth, and curcuminoid production in Curcuma longa. *Plants*, *13*(13), 1819. DOI: 10.3390/plants13131819 PMID: 38999659

Kasoju, N., & Sunilkumar, A. (2024). *Convergence of tissue engineering and sustainable development goals*. Biotechnology for Sustainable Materials., DOI: 10.1186/s44316-024-00021-y

Khan, M. A., Kumar, R., & Singh, P. (2024). Plant tissue culture and micropropagation as tools for sustainable agriculture and biodiversity conservation. *Sustainability*, *16*(3), 1120. DOI: 10.3390/su16031120

Khuder, J. K. H., & Mohammed, A. A. (2025). Isolation Endophytic Bacteria from Leaf Glands and Seeds of *Ardisia crenata* Plant and its Molecular Detection. *Egyptian Journal of Veterinary Science*, *56*(6), 1265–1270. DOI: 10.21608/ejvs.2024.282284.2009

Kumar, D., Chakradhar, P., Ranganna, G., Vimal, V., Raj, R., Anusha, C., Patra, S., & Nagaraju, V. (2024). Tissue Culture in Banana Cultivation: A Review of its Impact on Disease Management, Yield Improvement, and Sustainable Production. *Journal of Advances in Biology & Biotechnology*, *27*(9), 628–644. Advance online publication. DOI: 10.9734/jabb/2024/v27i91336

Mahdi, G.A. and Yahya, R.T. (2024). Effect of zinc oxide nanoparticles on soybean's biomolecules and oil content (*Glycine max* L.) seedlings tissues. Malaysian Journal of Microbiology. 20(Special issue 6): 1–5.

Opoku, A., & Fortune, C. (2011) The implementation of sustainable practices through leadershipin construction organizations In: Egbu C, Lou ECW (eds) Proceedings 27th annual ARCOM Bosch, C. (2009). Agenda 21 for Culture: An Agreement by Cities and Local Governments for Cultural Development. Available at SSRN: https://ssrn.com/abstract=2056382 or Packalén, S. (2010). Culture and sustainability. Corporate social responsibility and environmental management, 17(2), 118-121.

Parodi, O. (2011). Three steps towards a culture of sustainability. In Parodi, O., Ayestaran, I., & Banse, G. (Eds.), *Sustainable development-relationships to culture, knowledge and ethics* (pp. 76–92). KIT Scientific Publishing.

Parodi, O. (2011). Three steps towards a culture of sustainability. In Parodi, O., Ayestaran, I., & Banse, G. (Eds.), *Sustainable development-relationships to culture, knowledge and ethics* (pp. 76–92). KIT Scientific Publishing.

Patel, B., & Sahay, N. S. (2025). Strategies for the improvement of micropropagation of banana through in vitro culture. *Plant Tissue Culture & Biotechnology*, *35*(1), 185–198. DOI: 10.3329/ptcb.v35i1.82675

Pithiya, M., Sharma, S., Sharma, M., Sharma, M., & Kotwal, N. (2022). Advancements and Challenges in Plant Tissue Culture: a Comprehensive Overview. Journal of Plant Biota. DOI: 10.51470/JPB.2022.1.1.12

Rahmawati, D., Kartiwijaya, R., & Yuliawati, L. B. (2023). Micropropagation of banana plant Cavendish (*Musa acuminata* L.) using shoot-tip culture to support food security. *BIODIVERS–BIOTROP Science Magazine*, *2*(2), 16–21. DOI: 10.56060/bdv.2023.2.2.2092

Ramadan, E. A., Ramadan, N. A., & Mohammed, A. A. H. (2022). Aflatoxigenic fungi in nuts and dried fruits in Mosul and Duhok city. [Vojenske Zdravotnicke Listy]. *Vojenske Zdravotnicke Listy*, *91*(3), 224–234. DOI: 10.31482/mmsl.2021.049

Sultan, S. J., Qaddawi, Z. T., & Mohammed, A. A. H. (2024). Molecular diagnosis of bacteria isolated from *Trifolium repens* root nodules. *Journal of Applied and Natural Science*, *16*(3), 1153–1157. DOI: 10.31018/jans.v16i3.5404

Tamimi, S. M., & Othman, H. (2023). Silver nanoparticles for enhancing the efficiency of micropropagation of banana (Musa acuminata L.). *Tropical Life Sciences Research*, *34*(2), 161–175. DOI: 10.21315/tlsr2023.34.2.8 PMID: 38144380

Thakur, S. S., Hashmi, S., Mishra, S., Ekka, S., Kushwaha, A., & Kujur, R. (2024). *A Review on Plant Tissue Culture*. Asian Journal of Biology., DOI: 10.9734/ajob/2024/v20i2387

UNESCO. (2012a). *Culture: a driver and an enabler of sustainable development. UN systemtask team on the post-2015 development agenda, The United Nations Scientific, Educational and Cultural Organization*. UNESCO.

Yahya, R. T., Qaddawi, Z. T., Rasheed, A. A., & Sultan, S. J. (2024). Effect of diode laser rays in stimulating anthocyanin pigment levels in radish (*Raphanus sativus* L.) plant tissues. Journal of Applied and Natural Science., 16(3): 1158 – 1163. ISSN: 0974-9411 (Print), 2231-5209 (Online).

Yahya, R. T., Qaddawi, Z. T., Rasheed, A. A., & Sultan, S. J. (2024). Effect of diode laser rays in stimulating anthocyanin pigment levels in radish (*Raphanus sativus* L.) plant tissues. Journal of Applied and Natural Science., 16(3): 1158 – 1163. ISSN: 0974-9411 (Print), 2231-5209 (Online).

Zhang, C., Wang, S., Wang, Y., Wang, H., Qin, M., Dai, X., Yan, B., Guo, X., Zhou, L., Lin, H., & Guo, L. (2023).[Application of tissue culture technology of medicinal plants in sustainable development of Chinese medicinal resources] 48 5, 1186-1193. https://doi.org/DOI: 10.19540/j.cnki.cjcmm.20221017.104

Zuzarte, M., Salgueiro, L., & Canhoto, J. (2024). Plant Tissue Culture: Industrial Relevance and Future Directions. *Advances in Biochemical Engineering/Biotechnology*, *188*, 1–15. Advance online publication. DOI: 10.1007/10_2024_254 PMID: 38796640

Compilation of References

Abbas, K., Li, J., Gong, B., Lu, Y., Wu, X., Lü, G., & Gao, H. (2023). Drought stress tolerance in vegetables: The functional role of structural features, key gene pathways, and exogenous hormones. *International Journal of Molecular Sciences*, *24*(18), 13876. DOI: 10.3390/ijms241813876 PMID: 37762179

Abbas, M. M., & Hussain, W. S. (2020). Biostimulants of pepper and eggplant by using plants aqueous extract. *Plant Cell Biotechnology and Molecular Biology*, *21*(65&66), 78–82.

Abdalla, N., El-Ramady, H., Seliem, M. K., El-Mahrouk, M. E., Taha, N., Bayoumi, Y., Shalaby, T. A., & Dobránszki, J. (2022). An academic and technical overview on plant micropropagation challenges. *Horticulturae*, *8*(8), 677. DOI: 10.3390/horticulturae8080677

Abdul Aziz, M., Brini, F., Rouached, H., & Masmoudi, K. (2022, November 8). Genetically engineered crops for sustainably enhanced food production systems. *Frontiers in Plant Science*, *13*, 1027828. DOI: 10.3389/fpls.2022.1027828 PMID: 36426158

Abdullah, Z. Y., & Al-dulaimi, F. K. Y. (2024). Evaluating the role of Ultraviolet Radiation and Ozone gas in reducing Aflatoxin in Rice. *NTU Journal of Agricultural and Veterinary Sciences*, *4*(4), 249–253.

Abdullah, Z. Y., & Al-dulaimi, F. K. Y.Fahad K. Y. Al-dulaimiZainab Yahya Abdullah. (2024). Evaluating the role of Ultraviolet Radiation and Ozone gas in reducing Aflatoxin in Rice. (2024). *NTU Journal of Agriculture and Veterinary Science*, *4*(4). Advance online publication. DOI: 10.56286/npeme074

Abdulraheem, M. I., Xiong, Y., Moshood, A. Y., Cadenas-Pliego, G., Zhang, H., & Hu, J. (2024, January 7). Mechanisms of Plant Epigenetic Regulation in Response to Plant Stress: Recent Discoveries and Implications. *Plants*, *13*(2), 163. DOI: 10.3390/plants13020163 PMID: 38256717

Abiri, R., Rizan, N., Balasundram, S. K., Shahbazi, A. B., & Abdul-Hamid, H. (2023). Application of digital technologies for ensuring agricultural productivity. *Heliyon*, *9*(12), e22601. DOI: 10.1016/j.heliyon.2023.e22601 PMID: 38125472

Abley, K., Goswami, R., and Locke, J. C. (2024). Bet-hedging and variability in plant development: seed germination and beyond. *Philosophical Transactions of the Royal Society B, 379*(1900), 20230048.

Abobatta, W. F. (2018). Development growth and productivity of orange orchards (*Citrus sinensis* L) in Egypt (delta region). *Advances in Agricultural Technology & Plant Sciences*, *1*(1), 1–11. DOI: 10.63235/AATPS.180003

Abobatta, W. F. (2025). Importance of propagation techniques of fruit trees. *J Appl Biotechnol Bioeng*, *12*(4), 180–182.

Abobatta, W. F., & Ahmed, F. K. (2023). Effect of Urea and Nano-nitrogen Spray Treatments on Some Citrus Rootstock Seedlings. *Horticulture Research Journal*, *1*(1), 68–84. DOI: 10.21608/hrj.2023.306950

Abobatta, W. F., & Hussain, W. S. (2024). *Achieving food security through sustainable agriculture Achieving Food Security Through Sustainable Agriculture*. IGI Global. DOI: 10.4018/979-8-3693-4240-4

Absakine, S. I., Tchimbi, B., and Ousman, A. H. (2024). The Difficulty of Chadian Forest Tree Seed Germination: Overcoming Physical Dormancy. *African Journal of Advanced Pure and Applied Sciences (AJAPAS)*, 98-109.

Acemoglu, D., Autor, D., Dorn, D., Hanson, G. H., & Price, B. (2016). Import competition and the great US employment sag of the 2000s. *Journal of Labor Economics*, *34*(S1), S141–S198. DOI: 10.1086/682384

Adhikari, K., Bikundo, E., Chacko, X., Chapman, S., Humphries, F., Johnson, H., Keast, E., Lawson, C., Malbon, J., Robinson, D., Rourke, M., Sanderson, J., & Tranter, K. (2021). What Should Farmers' Rights Look Like? The Possible Substance of a Right. *Agronomy (Basel,* 11(2), 367. DOI:DOI: 10.3390/agronomy11020367

Adu Donyina, G., Szarvas, A., Opoku, V. A., Miko, E., Tar, M., Czóbel, S., & Monostori, T. (2025). Enhancing sweet potato production: A comprehensive analysis of the role of auxins and cytokinins in micropropagation. *Planta*, *261*(4), 1–14. DOI: 10.1007/s00425-025-04650-z PMID: 40035890

Ağaoğlu, Y. S., Çelik, H., Çelik, M., Fidan, Y., Gülşen, Y., Günay, A., & Yanmaz, R. (2010). *Genel bahçe bitkileri*. Ankara Üniversitesi Ziraat Fakültesi.

Aguilar, R., Ashworth, L., Galetto, L., & Aizen, M. A. (2006). Plant reproductive susceptibility to habitat fragmentation: Review and synthesis through a meta-analysis. *Ecology Letters*, *9*(8), 968–980.

Ahluwalia, O., Singh, P. C., & Bhatia, R. (2021). A review on drought stress in plants: Implications, mitigation, and the role of plant growth-promoting rhizobacteria. *Resour Environ Sustain*, *5*, 100032. DOI: 10.1016/j.resenv.2021.100032

Ahmad, A., Ahmad, N., Anis, M., Faisal, M., Alatar, A. A., Abdel-Salam, E. M., Meena, R. P., & Sivanesan, I. (2022). Biotechnological Advances in Pharmacognosy and In Vitro Manipulation of *Pterocarpus marsupium* Roxb. *Plants*, *11*(3), 247. DOI: 10.3390/plants11030247 PMID: 35161227

Ahmar, S., Gill, R. A., Jung, K. H., Faheem, A., Qasim, M. U., Mubeen, M., & Zhou, W. (2020). Conventional and molecular techniques from simple breeding to speed breeding in crop plants: Recent advances and future outlook. *International Journal of Molecular Sciences*, *21*(7), 2590. DOI: 10.3390/ijms21072590 PMID: 32276445

Ak, B. E. (2018). The Importance of in vitro micropropagation of fruit crops. In *1st International GAP agriculture and livestock congress proceedings book* (pp. 716-723).

Ak, B. E., Hatipoglu, I. H., & Dikmetas, B. (2021). Propagation of fruit trees. *Recent Headways in Pomology (Edited by: M. Pakyurek) Chapter, 3*, 55-92.

Ak, B. E., Hatipoğlu, İ. H., & Hatipoğlu, M. (2024). Modern Methods in Ornamental Plant Propagation: Improving Resistance and Sustainability. *Endangered Species*, *1*(2), 3.

Akbarinia, G., Sefidkonf, R., & Sharifia, M. B. (2003). Study on the effect of different rates of chemical fertilizer, manure and mixture of them on seed yield and main - composition of essential oil of Ajowan (*Trachyspermum copticum*). *Iranian Research and Development*, *61*, 32–41.

Akhtar, H., Usman, M., Binyamin, R., Hameed, A., Arshad, S. F., Aslam, H. M. U., Khan, I. A., Abbas, M., Zaki, H. E. M., Ondrasek, G., & Shahid, M. S. (2024). Traditional Strategies and Cutting-Edge Technologies Used for Plant Disease Management: A Comprehensive Overview. *Agronomy (Basel)*, *14*(9), 2175. DOI: 10.3390/agronomy14092175

Al Makhmari, S. M. (2016). Effect of substrates, IBA and types of cuttings on rooting. *Theses*. 722. https://scholarworks.uaeu.ac.ae/all_theses/722

Al-Aali, A. Y., & Teece, D. J. (2013). Towards the (strategic) management of intellectual property: Retrospective and prospective. *California Management Review*, *55*(4), 15–30. DOI: 10.1525/cmr.2013.55.4.15

Al-Dulaimi, F. K. Y., & Al-Tarjuman, J. K. (2024). Food security challenges. Achieving Food Security Through Sustainable Agriculture, igi-global. pp. 28–49. DOI: 10.4018/979-8-3693-4240-4.ch002

Ali, A. S., and Elozeiri, A. A. (2017). Metabolic processes during seed germination. *Advances in seed biology, 2017*, 141-166.

Ali, H. J.(1988). Fundamentals of Breeding and Genetics of Field Crops. Ministry of Higher Education and Scientific Research, University of Baghdad. Directorate of Dar Al-Kutub for Printing and Publishing, University of Mosul. PP:76.

Ali, J., Mukarram, M., Ojo, J., Dawam, N., Riyazuddin, R., Ghramh, H. A., Khan, K. A., Chen, R., Kurjak, D., & Bayram, A. (2024). Harnessing phytohormones: Advancing plant growth and defence strategies for sustainable agriculture. *Physiologia Plantarum, 176*(3), e14307. DOI: 10.1111/ppl.14307 PMID: 38705723

Alkhoraif, A. (2024). The impact of innovation governance and policies on government funding for emerging science and technology sectors in Saudi Arabia. *Journal of Infrastructure. Policy and Development, 8*(14), 8515. DOI: 10.24294/jipd8515

Alliance, O. S. (2022). *Putting the power of seed into the hands of growers*. Policy.

Al-Nuaimy Maha, M. T., & Abed Faten, N. A. (2024). Zearalenone toxins produced by some species of Fusarium. (2024). *NTU Journal of Agriculture and Veterinary Science, 4*(4). Advance online publication. DOI: 10.56286/bb44fh80

Al-Shabini, J. M. (2006). The phosphorus in the soil and plants. Egyptian Library for printing and publishing MSc. Thesis University of Muthanna.

Al-Tai, A.A. and Mohammed, A.A.-H. (2024). Germination of lavender (Lavandula angustifolia) plant seeds and production of its seedlings callus cultures. Malaysian Journal of Microbiology., 20(Special issue 6): 36–39.

Al-Tai, A.A., Mohammed, A.A.H. (2024). Germination of lavender (*Lavandula angustifolia*) plant seeds and production of its seedlings callus cultures. *Malaysian Journal of Microbiology.* 20 (Special issue 6): 36 - 39.

Al-Tarjuman, Janan K.,Abed, Faten N. Mula and Al-Dulaimi, Fahad K.Y. (2020). Study on fungal contamination of IRAQI paper currency. *Biochemical and Cellular Archives, 20*(1), 2201–2208.

Al-Tawaha, A. R., Turk, M. A., Abu-Zaitoon, Y. M., Aladaileh, S. H., Al-Rawashdeh, I. M., Alnaimat, S., & Wedyan, M. (2017). Plants adaptation to drought environment. *Bulgarian Journal of Agricultural Science, 23*(3), 381–388.

Altieri, M. A., Funes-Monzote, F. R., & Petersen, P. (2012). Agroecologically efficient agricultural systems for smallholder farmers: Contributions to food sovereignty. *Agronomy for Sustainable Development, 32*(1), 1–13. DOI: 10.1007/s13593-011-0065-6

Altpeter, F., Springer, N. M., Bartley, L. E., Blechl, A. E., Brutnell, T. P., Citovsky, V., Conrad, L., Gelvin, S. B., Jackson, D., Kausch, A. P., Lemaux, P. G., Medford, J. I., Orozo-Cardenas, M., Tricoli, D., VanEck, J., Voytas, D. F., Walbot, V., Wang, K., Zhang, Z. J., & Stewart, C. N.Jr. (2016). Advancing crop transformation in the era of genome editing. *The Plant Cell, 28*(7), 1510–1520. DOI: 10.1105/tpc.16.00196 PMID: 27335450

Amara, U., Khalid, R., & Hayat, R. (2015). In Maheshwari, D. K. (Ed.), *"Soil bacteria and phytohormones for sustainable crop production" in Bacterial metabolites in sustainable agroecosystem. Sustainable development and biodiversity. Vol. 12* (pp. 87–103). Springer.

Amiteye, S. (2021, September 30). Basic concepts and methodologies of DNA marker systems in plant molecular breeding. *Heliyon, 7*(10), e08093. DOI: 10.1016/j.heliyon.2021.e08093 PMID: 34765757

Andersen, R. (1996, June). The history of farmers' rights: A guide to central documents and literature. In *Conference on Plant Genetic Resources in Leipzig* (Vol. 17, p. 50).

Anderson, K. (2023). Agriculture's globalization: Endowments, technologies, tastes and policies. *Journal of Economic Surveys, 37*(4), 1314–1352. DOI: 10.1111/joes.12529

Anjanappa, R. B., & Gruissem, W. (2021). Current progress and challenges in crop genetic transformation. *Journal of Plant Physiology, 261*, 153411. DOI: 10.1016/j.jplph.2021.153411 PMID: 33872932

Ansori, A. N., Antonius, Y., Susilo, R. J., Hayaza, S., Kharisma, V. D., Parikesit, A. A., Zainul, R., Jakhmola, V., Saklani, T., Rebezov, M., Ullah, M. E., Maksimiuk, N., Derkho, M., & Burkov, P. (2023, August). Application of CRISPR-Cas9 genome editing technology in various fields: A review. *Narra J, 3*(2), e184. DOI: 10.52225/narra.v3i2.184 PMID: 38450259

Arcus, V. L., Prentice, E. J., Hobbs, J. K., Mulholland, A. J., Van der Kamp, M. W., Pudney, C. R., Parker, E. J., & Schipper, L. A. (2016). On the temperature dependence of enzyme-catalyzed rates. *Biochemistry, 55*(12), 1681–1688. DOI: 10.1021/acs.biochem.5b01094 PMID: 26881922

Aremu, A. O., Fawole, O. A., Makunga, N. P., Masondo, N. A., Moyo, M., Buthelezi, N. M. D., Amoo, S. O., Spíchal, L., & Doležal, K. (2020, August 22). Applications of Cytokinins in Horticultural Fruit Crops: Trends and Future Prospects. *Biomolecules, 10*(9), 1222. DOI: 10.3390/biom10091222 PMID: 32842660

Aronen, T., Varis, S., & Tikkinen, M. (2025). Somatic embryogenesis: concept, principles, and applications. In *Forest Microbiology* (pp. 373–388). Academic press. DOI: 10.1016/B978-0-443-21903-0.00022-9

Arpitha, K., & Navya, V. (2025). SEED DORMANCY AND IT'S MECHANISM. *Seed Biology and Biotechnology*, 18.

Arpitha, K., & Navya, V. (2025). Seed dormany and it's mechanism. *Seed Biology and Biotechnology*, 18.

Asmamaw, M., & Zawdie, B. (2021, August 21). Mechanism and Applications of CRISPR/Cas-9-Mediated Genome Editing. *Biologics, 15*, 353–361. DOI: 10.2147/BTT.S326422 PMID: 34456559

Atzori, G., & Caparrotta, S. (2023). Different salt stress response in Kalanchoe plants during vegetative and asexual propagation. *South African Journal of Botany, 154*, 282–288. DOI: 10.1016/j.sajb.2023.01.027

Awasthi, P. (2023). Review on dormancy, causes, uses, and measures of overcoming it. *Science, 7*(2), 24–27.

Awazade, A. S., & Verma, D. (2024). Advancing Vegetable Grafting: A comprehensive review of techniques, challenges, and the future of automated solutions. *Journal of Scientific Research and Reports, 30*(11), 517–530. DOI: 10.9734/jsrr/2024/v30i112580

Awotedu, B. F., Omolola, T. O., Akala, A. O., Awotedu, O. L., & Olaoti-Laaro, S. O. (2021). Vegetative propagation: A unique technique of improving plants growth. *World News Nat Sci., 35*, 83–101.

Awotedu, B. F., Omolola, T. O., Akala, A. O., Awotedu, O. L., & Olaoti-Laaro, S. O. (2021). Vegetative propagation: A unique technique of improving plants growth. *World News of Natural Sciences, 35*, 83–101.

Ayuso-Calles M, García-Estévez I, Jiménez-Gómez A, Flores-Félix JD, Escribano-Bailón MT, Rivas R (2020) *Rhizobium laguerreae* improves productivity and phenolic compound content of lettuce (*Lactuca sativa* L.) under saline stress conditions. Foods 9:1166. 10.3390 .

Azadi, H., Samiee, A., Mahmoudi, H., Jouzi, Z., Rafiaani Khachak, P., De Maeyer, P., & Witlox, F. (2016). Genetically modified crops and small-scale farmers: Main opportunities and challenges. *Critical Reviews in Biotechnology, 36*(3), 434–446. PMID: 25566797

Bag, T. K., Srivastava, A. K., Yadav, S. K., Gurjar, M. S., Diengdoh, L. C., Rai, R., & Singh, S. (2015). Potato (Solanum tuberosum) aeroponics for quality seed production in north eastern Himalayan region of India. *Indian Journal of Agricultural Sciences, 85*(10), 1360–1364. DOI: 10.56093/ijas.v85i10.52303

Baharudin, N. F., & Osman, N. I. (2023). Plant development, stress responses, and secondary metabolism under ethylene regulation. *Plant Stress, 7*, 100146. DOI: 10.1016/j.stress.2023.100146

Bahuguna, R. N., & Jagadish, K. S. (2015). Temperature regulation of plant phenological development. *Environmental and Experimental Botany, 111*, 83–90. DOI: 10.1016/j.envexpbot.2014.10.007

Balliu, A., Zheng, Y., Sallaku, G., Fernández, J. A., Gruda, N. S., & Tuzel, Y. (2021). Environmental and cultivation factors affect the morphology, architecture and performance of root systems in soilless grown plants. *Horticulturae, 7*(8), 243. DOI: 10.3390/horticulturae7080243

Banadka, A., Toppo, V., Kashyap, K., Nandhini, M., Akshita, B., Sudheer, W. N., & Nagella, P. (2025). Revolutionizing Ashwagandha Breeding: From Traditional to Modern Breeding Techniques for Indian Ginseng, Withania somnifera (L.) Dunal. In *Biodiversity and Genetic Improvement of Medicinal and Aromatic Plants II* (pp. 1–34). Springer. DOI: 10.1007/978-3-031-81857-8_1

Bano, C., Amist, N., & Singh, N. B. (2019). Morphological and anatomical modifications of plants for environmental stresses. *Molecular plant abiotic stress: Biology and biotechnology*, 29-44.

Baron, J., Contreras, J. L., Husovec, M., Larouche, P., & Thumm, N. (2019). Making the rules: The governance of standard development organizations and their policies on intellectual property rights. *JRC Science for Policy Report. EUR*, 29655.

Barrero, J. M., Buitink, J., Chen, S. C., Da Silva, E. A. A., Hay, F. R., Kawakami, N., Leprince, O., Ojiewo, C. O., Pawłowski, T. A., Pérez, H. E., & Vijay, D. (2025). Celebrating the 25th Anniversary of the International Society for Seed Science. *Seed Science Research, 35*(1), 1–8. DOI: 10.1017/S0960258525000029

Barrett, C. B. (2021). Overcoming global food security challenges through science and solidarity. *American Journal of Agricultural Economics, 103*(2), 422–447. DOI: 10.1111/ajae.12160

Bashir, M. A., Rehim, A., Qurat-Ul-Ain Raza, H. M., Raza, A., Zhai, L., Liu, H., and Wang, H. (2021). Biostimulants as Plant Growth Stimulators in Modernized Agriculture and Environmental Sustainability. Journal of Technology in Agriculture, 311-322.

Bekalu, Z. E., Panting, M., Bæksted Holme, I., & Brinch-Pedersen, H. (2023, July 25). Opportunities and Challenges of In Vitro Tissue Culture Systems in the Era of Crop Genome Editing. *International Journal of Molecular Sciences, 24*(15), 11920. DOI: 10.3390/ijms241511920 PMID: 37569295

Ben Mahmoud, M. E.-T. (2016). The effect of inoculation with Burkholderia Burkholdriaspp. on tomato plant productivity. *Syrian Journal of Agricultural Research, 3*(2), 235–242.

Ben Mahmoud, M. E.-T. (2019). Study of the interaction effect between organic fertilizer (cow manure) and biofertilizer (Burkholderia spp.) on wheat plant productivity. *Syrian Journal of Agricultural Research, 6*(1), 314–320.

Bennur, P. L., O'Brien, M., Fernando, S. C., & Doblin, M. S. (2025). Improving transformation and regeneration efficiency in medicinal plants: Insights from other recalcitrant species. *Journal of Experimental Botany, 76*(1), 52–75. DOI: 10.1093/jxb/erae189 PMID: 38652155

Berri, N., Moise, S., Keirouz, A., Jennings, A., Castro-Dominguez, B., & Leese, H. (2025). Repurposing Laboratory Plastic into Functional Fibrous Scaffolds via Green Electrospinning for Cell Culture and Tissue Engineering Applications. *ACS Biomaterials Science & Engineering, 11*(6), 3573–3585. DOI: 10.1021/acsbiomaterials.5c00146 PMID: 40207865

Beshara, S., Kassem, A., Fors, H., & Harraz, N. (2024). A Comprehensive Review and Mapping Citrus Supply Chains from a Sustainability Perspective across the European Union, Middle East, and Africa. *Sustainability, 16*(19), 8582. DOI: 10.3390/su16198582

Beyl, C. A. (2018). Getting started with tissue culture—media preparation, sterile technique, and laboratory equipment. In *Plant tissue culture concepts and laboratory exercises* (pp. 21–38). Routledge.

Beyl, C. A., & Trigiano, R. N. (Eds.). (2016). *Plant propagation concepts and laboratory exercises*. CRC press. DOI: 10.1201/b17340

Bhatla, S. C., & Kathpalia, R. (2023). Seed dormancy and germination. In *Plant physiology, development and metabolism* (pp. 625–640). Springer Nature Singapore. DOI: 10.1007/978-981-99-5736-1_28

Bhatla, S. C., & Lal, M. A. (2023). Auxins. In *Plant Physiology, Development and Metabolism*. Springer., DOI: 10.1007/978-981-99-5736-1_15

Bhattacharya, A. (2022). Effect of low-temperature stress on germination, growth, and phenology of plants: A review. Physiological processes in plants under low temperature stress, 1-106.

Bhattacharya, A. (2022). Effect of low-temperature stress on germination, growth, and phenology of plants: A review. *Physiological processes in plants under low temperature stress*, 1-106.

Bidabadi, S. S., & Jain, S. M. (2020). Cellular, Molecular, and Physiological Aspects of In Vitro Plant Regeneration. *Plants*, *9*(6), 702. DOI: 10.3390/plants9060702 PMID: 32492786

Bidyananda, N., Jamir, I., Nowakowska, K., Varte, V., Vendrame, W. A., Devi, R. S., & Nongdam, P. (2024). Plant Genetic Diversity Studies: Insights from DNA Marker Analyses. *International Journal of Plant Biology*, *15*(3), 607–640. DOI: 10.3390/ijpb15030046

Biswas, P., and Kumar, N. (2023). Application of molecular markers for the assessment of genetic fidelity of in vitro raised plants: Current status and future prospects. *Molecular Marker Techniques: A Potential Approach of Crop Improvement*, 233-256.

Blicharska, M., Smithers, R. J., Mikusiński, G., Rönnbäck, P., Harrison, P. A., Nilsson, M., & Sutherland, W. J. (2019). Biodiversity's contributions to sustainable development. *Nature Sustainability*, *2*(12), 1083–1093.

Boccaletti, G. (2022). *Water: A biography*. Vintage.

Boes, J., Boettcher, P., & Honkatukia, M. (2023). *Innovations in cryoconservation of animal genetic resources: practical guide*. Food and Agriculture Org.

Bonnet, M., Lagier, J. C., Raoult, D., & Khelaifia, S. (2019, November 30). Bacterial culture through selective and non-selective conditions: The evolution of culture media in clinical microbiology. *New Microbes and New Infections*, *34*, 100622. DOI: 10.1016/j.nmni.2019.100622 PMID: 31956419

Borowiak, C. (2004). Farmers' rights: Intellectual property regimes and the struggle over seeds. *Politics & Society*, *32*(4), 511–543. DOI: 10.1177/0032329204269979

Boudadi, I., El Merzougui, S., Lachguer, K., Lagram, K., El Boullani, R., & Serghini, M. A. (2025). Enhancing in vitro rooting and acclimatization of Spathiphyllum wallisii through explant size, NAA, and substrate selection. [PCTOC]. *Plant Cell, Tissue and Organ Culture, 160*(2), 38. DOI: 10.1007/s11240-025-02987-7

Brush, S. B. (2007). Farmers' rights and protection of traditional agricultural knowledge. *World Development, 35*(9), 1499–1514. DOI: 10.1016/j.worlddev.2006.05.018

Burak, Ç., & Meltem, I. Y. (2021). The Role of Culture in Environmental Sustainability. *Present Environment and Sustainable Development., 15*(2), 259–272. Advance online publication. DOI: 10.15551/pesd2021152021

Burger, J. (2014). Indigenous peoples, extractive industries and human rights.

Butler, T. J., Celen, A. E., Webb, S. L., Krstic, D., & Interrante, S. M. (2014). Temperature affects the germination of forage legume seeds. *Crop Science, 54*(6), 2846–2853. DOI: 10.2135/cropsci2014.01.0063

Byrne, P. (2014). Genetically Modified (GM) Crops: Techniques and Applications. *Colorado State.*

Cabahug, R. A. M., Nam, S. Y., Lim, K. B., Jeon, J. K., & Hwang, Y. J. (2018). Propagation techniques for ornamental succulents. *Flower Research Journal, 26*(3), 90–101. DOI: 10.11623/frj.2018.26.3.02

Calfapietra, C., Peñuelas, J., & Niinemets, Ü. (2015). Urban plant physiology: Adaptation-mitigation strategies under permanent stress. *Trends in Plant Science, 20*(2), 72–75. DOI: 10.1016/j.tplants.2014.11.001 PMID: 25476199

Calicioglu, O., Flammini, A., Bracco, S., Bellù, L., & Sims, R. (2019). The future challenges of food and agriculture: An integrated analysis of trends and solutions. *Sustainability (Basel), 11*(1), 222. DOI: 10.3390/su11010222

Campi, M., & Nuvolari, A. (2021). Intellectual property rights and agricultural development: Evidence from a worldwide index of IPRs in agriculture (1961-2018). *The Journal of Development Studies, 57*(4), 650–668. DOI: 10.1080/00220388.2020.1817395

Carrera-Castaño, G., Calleja-Cabrera, J., Pernas, M., Gómez, L., & Oñate-Sánchez, L. (2020). An updated overview on the regulation of seed germination. *Plants, 9*(6), 703. DOI: 10.3390/plants9060703 PMID: 32492790

Castro-Camba, R., Sánchez, C., Vidal, N., & Vielba, J. M. (2022). Plant development and crop yield: The role of gibberellins. *Plants, 11*(19), 2650. DOI: 10.3390/plants11192650 PMID: 36235516

Chandrasekhar, S., & Mehrotra, N. (2016). Doubling Farmers' Incomes by 2022: What Would It Take? *Economic and Political Weekly*, 10–13.

Chaparro, J. M., Sheflin, A. M., Manter, D. K., & Vivanco, J. M. (2012). Manipulating the soil microbiome to increase soil health and plant fertility. *Biology and Fertility of Soils*, *48*(5), 489–499. DOI: 10.1007/s00374-012-0691-4

Chauhan, A., Sharma, D., Kumar, R., Shiwani, K., & Sharma, N. (2021). Methods of propagation in vegetable crops. In *Recent trends in propagation of forest and horticultural crops* (pp. 270–281). Taran Publication.

Chauhan, A., Sharma, D., Kumar, R., Shiwani, K., & Sharma, N. (2021). *Methods of propagation in vegetable crops. Recent trends in propagation of forest and horticultural crops*. Taran Publication.

Chen, F., Chen, L., Yan, Z., Xu, J., Feng, L., He, N., Guo, M., Zhao, J., Chen, Z., Chen, H., Yao, G., & Liu, C. (2024, September 23). Recent advances of CRISPR-based genome editing for enhancing staple crops. *Frontiers in Plant Science*, *15*, 1478398. DOI: 10.3389/fpls.2024.1478398 PMID: 39376239

Chen, J., & Stamps, R. H. (2006). *Cutting propagation of foliage plants. Cutting propagation: A guide to propagating and producing floriculture crops*. Ball Publishing.

Chen, Y. M., Huang, J. Z., Hou, T. W., & Pan, I. C. (2019). Effects of light intensity and plant growth regulators on callus proliferation and shoot regeneration in the ornamental succulent Haworthia. *Botanical Studies*, *60*(1), 10. DOI: 10.1186/s40529-019-0257-y PMID: 31267253

Chin, C. F., & Tan, H. S. (2018). The Use of Proteomic Tools to Address Challenges Faced in Clonal Propagation of Tropical Crops through Somatic Embryogenesis. *Proteomes*, *6*(2), 21. DOI: 10.3390/proteomes6020021 PMID: 29734680

Chon, M. (2011). Global intellectual property governance (under construction). *Theoretical Inquiries in Law*, *12*(1), 349–380. DOI: 10.2202/1565-3404.1271

Chung, K. P. (2025, April 30). Cytoplasmic inheritance: The transmission of plastid and mitochondrial genomes across cells and generations. *Plant Physiology*, *198*(1), kiaf168. Advance online publication. DOI: 10.1093/plphys/kiaf168 PMID: 40304456

Churchman, J. C., & Lowe, D. J. (2012). Alteration, formation, and occurrence of minerals in soils. In Huang, P. M., Li, Y., & Sumner, M. E. (Eds.), *Handbook of Soil Sciences — Properties and Processes* (pp. 20-1–20-72). CRC Press.

Clapp, J. (2022). The rise of big food and agriculture: corporate influence in the food system. In *A research agenda for food systems* (pp. 45–66). Edward Elgar Publishing.

Clark, H., & Wu, H. (2016). The sustainable development goals: 17 goals to transform our world. Furthering the work of the United Nations, 36-54.

Clements, D. R., & Jones, V. L. (2021). Ten ways that weed evolution defies human management efforts amidst a changing climate. *Agronomy (Basel)*, *11*(2), 284. DOI: 10.3390/agronomy11020284

Cofer, C. M. (2015). *The Ara Pacis Augustae and the Ancient Understanding of Grafting* (Doctoral dissertation, Bryn Mawr College)

Collard, B. C., & Mackill, D. J. (2008, February 12). Marker-assisted selection: An approach for precision plant breeding in the twenty-first century. *Philosophical Transactions of the Royal Society of London. Series B, Biological Sciences*, *363*(1491), 557–572. DOI: 10.1098/rstb.2007.2170 PMID: 17715053

Corbineau, F. (2022). Oxygen, a key signalling factor in the control of seed germination and dormancy. *Seed Science Research*, *32*(3), 126–136. DOI: 10.1017/S096025852200006X

Correa, C. M. (2006). Considerations on the standard material transfer agreement under the fao treaty on plant genetic resources for food and agriculture. *The Journal of World Intellectual Property*, *9*(2), 137–165. DOI: 10.1111/j.1422-2213.2006.00272.x

Corredoira, E., Merkle, S. A., Martínez, M. T., Toribio, M., Canhoto, J. M., Correia, S. I., Ballester, A., & Vieitez, A. (2019). Non-zygotic embryogenesis in hardwood species. *Critical Reviews in Plant Sciences*, *38*(1), 29–97. DOI: 10.1080/07352689.2018.1551122

Correia, S., Matos, M., & Leal, F. (2024). Advances in blueberry (Vaccinium spp.) in vitro culture: A review. *Horticulturae*, *10*(6), 533. DOI: 10.3390/horticulturae10060533

Cortese, A. D. (2003). The critical role of higher education in creating a sustainable future. *Planning for Higher Education*, *31*(3), 15–22.

COST, R. (2025). Integrative approaches to enhance reproductive resilience of crops for climate-proof agriculture. *Plant Stress*, •••, 15.

Crawford, A. J., McLachlan, D. H., Hetherington, A. M., & Franklin, K. A. (2012). High temperature exposure increases plant cooling capacity. *Current Biology*, *22*(10), R396–R397. DOI: 10.1016/j.cub.2012.03.044 PMID: 22625853

Cucu, M. A., Choudhary, R., Trkulja, V., Garg, S., & Matić, S. (2025). Utilizing Environmentally Friendly Techniques for the Sustainable Control of Plant Pathogens: A Review. *Agronomy (Basel)*, *15*(7), 1551. DOI: 10.3390/agronomy15071551

Cullet, P. (2004). *Food security and intellectual property rights in developing countries.* RIBios (Réseau interdisciplinaire biosécurité): Institut universitaire d'études du développement (IUED).

Cullet, P. (1999). Revision of the TRIPS Agreement concerning the protection of plant varieties: Lessons from India concerning the development of a sui generis system. *The Journal of World Intellectual Property, 2*(4), 617–656. DOI: 10.1111/j.1747-1796.1999.tb00081.x

Cuthill, M. (2010). Strengthening the 'social' in sustainable development: Developing aconceptual framework for social sustainability in a rapid urban growth region in Australia. *Sustainable Development (Bradford), 18*(6), 362–373. DOI: 10.1002/sd.397

Dai, K., Bergot, A., Liang, C., Xiang, W. N., & Huang, Z. (2015). Environmental issues associated with wind energy–A review. *Renewable Energy, 75*, 911–921. DOI: 10.1016/j.renene.2014.10.074

Dal Monico, S. (2019). Biopiracy, or the misappropriation of traditional knowledge for profit: a human rights perspective.

Das, R., & Biswas, S. (2022). Influence of abiotic stresses on seed production and quality. In *Seed biology updates.* IntechOpen., DOI: 10.5772/intechopen.106045

Daubenmire, R. (2012). *Plant geography: with special reference to North America.* Elsevier.

De Almeida, M. R., Aumond, M.Jr, Da Costa, C. T., Schwambach, J., Ruedell, C. M., Correa, L. R., & Fett-Neto, A. G. (2017). Environmental control of adventitious rooting in Eucalyptus and Populus cuttings. *Trees (Berlin), 31*(5), 1377–1390. DOI: 10.1007/s00468-017-1550-6

De Carlo, A., Tarraf, W., Lambardi, M., & Benelli, C. (2021). Temporary Immersion System for Production of Biomass and Bioactive Compounds from Medicinal Plants. *Agronomy (Basel), 11*(12), 2414. DOI: 10.3390/agronomy11122414

de Carvalho, D. U., Boakye, D. A., Gast, T., Leite Junior, R. P., & Alferez, F. (2021). Determining seed viability during fruit maturation to improve seed production and availability of new citrus rootstocks. *Frontiers in Plant Science, 12*, 777078. DOI: 10.3389/fpls.2021.777078

De Frenne, P., Graae, B. J., Rodríguez-Sánchez, F., Kolb, A., Chabrerie, O., Decocq, G., De Kort, H., De Schrijver, A., Diekmann, M., Eriksson, O., Gruwez, R., Hermy, M., Lenoir, J., Plue, J., Coomes, D. A., & Verheyen, K. (2013). Latitudinal gradients as natural laboratories to infer species' responses to temperature. *Journal of Ecology, 101*(3), 784–795. DOI: 10.1111/1365-2745.12074

De Klerk, G. J. (2012). Micropropagation of bulbous crops: Technology and present state. *Floriculture & Ornamental Biotechnology*, *6*(1), 1–8.

De Schutter, O. (2009). *Report of the Special Rapporteur on the right to food*. UN.

Debnath, S. C., Bhatt, D., & Goyali, J. C. (2023). DNA-Based Molecular Markers and Antioxidant Properties to Study Genetic Diversity and Relationship Assessment in Blueberries. *Agronomy (Basel)*, *13*(6), 1518. DOI: 10.3390/agronomy13061518

Dehaven, C. D., Evans, A. M., Dai, H., & Lawton, K. A. (2010, October 18). Organization of GC/MS and LC/MS metabolomics data into chemical libraries. *Journal of Cheminformatics*, *2*(1), 9. DOI: 10.1186/1758-2946-2-9 PMID: 20955607

De-la-Peña, C., Nic-Can, G. I., Galaz-Ávalos, R. M., Avilez-Montalvo, R., & Loyola-Vargas, V. M. (2015). The role of chromatin modifications in somatic embryogenesis in plants. *Frontiers in Plant Science*, *6*, 635. DOI: 10.3389/fpls.2015.00635 PMID: 26347757

Delgado, A., & Gómez, J. A. (2024). The soil: Physical, chemical, and biological properties. In *Principles of agronomy for sustainable agriculture* (pp. 15–30). Springer International Publishing. DOI: 10.1007/978-3-031-69150-8_2

Delmer, D. P., Nottenburg, C., Graff, G. D., & Bennett, A. B. (2003). Intellectual property resources for international development in agriculture. *Plant Physiology*, *133*(4), 1666–1670. DOI: 10.1104/pp.103.028993 PMID: 14681529

Delporte, F., Jacquemin, J. M., Masson, P., & Watillon, B. (2012, December). Insights into the regenerative property of plant cells and their receptivity to transgenesis: Wheat as a research case study. *Plant Signaling & Behavior*, *7*(12), 1608–1620. DOI: 10.4161/psb.22424 PMID: 23072995

Dennis, E. S., Ellis, J., Green, A., Llewellyn, D., Morell, M., Tabe, L., & Peacock, W. J. (2008, February 12). Genetic contributions to agricultural sustainability. *Philosophical Transactions of the Royal Society of London. Series B, Biological Sciences*, *363*(1491), 591–609. DOI: 10.1098/rstb.2007.2172 PMID: 17656342

Desai, P., Desai, S., Rafaliya, R., & Patil, G. (2022). Plant tissue culture: Somatic embryogenesis and organogenesis. In *Advances in plant tissue culture* (pp. 109–130). Academic Press., DOI: 10.1016/B978-0-323-90795-8.00006-0

Devi, K. V., Raj, S., Alavekar, M. S., & Bhrihavasi, G. (2025). Advancements and Future Trends in Horticultural Propagation. *Horticulture Crops*, 204.

Didi, D. A., Su, S., Sam, F. E., Tiika, R. J., & Zhang, X. (2022, September 28). Effect of Plant Growth Regulators on Osmotic Regulatory Substances and Antioxidant Enzyme Activity of *Nitraria tangutorum*. *Plants*, *11*(19), 2559. DOI: 10.3390/plants11192559 PMID: 36235426

Din, S., Wani, R. A., Ab Waheed Wani, F. N., Farwah, S., Rizvi, S., & Nisar, T. F. W. S. (2019). Enhancement in temperate fruit crops. *Journal of Pharmacognosy and Phytochemistry*, *8*(2), 1203–1216.

Dobrzyński, J., Wróbel, B., & Górska, E. B. (2023). Taxonomy, ecology, and cellulolytic properties of the genus *Bacillus* and related genera. *Agriculture*, *13*(10), 1979. DOI: 10.3390/agriculture13101979

Dormatey, R., Sun, C., Ali, K., Fiaz, S., Xu, D., Calderón-Urrea, A., Bi, Z., Zhang, J., & Bai, J. (2021). ptxD/Phi as alternative selectable marker system for genetic transformation for bio-safety concerns: A review. *PeerJ*, *9*, e11809. DOI: 10.7717/peerj.11809 PMID: 34395075

Driesen, E., Van den Ende, W., De Proft, M., & Saeys, W. (2020). Influence of environmental factors light, CO2, temperature, and relative humidity on stomatal opening and development: A review. *Agronomy (Basel)*, *10*(12), 1975. DOI: 10.3390/agronomy10121975

Druege, U., Hilo, A., Pérez-Pérez, J. M., Klopotek, Y., Acosta, M., Shahinnia, F., Zerche, S., Franken, P., & Hajirezaei, M. R. (2019). Molecular and physiological control of adventitious rooting in cuttings: Phytohormone action meets resource allocation. *Annals of Botany*, *123*(6), 929–949. DOI: 10.1093/aob/mcy234 PMID: 30759178

Dubois, M., Van den Broeck, L., & Inzé, D. (2018). The pivotal role of ethylene in plant growth. *Trends in Plant Science*, *23*(4), 311–323. DOI: 10.1016/j.tplants.2018.01.003 PMID: 29428350

Dubouzet, J. G., Strabala, T. J., & Wagner, A. (2013). Potential transgenic routes to increase tree biomass. *Plant Science*, *212*, 72–101. DOI: 10.1016/j.plantsci.2013.08.006 PMID: 24094056

Duca, D. R., & Glick, B. R. (2020). Indole-3-acetic acid biosynthesis and its regulation in plant-associated bacteria. *Applied Microbiology and Biotechnology*, *104*(20), 8607–8619. DOI: 10.1007/s00253-020-10869-5 PMID: 32875364

Duke, S. O., Kigel, J., & Koller, D. (2018). Asexual reproduction of weeds. In *Weed physiology* (pp. 65–100). CRC Press. DOI: 10.1201/9781351077743-3

Du, Q., Liu, T., Jiao, X., Song, X., Zhang, J., & Li, J. (2019). Leaf anatomical adaptations have central roles in photosynthetic acclimation to humidity. *Journal of Experimental Botany*, *70*(18), 4949–4962. DOI: 10.1093/jxb/erz238 PMID: 31145790

Dusenge, M. E., Duarte, A. G., & Way, D. A. (2019). Plant carbon metabolism and climate change: Elevated CO 2 and temperature impacts on photosynthesis, photorespiration and respiration. *The New Phytologist*, *221*(1), 32–49. DOI: 10.1111/nph.15283 PMID: 29983005

Ebrahimi, S., Khosravi, M. A., Raz, A., Karimipoor, M., & Parvizi, P. (2023, September 1). CRISPR-Cas Technology as a Revolutionary Genome Editing tool: Mechanisms and Biomedical Applications. *Iranian Biomedical Journal*, *27*(5), 219–246. DOI: 10.61186/ibj.27.5.219 PMID: 37873636

Edler, J., & Fagerberg, J. (2017). Innovation policy: What, why, and how. *Oxford Review of Economic Policy*, *33*(1), 2–23. DOI: 10.1093/oxrep/grx001

Efferth, T. (2019). Biotechnology applications of plant callus cultures. *Engineering (Beijing)*, *5*(1), 50–59. DOI: 10.1016/j.eng.2018.11.006

Egertsdotter, U., Ahmad, I., & Clapham, D. (2019). Automation and scale up of somatic embryogenesis for commercial plant production, with emphasis on conifers. *Frontiers in Plant Science*, *10*, 109. DOI: 10.3389/fpls.2019.00109 PMID: 30833951

Egwu, C. C. N., & Egwu, C. (2018). Genetically Modified Crops Controversy: The Actualization of Intellectual Property Rights Regime.

Eid, A. A. M., Nomier, S. A., Ibrahim, M. M., & Gad, M. M. (2018). Effect of some natural extracts, indolbutiric acid and naphthalene acetic acid on rooting of picual olive cuttings. *Zagazig Journal of Agricultural Research*, *45*(1), 119–136. DOI: 10.21608/zjar.2018.49816

El Bilali, H., Strassner, C., & Ben Hassen, T. (2021). Sustainable agri-food systems: Environment, economy, society, and policy. *Sustainability (Basel)*, *13*(11), 6260. DOI: 10.3390/su13116260

El-Akabawy, M. A. (2000). Effect of some bio fertilizers and farmyard manure on yield and nutrient uptake of Egyptian clover grown on lomy sand soil. *Egypt. J. Agric. Res.*, *78*(5).

ELASI. M. F., Ak, B. E., & Ekinci, H. (2024). Some researches on micropropagation of Orchid (Phalaenopsis spp.) in vitro conditions. In *5 th International Conference on Engineering and Applied Natural Sciences August* (pp. 25-26).

Elçi, A. (2000). Türkiye'de tohumculuğun durumu ve gelişimi. In *Türkiye Ziraat Mühendisliği V* (Vol. 2, pp. 859–870). Teknik Kongresi.

Eliwa, G. I., El-Dengawy, E. F., Gawish, M. S., & Yamany, M. M. (2024, March 7). Comprehensive study on in vitro propagation of some imported peach rootstocks: In vitro explant surface sterilization and bud proliferation. *Scientific Reports*, *14*(1), 5586. DOI: 10.1038/s41598-024-55685-3 PMID: 38454056

El-Sherif, N. (2018). Impact of Plant Tissue Culture on Agricultural Sustainability. 93-107.DOI: 10.1007/698_2017_160

Erdoğan, İ., Cevher-Keskin, B., Bilir, Ö., Hong, Y., & Tör, M. (2023, July 22). Recent Developments in CRISPR/Cas9 Genome-Editing Technology Related to Plant Disease Resistance and Abiotic Stress Tolerance. *Biology (Basel)*, *12*(7), 1037. DOI: 10.3390/biology12071037 PMID: 37508466

Erol, M. H., Dönmez, D., Biçen, B., Şimşek, Ö., & Aka Kaçar, Y. (2023). Modern approaches to In Vitro clonal banana production: Next-generation tissue culture systems. *Horticulturae*, *9*(10), 1154. DOI: 10.3390/horticulturae9101154

Espinosa-Leal, C. A., Puente-Garza, C. A., & García-Lara, S. (2018). In vitro plant tissue culture: Means for production of biological active compounds. *Planta*, *248*(1), 1–18. DOI: 10.1007/s00425-018-2910-1 PMID: 29736623

Evstropov, V., Trushkova, E., & Nikhayeva, A. (2019). Recent research questions of ecological aspects of the interaction of the organism and the environment. *Научный альманах стран Причерноморья*, (3 (19)), 41-51.

Ezell, S., & Cory, N. (2019). *The way forward for intellectual property internationally*. Information Technology and Innovation Foundation.

Faegri, K., & Van Der Pijl, L. (2013). Principles of pollination ecology.

Farooq, M. A., Ali, S., Hassan, A., Tahir, H. M., Mumtaz, S., & Mumtaz, S. (2021). Biosynthesis and industrial applications of α-amylase: A review. *Archives of Microbiology*, *203*(4), 1281–1292. DOI: 10.1007/s00203-020-02128-y PMID: 33481073

Fehér, A. (2019, April 26). Callus, Dedifferentiation, Totipotency, Somatic Embryogenesis: What These Terms Mean in the Era of Molecular Plant Biology? *Frontiers in Plant Science*, *10*, 536. DOI: 10.3389/fpls.2019.00536 PMID: 31134106

Ferrante, A., & Mariani, L. (2018). Agronomic management for enhancing plant tolerance to abiotic stresses: High and low values of temperature, light intensity, and relative humidity. *Horticulturae*, *4*(3), 21. DOI: 10.3390/horticulturae4030021

Fithian, C., & Powell, A. (2009). *Cultural aspects of sustainable development. Student paperfor sustainable design seminar/studio, fall 2009, seminar in sustainable architecture*. University of Texas School of Architecture.

Fjellheim, S., Boden, S., & Trevaskis, B. (2014). The role of seasonal flowering responses in adaptation of grasses to temperate climates. *Frontiers in Plant Science*, *5*, 431. DOI: 10.3389/fpls.2014.00431 PMID: 25221560

Flandrin, J. L., Montanari, M., & Sonnenfeld, A. (Eds.). (1999). *Food: a culinary history from antiquity to the present*. Columbia University Press.

Frankel, R., & Galun, E. (2012). *Pollination mechanisms, reproduction and plant breeding* (Vol. 2). Springer Science & Business Media.

Fuller, E., Germaine, K. J., & Rathore, D. S. (2023). The good, the bad, and the useable microbes within the common Alder (Alnus glutinosa) microbiome—Potential bio-agents to combat Alder dieback. *Microorganisms*, *11*(9), 2187. DOI: 10.3390/microorganisms11092187 PMID: 37764031

Fussy, A., & Papenbrock, J. (2022). An Overview of Soil and Soilless Cultivation Techniques—Chances, Challenges and the Neglected Question of Sustainability. *Plants*, *11*(9), 1153. DOI: 10.3390/plants11091153 PMID: 35567154

Gad, M. M., & Ibrahim, M. M. (2018). Effect of IBA and some natural extracts on rooting and vegetative growth of Picual olive sucker and shoot cuttings. *Current Science International*, *7*(2), 191–203.

Ganie, S. H., Upadhyay, P., Das, S., & Prasad Sharma, M. (2015, December). Authentication of medicinal plants by DNA markers. *Plant Gene*, *4*, 83–99. DOI: 10.1016/j.plgene.2015.10.002 PMID: 32289060

Garban, Z., & Ilia, G. (2024). Structure-Activity of Plant Growth Bioregulators and Their Effects on Mammals. *Molecules (Basel, Switzerland)*, *29*(23), 5671. DOI: 10.3390/molecules29235671 PMID: 39683830

Gardarin, A., Dürr, C., & Colbach, N. (2011). Prediction of germination rates of weed species: Relationships between germination speed parameters and species traits. *Ecological Modelling*, *222*(3), 626–636. DOI: 10.1016/j.ecolmodel.2010.10.005

Gardiner, B., Berry, P., & Moulia, B. (2016). Wind impacts on plant growth, mechanics and damage. *Plant Science*, *245*, 94–118. DOI: 10.1016/j.plantsci.2016.01.006 PMID: 26940495

Gebresenbet, G., Bosona, T., Patterson, D., Persson, H., Fischer, B., Mandaluniz, N., Chirici, G., Zacepins, A., Komasilovs, V., Pitulac, T., & Nasirahmadi, A. (2023). A concept for application of integrated digital technologies to enhance future smart agricultural systems. *Smart Agricultural Technology*, *5*, 100255. DOI: 10.1016/j.atech.2023.100255

Gentili, R., Ambrosini, R., Montagnani, C., Caronni, S., & Citterio, S. (2018). Effect of soil pH on the growth, reproductive investment and pollen allergenicity of Ambrosia artemisiifolia L. *Frontiers in Plant Science*, *9*, 1335. DOI: 10.3389/fpls.2018.01335 PMID: 30294333

George, E. F., Hall, M. A., & Klerk, G. J. D. (2008). The components of plant tissue culture media I: macro-and micro-nutrients. In *Plant Propagation by Tissue Culture: Volume 1. The Background* (pp. 65-113). Dordrecht: Springer Netherlands

George, E. F., Hall, M. A., & De Klerk, G. J. (2008). *Plant propagation by tissue culture* (3rd ed.). The Back Ground Springer.

George, P., & Manuel, J. (2013). Low cost tissue culture technology for the regeneration of some economically important plants for developing countries. *International Journal of Agriculture Environment and Biotechnology*, *6*(Special Issue), 703–711.

Gerace, E., Mancuso, G., Midiri, A., Poidomani, S., Zummo, S., & Biondo, C. (2022, June 8). Recent Advances in the Use of Molecular Methods for the Diagnosis of Bacterial Infections. *Pathogens (Basel, Switzerland)*, *11*(6), 663. DOI: 10.3390/pathogens11060663 PMID: 35745518

Getabalew, M., Alemneh, T., & Zewdie, D. (2020). Effect of rangeland vegetation cover on climate change mitigation.

Ghimire, B. K., Yu, C. Y., Kim, W.-R., Moon, H.-S., Lee, J., Kim, S. H., & Chung, I. M. (2023). Assessment of Benefits and Risk of Genetically Modified Plants and Products: Current Controversies and Perspective. *Sustainability (Basel)*, *15*(2), 1722. DOI: 10.3390/su15021722

Ghosh, T., Samai, I., Debnath, S., Singh, A. P., & Atta, K. (2025). Synthetic Seeds: Innovation in Plant Propagation and Conservation for Sustainable Agriculture. In *Advances in Seed Quality Evaluation and Improvement* (pp. 431–446). Springer Nature Singapore. DOI: 10.1007/978-981-96-5178-8_17

Gibson, A. K. (2022). Genetic diversity and disease: The past, present, and future of an old idea. *Evolution; International Journal of Organic Evolution*, *76*(S1, s1), 20–36. DOI: 10.1111/evo.14395 PMID: 34796478

Gilliham, M., Able, J. A., & Roy, S. J. (2017). Translating knowledge about abiotic stress tolerance to breeding programmes. *The Plant Journal, 90*(5), 898–917. DOI: 10.1111/tpj.13456 PMID: 27987327

Giordano, M., Petropoulos, S. A., & Rouphael, Y. (2021). Response and Defence Mechanisms of Vegetable Crops against Drought, Heat and Salinity Stress. *Agriculture, 11*(5), 463. DOI: 10.3390/agriculture11050463

Gollin, M. A. (2008). *Driving innovation: intellectual property strategies for a dynamic world.* Cambridge University Press., DOI: 10.1017/CBO9780511805394

Gómez, J. M., Schupp, E. W., & Jordano, P. (2019). Synzoochory: The ecological and evolutionary relevance of a dual interaction. *Biological Reviews of the Cambridge Philosophical Society, 94*(3), 874–902. DOI: 10.1111/brv.12481 PMID: 30467946

Gopal, G. V., & Sahadevan, N. (2025). *Plant Tissue Culture basic Laboratory Techniques and Advances in Biotechnological Standards for GM Crops.* Crown Publishing.

GRAIN. L. V. C. (2015). *Seed laws that criminalize farmers: resistance and fight back.* https://grain.org/en/article/5142-seed -laws -that -criminalise -farmers -resistance-and-fightback

Granieri, M. (2015). Genetically modified seeds, intellectual property protection and the role of law in transnational perspective. In *Law, Development and Innovation (pp. 89–113). Springer International Publishing.*

Gratani, L. (2014). Plant phenotypic plasticity in response to environmental factors. *Advances in Botany, 2014*(1), 208747. DOI: 10.1155/2014/208747

Grath, A., & Dai, G. (2019). Direct cell reprogramming for tissue engineering and regenerative medicine. *Journal of Biological Engineering, 13*(1), 14. DOI: 10.1186/s13036-019-0144-9 PMID: 30805026

Grech, A., Howse, E., & Boylan, S. (2020). A scoping review of policies promoting and supporting sustainable food systems in the university setting. *Nutrition Journal, 19*(1), 1–13. DOI: 10.1186/s12937-020-00617-w PMID: 32912299

Green, A. J., & Wilkinson, D. M. (2024). Darwin's digestion myth: Historical and modern perspectives on our understanding of seed dispersal by waterbirds. *Seeds (New York, N.Y.), 3*(4), 505–527. DOI: 10.3390/seeds3040034

Grega, L., Anderson, S., Cheetham, M., Clemente, M., Colletti, A., Moy, W., Talarico, D., Thatcher, S. L., & Osborn, J. M. (2013). Aerodynamic characteristics of saccate pollen grains. *International Journal of Plant Sciences, 174*(3), 499–510. DOI: 10.1086/668694

Grierson, D. (2009). Towards a sustainable built environment. *CIC Start Online Innovation Rev, 1*, 70–77.

Grigoriev, S. N., Fedorov, S. V., & Hamdy, K. (2019). Materials, properties, manufacturing methods and cutting performance of innovative ceramic cutting tools– a review. *Manufacturing Review, 6*, 19. DOI: 10.1051/mfreview/2019016

Grossnickle, S. C., & MacDonald, J. E. (2018). Seedling quality: History, application, and plant attributes. *Forests, 9*(5), 283. DOI: 10.3390/f9050283

Grzelak, M., Pacholczak, A., & Nowakowska, K. (2024). Challenges and insights in the acclimatization step of micropropagated woody plants. [PCTOC]. *Plant Cell, Tissue and Organ Culture, 159*(3), 72. DOI: 10.1007/s11240-024-02923-1

Gudynaitė-Franckevičienė, V., & Pliūra, A. (2021). The Impact of different environmental conditions during vegetative propagation on growth, survival, and biochemical characteristics in Populus hybrids in clonal field trial. *Forests, 12*(7), 892. DOI: 10.3390/f12070892

Gulati, R. (2019). BIOPIRACY, A BIOLOGICAL THEFT? [IJOLS]. *International Journal of Legal Studies, 5*(1), 317–350.

Guo, Y. W., Liu, Y., Huang, P. C., Rong, M., Wei, W., Xu, Y. H., & Wei, J. H. (2025). Adaptive Changes and Genetic Mechanisms in Organisms Under Controlled Conditions: A Review. *International Journal of Molecular Sciences, 26*(5), 2130. DOI: 10.3390/ijms26052130 PMID: 40076752

Gupta, A. K. (2004). *WIPO-UNEP study on the role of intellectual property rights in the sharing of benefits arising from the use of biological resources and associated traditional knowledge* (Vol. 769). WIPO.

Gupta, S., Schillaci, M., & Roessner, U. (2022, April 15). Metabolomics as an emerging tool to study plant-microbe interactions. *Emerging Topics in Life Sciences, 6*(2), 175–183. DOI: 10.1042/ETLS20210262 PMID: 35191478

Halewood, M., & Lapeña, I. (2016). Farmers' varieties and farmers' rights: Challenges at the crossroads of agriculture, taxonomy and law. In *Farmers' Crop Varieties and Farmers' Rights* (pp. 15-38). Routledge.

Hallman, W. K., Adelaja, A. O., & Schilling, B. J. (2002, August). Public perceptions of agricultural biotechnology in the United States in 2001. In *XXVI International Horticultural Congress: Biotechnology in Horticultural Crop Improvement: Achievements, Opportunities and 625* (pp. 445-455).

Hamany Djande, C. Y., Pretorius, C., Tugizimana, F., Piater, L. A., & Dubery, I. A. (2020). Metabolomics: A Tool for Cultivar Phenotyping and Investigation of Grain Crops. *Agronomy (Basel), 10*(6), 831. DOI: 10.3390/agronomy10060831

Hanapi, S. Z., Awad, H. M., Ali, S. I. S., Sarip, S. H. M., Sarmidi, M. R., & Aziz, R. (2013). Agriculture wastes conversion for biofertilizer production using beneficial microorganisms for sustainable agriculture applications. Malaysian Journal of Microbiology, 60-67.

Hanson, B. (Ed.). (2007). *Buried Treasures: Tasty Tubers of the World: how to Grow and Enjoy Root Vegetables, Tubers, Rhizomes, and Corms* (Vol. 188). Brooklyn Botanic Garden.

Hardinsyah, H., Dwikurnia, F., & Kurniawan, A. (2007). Faktor-faktor penentu pola konsumsi pangan di Indonesia. [Factors determining food consumption patterns in Indonesia] In Thaha, R., Jalal, A., & Setiawan, B. (Eds.), *Pembangunan gizi dan pangan dalam perspektif kemandirian lokal* (pp. 45–62). IPB Press.

Hari, M., Seshadri, S., & Perumal, K. (2010). Booklet on Bio-fertilizer (Phospho-Bacteria).Shri AMM Murugappa Chettiar Research Center. *Taramani, Chennai, 600*, 113.

Hart, J. W. (2012). *Light and plant growth* (Vol. 1). Springer Science & Business Media.

Hasan, N., Choudhary, S., Naaz, N., Sharma, N., & Laskar, R. A. (2021, August 27). Recent advancements in molecular marker-assisted selection and applications in plant breeding programmes. *Journal of Genetic Engineering and Biotechnology, 19*(1), 128. DOI: 10.1186/s43141-021-00231-1 PMID: 34448979

Hasnain, A., Naqvi, S. A. H., Ayesha, S. I., Khalid, F., Ellahi, M., Iqbal, S., Zeeshan, H. M.,Abbas, A., Adamski, R., Markowska, D., Baazeem, A., Mustafa, G., MOustafa, M., Hasan, M.E., & Abdelhamid, M. M. (2022). Plants in vitro propagation with its applications in food, pharmaceuticals and cosmetic industries; current scenario and future approaches. *Frontiers in plant science, 13*, 1009395. DOI: 10.3389/fpls.2022.1009395

Hauser, R., & Banse, G. (2011). Culture and culturality: approaching a multi-faceted concept. In *Parodi, O., Ayestaran, I. Banse, G.B., Sustainable Development- Relationships to Culture, Knowledge and Ethics*. KIT Scientific Publishing. DOI: 10.5771/9783845271156-31

Hawkes, J. (2001). *The fourth pillar of sustainability: culture's essential role in public planning*. Common Ground Press.

Hayat, M. K., Khan, T. Y., Sial, T. N., Khan, H. A., Ashraf, Z. U., Sheraz, M., & Belizaire, P. J. (2025). The Biotech And Precision Agriculture Revolution: Feeding the World Sustainably. *Policy Research Journal, 3*(3), 293–316.

He, Y. (2023). Unraveling the interplay between food security, agriculture, trade policy, and energy consumption: An environmental sustainability insight. *Energy & Environment,* 0958305X231195604.

Henao-Ramírez, A. M., & Urrea-Trujillo, A. I. (2020). Somatic Embryogenesis for Clonal Propagation and Associated Molecular Studies in Cacao (*Theobroma cacao L.*). In Chong, P., Newman, D., & Steinmacher, D. (Eds.), *Agricultural, Forestry and Bioindustry Biotechnology and Biodiscovery*. Springer., DOI: 10.1007/978-3-030-51358-0_5

Herrera-Estrella, L., & Alvarez-Morales, A. (2001). Genetically modified crops: Hope for developing countries? *EMBO Reports, 2*(4), 256–258. DOI: 10.1093/embo-reports/kve075 PMID: 11306538

Hesami, M., Baiton, A., Alizadeh, M., Pepe, M., Torkamaneh, D., & Jones, A. M. P. (2021). Advances and Perspectives in Tissue Culture and Genetic Engineering of Cannabis. *International Journal of Molecular Sciences, 22*(11), 5671. DOI: 10.3390/ijms22115671 PMID: 34073522

Hillel, D. (2013). *Introduction to soil physics*. Academic press.

Hosseini, N., & Rezanejad, F., & ZamaniBahramabadi, E. (2022). Effects of soil texture, irrigation intervals, and cultivar on some nut qualities and different types of fruit blankness in pistachio (Pistacia vera L.). *International Journal of Horticultural Science and Technology, 9*(1), 41–53.

Howitt, C. A., & Miskelly, D. (2017). Identification of grain variety and quality type. In *Cereal Grains* (pp. 453–492). Woodhead Publishing., DOI: 10.1016/B978-0-08-100719-8.00017-6

Huang, W., & Wang, X. (2024). The Impact of Technological Innovations on Agricultural Productivity and Environmental Sustainability in China. *Sustainability (Basel), 16*(19), 8480. DOI: 10.3390/su16198480

Hussain, W. S., & Abbas, M. M. (2025). Importance of Aeroponics in Vegetable Crops. In *Utilizing Aeroponics Techniques for Improved Farming* (pp. 57–70). IGI Global Scientific Publishing. DOI: 10.4018/979-8-3693-2320-5.ch003

Hussain, W. S., Ahmed, H. F., Al-Saffar, R. S., & Taher, N. A. H. A. (2023). Effect of bio fertilization using banana and apple peels on flax plant's vegetative characteristics. *Al-Azhar Journal of Agricultural Research*, *48*(3), 395–400.

Hussain, W. S., & Mohammad, A. K. (2024). Sustainable Agriculture Is a Way to Increase Wheat Production. In *Achieving Food Security Through Sustainable Agriculture* (pp. 252–267). IGI Global. DOI: 10.4018/979-8-3693-4240-4.ch010

Hwang, H. H., Yu, M., & Lai, E. M. (2017, October 20). Agrobacterium-mediated plant transformation: Biology and applications. *The Arabidopsis Book / American Society of Plant Biologists*, *15*, e0186. DOI: 10.1199/tab.0186 PMID: 31068763

Ievinsh, G. (2023). Water content of plant tissues: So simple that almost forgotten? *Plants*, *12*(6), 1238. DOI: 10.3390/plants12061238 PMID: 36986926

Ikeuchi, M., Sugimoto, K., & Iwase, A. (2013, September). Plant callus: Mechanisms of induction and repression. *The Plant Cell*, *25*(9), 3159–3173. DOI: 10.1105/tpc.113.116053 PMID: 24076977

Imran, Y., Wijekoon, N., Gonawala, L., Chiang, Y. C., & De Silva, K. R. D. (2021). Biopiracy: Abolish Corporate Hijacking of Indigenous Medicinal Entities. *TheScientificWorldJournal*, *2021*(1), 8898842. DOI: 10.1155/2021/8898842 PMID: 33679261

Ingold, T. (2024). Growing plants and raising animals: an anthropological perspective on domestication. In *The origins and spread of agriculture and pastoralism in Eurasia* (pp. 12–24). Routledge. DOI: 10.4324/9781003579861-3

Ingram, J., & Maye, D. (2020). What are the implications of digitalisation for agricultural knowledge? *Frontiers in Sustainable Food Systems*, *4*, 66. DOI: 10.3389/fsufs.2020.00066

Iqbal, M., & Singh, K. K. (2020). Propagation of temperate fruit crops. *Innovative agriculture and botany*, 119-135.

Iqbal, M., Aftab, Z.-H., Anjum, T., Rizwana, H., Akram, W., Aftab, A., Sajid, Z. A., & Li, G. (2024). Nano-Integrated plant tissue culture to increase the rate of callus induction, growth, and curcuminoid production in Curcuma longa. *Plants*, *13*(13), 1819. DOI: 10.3390/plants13131819 PMID: 38999659

Iskra, M., Voigt, L., & Raab, M. (2024). Accounting for dynamic cognition–action interaction in decision-making tasks in sports: A scoping review. *Sport, Exercise, and Performance Psychology*. Advance online publication. DOI: 10.1037/spy0000345

Istiandari, P., & Faizal, A. (2025). Integrating In Vitro Cultivation and Sustainable Field Practices of Sacha Inchi (*Plukenetia volubilis* L.) for Enhanced Oil Yield and Quality: A Review. *Horticulturae*, *11*(2), 194. DOI: 10.3390/horticulturae11020194

Jain, S., Sharma, R., Nidhi, N., Behera, S. D., Baksh, H., Sharma, R., & Kumar, V. (2023). Advancement in tissue culture techniques for fruit crops. *Int J Environ Clim Change*, *13*(11), 4396–4407. DOI: 10.9734/ijecc/2023/v13i113620

Jain, S., Singh, H., Rathod, M., Meena, R., Deshmukh, R. N., Mohapatra, A., Sharma, R., & Anushi, . (2023). Preserving for the future: The critical role of germplasm conservation in fruit crop resilience. *Int J Environ Clim Change*, *13*(11), 4651–4661. DOI: 10.9734/ijecc/2023/v13i113644

Jajoriya, R., Sharma, R., Meena, V. K., Vadithya, A. S., Singh, K., Kumar, P., & Nathawat, T. S. (2025). EMERGENCE OF MOLECULAR MARKER FROM RFLP TO SNP: A REVIEW. *Plant Archives*, *25*(1), 2062–2074.

Jakubowska, Z., Gradowski, M., & Dobrzyński, J. (2025). Role of plant growth-promoting bacteria (PGPB) in enhancing phenolic compounds biosynthesis and its relevance to abiotic stress tolerance in plants: A review. *Antonie van Leeuwenhoek*, *118*(9), 1–17. DOI: 10.1007/s10482-025-02130-8 PMID: 40707699

Jaskani, M. J., & Khan, I. A. (2021). Horticulture: an overview. *University of Agriculture Faisalabad*, 3-22.

Javed, A., Ali, E., Afzal, K. B., Osman, A., & Riaz, S. (2022). Soil fertility: Factors affecting soil fertility, and biodiversity responsible for soil fertility. *International Journal of Plant. Animal and Environmental Sciences*, *12*(1), 21–33.

Jhanji, S., Goyal, E., Chumber, M., & Kaur, G. (2024). Exploring fine tuning between phytohormones and ROS signaling cascade in regulation of seed dormancy, germination and seedling development. *Plant Physiology and Biochemistry*, *207*, 108352. DOI: 10.1016/j.plaphy.2024.108352 PMID: 38266558

Jiang, Z., Li, Y. P., Gai, P. Z., Gao, J., & Xu, L. (2025). Exogenously applied ABA alleviates dysplasia of maize (Zea mays L.) ear under drought stress by altering photosynthesis and sucrose transport. *Plant Signaling & Behavior*, *20*(1), 2462497. DOI: 10.1080/15592324.2025.2462497 PMID: 39907639

Jie, L., & Haoran, M. (2025). Deciphering popular routes in urban parks: The impact of environmental factors on the amount, intensity and diversity of physical activity. *Urban Forestry & Urban Greening, 105*, 128684. DOI: 10.1016/j.ufug.2025.128684

Jin, L., Sharpe, A. G., Ji, Y., Hewavithana, T., Sharpe, A. G., and Jin, L. (2025). Yuanyuan Ji1, 2, Thulani Hewavithana1, Andrew G. Sharpe2* and Lingling Jin1. *Omics in Seed Development: Challenges and Opportunities for Improving of Seed Quality and Yield in Model and Crop Plants.*

Jin, X., Cheng, Q., Zhao, B., Ji, J., & Li, M. (2020). Design and test of 2ZYM-2 potted vegetable seedlings transplanting machine. *International Journal of Agricultural and Biological Engineering, 13*(1), 101–110. DOI: 10.25165/j.ijabe.20201301.5494

Johari-Pireivatlou, M. (2010). Effect of soil water stress on yield and proline content of four wheat lines. *African Journal of Biotechnology, 9*(1).

Johnston, S. E. (2024). Understanding the genetic basis of variation in meiotic recombination: Past, present, and future. *Molecular Biology and Evolution, 41*(7), msae112. DOI: 10.1093/molbev/msae112 PMID: 38959451

Johri, B. M., & Srivastava, P. S. (2012). *Reproductive Biology of Plants.* Narosa Publishing House.

Jones, J. R. (2012). *Plant nutrition and soil fertility manual.* CRC press. DOI: 10.1201/b11577

Jordano, P. (2017). What is long-distance dispersal? And a taxonomy of dispersal events. *Journal of Ecology, 105*(1), 75–84. DOI: 10.1111/1365-2745.12690

Joshi, B. K., Ghimire, K. H., Neupane, S. P., Gauchan, D., & Mengistu, D. K. (2023). Approaches and advantages of increased crop genetic diversity in the fields. *Diversity, 15*(5), 603. DOI: 10.3390/d15050603

Kaiser, N., Douches, D., Dhingra, A., Glenn, K. C., Herzig, P. R., Stowe, E. C., & Swarup, S. (2020). The role of conventional plant breeding in ensuring safe levels of naturally occurring toxins in food crops. *Trends in Food Science & Technology, 100*, 51–66. DOI: 10.1016/j.tifs.2020.03.042

Kakoulidou, I., Avramidou, E. V., Baránek, M., Brunel-Muguet, S., Farrona, S., Johannes, F., Kaiserli, E., Lieberman-Lazarovich, M., Martinelli, F., Mladenov, V., Testillano, P. S., Vassileva, V., & Maury, S. (2021, August 11). Epigenetics for Crop Improvement in Times of Global Change. *Biology (Basel), 10*(8), 766. DOI: 10.3390/biology10080766 PMID: 34439998

Kandar, C. C., & Pal, D. (2024). Relation Between Seed Life Cycle and Cell Proliferation. Metabolic Changes in Seed Germination. In *Seeds: anti-proliferative storehouse for bioactive secondary metabolites* (pp. 49–79). Springer Nature Singapore. DOI: 10.1007/978-981-97-3014-8_2

Kapazoglou, A., Tani, E., Papasotiropoulos, V., Letsiou, S., Gerakari, M., Abraham, E., & Bebeli, P. J. (2025). Enhancing abiotic stress resilience in Mediterranean woody perennial fruit crops: Genetic, epigenetic, and microbial molecular perspectives in the face of climate change. *International Journal of Molecular Sciences, 26*(7), 3160. DOI: 10.3390/ijms26073160 PMID: 40243927

Karami, O., Aghavaisi, B., & Mahmoudi Pour, A. (2009, November). Molecular aspects of somatic-to-embryogenic transition in plants. *Journal of Chemical Biology, 2*(4), 177–190. DOI: 10.1007/s12154-009-0028-4 PMID: 19763658

Kasoju, N., & Sunilkumar, A. (2024). *Convergence of tissue engineering and sustainable development goals*. Biotechnology for Sustainable Materials., DOI: 10.1186/s44316-024-00021-y

Kaur, S., Ijaz, S., & Nasir, B. (2024). Tissue Culture and Somatic Fusion in Plants. In Ijaz, S., Ul Haq, I., & Mohamed Ali, H. (Eds.), *Trends in Plant Biotechnology*. Springer., DOI: 10.1007/978-981-97-0814-7_1

Kauth, P. J., & Biber, P. D. (2015). Moisture content, temperature, and relative humidity influence seed storage and subsequent survival and germination of Vallisneria americana seeds. *Aquatic Botany, 120*, 297–303. DOI: 10.1016/j.aquabot.2014.09.009

Kavhiza, N. J., Zargar, M., Prikhodko, S. I., Pakina, E. N., Murtazova, K. M.-S., & Nakhaev, M. R. (2022). Improving Crop Productivity and Ensuring Food Security through the Adoption of Genetically Modified Crops in Sub-Saharan Africa. *Agronomy (Basel), 12*(2), 439. DOI: 10.3390/agronomy12020439

Kendrick, R. E., & Kronenberg, G. H. (Eds.). (2012). *Photomorphogenesis in plants*. Springer Science & Business Media.

Khadr, A., Wang, G. L., Wang, Y. H., Zhang, R. R., Wang, X. R., Xu, Z. S., Tian, Y. S., & Xiong, A. S. (2020, December 8). Effects of auxin (indole-3-butyric acid) on growth characteristics, lignification, and expression profiles of genes involved in lignin biosynthesis in carrot taproot. *PeerJ, 8*, e10492. DOI: 10.7717/peerj.10492 PMID: 33354430

Khaitov, M., Rasulov, I., & Karshiyev, J. (2024). Technology and methodology of propagation of fruit seedlings. *Science and innovation, 3*(D9), 276-279.

Khalil, M., Sarwar, M. S., Tariq, H., Farooq, M., & Ullah, M. (2023). A new method for plant breeding through the use of plant tissue culture. *Journal of Life and Social Sciences, 2023*(1), 20–20. DOI: 10.64013/bbasrjlifess.v2023i1.20

Khalimova, Y. S. (2024). Features of Sperm Development: Spermatogenesis and Fertilization. *American Journal of Bioscience and Clinical Integrity, 1*(11), 90–98.

Khan, A., & Korban, S. S. (2022). Breeding and genetics of disease resistance in temperate fruit trees: challenges and new opportunities. *Theoretical and Applied Genetics, 135*(11), 3961-3985. DOI: 10.1007/s00122-022-04093-0

Khan, M. A., Kumar, R., & Singh, P. (2024). Plant tissue culture and micropropagation as tools for sustainable agriculture and biodiversity conservation. *Sustainability, 16*(3), 1120. DOI: 10.3390/su16031120

Khan, S., Ullah, M. W., Siddique, R., Nabi, G., Manan, S., Yousaf, M., & Hou, H. (2016). Role of Recombinant DNA Technology to Improve Life. *International Journal of Genomics, 2016*, 2405954. DOI: 10.1155/2016/2405954 PMID: 28053975

Khatik, N. (2023). Role of Plant Biotechnology in Agricultural Sustainability and Improvement of Crops. *Journal of Agricultural Research, 11*(6), 929–936.

Khezri, M., Asghari-Zakaria, R., & Zare, N. (2024). Plant Cell and Tissue Culture: Propagation, Improvement, and Conservation of Medicinal Plants. In Kumar, N. (Ed.), *Biosynthesis of Natural Products in Plants*. Springer., DOI: 10.1007/978-981-97-2166-5_11

Khuder, J. K. H., & Mohammed, A. A. (2025). Isolation Endophytic Bacteria from Leaf Glands and Seeds of *Ardisia crenata* Plant and its Molecular Detection. *Egyptian Journal of Veterinary Science, 56*(6), 1265–1270. DOI: 10.21608/ejvs.2024.282284.2009

Kim, E., & Donohue, K. (2013). Local adaptation and plasticity of Erysimum capitatum to altitude: Its implications for responses to climate change. *Journal of Ecology, 101*(3), 796–805. DOI: 10.1111/1365-2745.12077

Kim, K. D., Kang, Y., & Kim, C. (2020, October 28). Application of Genomic Big Data in Plant Breeding:Past, Present, and Future. *Plants, 9*(11), 1454. DOI: 10.3390/plants9111454 PMID: 33126607

Kirkby C A & Fattore, A. (2006). Effect of Rice Stubble Burning on Soil Health . A report for the Rural Industries Research and Development Corporation. RIRDC Publication

Kirkby C A and Fattore, A. (2006). Effect of Rice Stubble Burning on Soil Health . A report for the Rural Industries Research and Development Corporation. RIRDC Publication

Klimek-Chodacka, M., Kadluczka, D., Lukasiewicz, A., Malec-Pala, A., Baranski, R., & Grzebelus, E. (2020). Effective callus induction and plant regeneration in callus and protoplast cultures of *Nigella damascena* L. *Plant Cell, Tissue and Organ Culture*, *143*(3), 693–707. DOI: 10.1007/s11240-020-01953-9

Klupczyńska, E. A., & Pawłowski, T. A. (2021). Regulation of seed dormancy and germination mechanisms in a changing environment. *International Journal of Molecular Sciences*, *22*(3), 1357. DOI: 10.3390/ijms22031357 PMID: 33572974

Kolb, M. (2018). What is globalization? And how has the global economy shaped the United States. *Peterson Institute for International Economics*.https://www. piie .com/microsites/globalization/what-is-globalization

Körner, C. (2016). Plant adaptation to cold climates. [-Faculty.]. *F1000 Research*, *5*, F1000. DOI: 10.12688/f1000research.9107.1 PMID: 27990251

Kovincic, A., Markovic, K., Ristic, D., Babic, V., Petrovic, T., Zivanovic, T., & Kravic, N. (2023). Efficiency of biological typing methods in maize hybrid genetic purity estimation. *Genes*, *14*(6), 1195. DOI: 10.3390/genes14061195 PMID: 37372375

Krasteva, G., Georgiev, V., & Pavlov, A. (2020, December 18). Recent applications of plant cell culture technology in cosmetics and foods. *Engineering in Life Sciences*, *21*(3-4), 68–76. DOI: 10.1002/elsc.202000078 PMID: 33716606

Krishnamoorthy, M., Palanisamy, K., Francis, A. P., & Gireesan, K. (2016). Impact of environmental factors and altitude on growth and reproductive characteristics of teak (Tectona grandis Linn. f.) in Southern India. *Journal of Forest and Environmental Science*, *32*(4), 353–366. DOI: 10.7747/JFES.2016.32.4.353

Krömer, T., Acebey, A., Kluge, J., & Kessler, M. (2013). Effects of altitude and climate in determining elevational plant species richness patterns: A case study from Los Tuxtlas, Mexico. Flora-Morphology, Distribution. *Flora (Jena)*, *208*(3), 197–210. DOI: 10.1016/j.flora.2013.03.003

Kruft, D. (2001). Impacts of genetically-modified crops and seeds on farmers. *The Agricultural Law Resource and Reference Center,* 1-11.

Kubota, C., McClure, M. A., Kokalis-Burelle, N., Bausher, M. G., & Rosskopf, E. N. (2008). Vegetable grafting: History, use, and current technology status in North America. *HortScience*, *43*(6), 1664–1669. DOI: 10.21273/HORTSCI.43.6.1664

Kulus, D., & Tymoszuk, A. (2024). Advancements in In Vitro Technology: A Comprehensive Exploration of Micropropagated Plants. *Horticulturae*, *10*(1), 88. DOI: 10.3390/horticulturae10010088

Kumar, D., Chakradhar, P., Ranganna, G., Vimal, V., Raj, R., Anusha, C., Patra, S., & Nagaraju, V. (2024). Tissue Culture in Banana Cultivation: A Review of its Impact on Disease Management, Yield Improvement, and Sustainable Production. *Journal of Advances in Biology & Biotechnology*, *27*(9), 628–644. Advance online publication. DOI: 10.9734/jabb/2024/v27i91336

Kumar, R., Das, S. P., Choudhury, B. U., Kumar, A., Prakash, N. R., Verma, R., Chakraborti, M., Devi, A. G., Bhattacharjee, B., Das, R., Das, B., Devi, H. L., Das, B., Rawat, S., & Mishra, V. K. (2024, November 7). Advances in genomic tools for plant breeding: Harnessing DNA molecular markers, genomic selection, and genome editing. *Biological Research*, *57*(1), 80. DOI: 10.1186/s40659-024-00562-6 PMID: 39506826

Kumar, S., Kumar, R., Bibwe, B. R., Nath, P., Singh, R. K., Mandhania, S., Pal, A., Soni, R., & Kumar, A. (2024). Postharvest handling of ethylene with oxidative and absorptive means. *Journal of Food Science and Technology*, *61*(5), 813–832. DOI: 10.1007/s13197-023-05777-1 PMID: 38487289

Kumlehn, J., Pietralla, J., Hensel, G., Pacher, M., & Puchta, H. (2018). The CRISPR/Cas revolution continues: From efficient gene editing for crop breeding to plant synthetic biology. *Journal of Integrative Plant Biology*, *60*(12), 1127–1153. DOI: 10.1111/jipb.12734 PMID: 30387552

Kutschera, U., & Ray, P. M. (2022, January). Forever young: Stem cell and plant regeneration one century after Haberlandt 1921. *Protoplasma*, *259*(1), 3–18. DOI: 10.1007/s00709-021-01683-5 PMID: 34292403

Lakehal, A., & Bellini, C. (2019). Control of adventitious root formation: Insights into synergistic and antagonistic hormonal interactions. *Physiologia Plantarum*, *165*(1), 90–100. DOI: 10.1111/ppl.12823 PMID: 30159890

Lamont, B. B., & Pausas, J. G. (2023). Seed dormancy revisited: Dormancy-release pathways and environmental interactions. *Functional Ecology*, *37*(4), 1106–1125. DOI: 10.1111/1365-2435.14269

Lange, O. L., Kappen, L., & Schulze, E. D. (Eds.). (2012). *Water and plant life: problems and modern approaches* (Vol. 19). Springer Science & Business Media.

Lavall, T. P., Fontanela, C., & Marocco, A. D. A. L. (2022). Perspectives for the Effectiveness of the National Intellectual Property Strategy: The 2021-2023 Biannal Plan and the Role of the Technological Innovation Centers. *Direito e Desenvolvimento, 13*, 70.

Lawson, C., & Bikundo, E. (2019). The Substance and Content of Farmers' Rights–A Framework? *Griffith University Law School Research Paper.*

Lawson, C. (2015). Implementing Farmers' Rights: Finding Meaning and Purpose for the International Treaty on Plant Genetic Resources for Food and Agriculture Commitments? *Available at SSRN* 2626668.

Legris, M., Nieto, C., Sellaro, R., Prat, S., & Casal, J. J. (2017). Perception and signalling of light and temperature cues in plants. *The Plant Journal, 90*(4), 683–697. DOI: 10.1111/tpj.13467 PMID: 28008680

Leljak-Levanić, D., Mihaljević, S., & Bauer, N. (2015). Somatic and zygotic embryos share common developmental features at the onset of plant embryogenesis. *Acta Physiologiae Plantarum, 37*(7), 127. DOI: 10.1007/s11738-015-1875-y

Lesser, W. (1987). The impacts of seed patents. *North Central Journal of Agricultural Economics, 9*(1), 37–48. DOI: 10.2307/1349340

Lewsey, M. G., Bassel, G. W., & Whelan, J. (2025). Dynamic and spatial control of cellular activity during seed germination. *Current Opinion in Plant Biology, 86*, 102754. DOI: 10.1016/j.pbi.2025.102754 PMID: 40555155

Li, B., Lecourt, J., & Bishop, G. (2018). Advances in non-destructive early assessment of fruit ripeness towards defining optimal time of harvest and yield prediction—A review. *Plants, 7*(1), 3. DOI: 10.3390/plants7010003 PMID: 29320410

Li, B., Sun, C., Li, J., & Gao, C. (2024). Targeted genome-modification tools and their advanced applications in crop breeding. *Nature Reviews. Genetics, 25*(9), 603–622. DOI: 10.1038/s41576-024-00720-2 PMID: 38658741

Li, H., Chang, L., Wei, Y., & Li, Y. (2023). Interacting effects of land use type, soil attributes, and environmental factors on aggregate stability. *Land (Basel), 12*(7), 1286. DOI: 10.3390/land12071286

Liu, Q., Jin, X., Cheng, J., Zhou, H., Zhang, Y., & Dai, Y. (2023, May). Advances in the application of molecular diagnostic techniques for the detection of infectious disease pathogens [Review]. *Molecular Medicine Reports, 27*(5), 104. DOI: 10.3892/mmr.2023.12991 PMID: 37026505

Liu, S., Li, X., Xu, L., & Zhang, G. (2024). Hormone functions in adventitious root formation during cutting propagation of woody plants. *Journal of Plant Research*, •••, 1–8. DOI: 10.1007/s10265-024-01602-8 PMID: 39656343

Liu, W., Liu, J., Yin, D., & Zhao, X. (2015). Influence of ecological factors on the production of active substances in the anti-cancer plant Sinopodophyllum hexandrum (Royle) TS Ying. *PLoS One*, *10*(4), e0122981. DOI: 10.1371/journal.pone.0122981 PMID: 25874701

Li, W., Li, W. B., Xing, L. J., & Guo, S. X. (2023). Effect of arbuscular mycorrhizal fungi (AMF) and plant growth-promoting rhizobacteria (PGPR) on microorganism of phenanthrene and pyrene contaminated soils. *International Journal of Phytoremediation*, *25*(2), 240–251. DOI: 10.1080/15226514.2022.2071832 PMID: 35549569

Li, Y. Y., Hao, Z. G., Miao, S., Zhang, X., Li, J. Q., Guo, S. X., & Lee, Y. I. (2022). Profiles of cytokinins metabolic genes and endogenous cytokinins dynamics during shoot multiplication in vitro of Phalaenopsis. *International Journal of Molecular Sciences*, *23*(7), 3755. DOI: 10.3390/ijms23073755 PMID: 35409120

Li, Z. G., & Gong, M. (2011). Mechanical stimulation-induced cross-adaptation in plants: An overview. *Journal of Plant Biology*, *54*(6), 358–364. DOI: 10.1007/s12374-011-9178-3

Lohani, N., Singh, M. B., & Bhalla, P. L. (2020). High temperature susceptibility of sexual reproduction in crop plants. *Journal of Experimental Botany*, *71*(2), 555–568. DOI: 10.1093/jxb/erz426 PMID: 31560053

Lokare, P. D., & Pandya, J. B. (2024). *Entrepreneurial Botany*. Book Saga Publications. DOI: 10.60148/978-8197603839

Long, Y., Yang, Y., Pan, G., & Shen, Y. (2022). New insights into tissue culture plant-regeneration mechanisms. *Frontiers in Plant Science*, *13*, 926752. DOI: 10.3389/fpls.2022.926752 PMID: 35845646

Lordanis, C., Io, K., Anastassia, V., & Wilhelm, H. (2013) Commissionnon genetic resources for food and agriculture, Background study paper. FAO, 64, 2-23.

Lor, P. (2004). Storehouses of knowledge? The role of libraries in preserving and promoting indigenous knowledge. *Indilinga*, *3*(1), 45–56.

Lowry, W. P. (2013). *Weather and life: An introduction to biometeorology*. Elsevier.

Lucas, J. A., Garcia-Villaraco, A., Montero-Palmero, M. B., Montalban, B., Ramos Solano, B., & Gutierrez-Mañero, F. J. (2023). Physiological and genetic modifications induced by plant-growth-promoting rhizobacteria (PGPR) in tomato plants under moderate water stress. *Biology (Basel)*, *12*(7), 901. DOI: 10.3390/biology12070901 PMID: 37508334

Madzivire, P. K. (2021). *In Vitro Propagation of Haemanthus pumilio and H. albiflos (Amaryllidaceae) and the population genetics of H. pumilio* (Doctoral dissertation, Stellenbosch: Stellenbosch University).

Mahdi, G.A. and Yahya, R.T. (2024). Effect of zinc oxide nanoparticles on soybean's biomolecules and oil content (*Glycine max* L.) seedlings tissues. Malaysian Journal of Microbiology. 20(Special issue 6): 1–5.

Mahoney, S. E. (2018). Owning the World's Seed Supply: How Seed Industry Mergers Threaten Global Food Security. *Geo. Envtl. L. Rev.*, *31*, 563.

Maity, A., Paul, D., Lamichaney, A., Sarkar, A., Babbar, N., Mandal, N., & Chakrabarty, S. K. (2023). Climate change impacts on seed production and quality: Current knowledge, implications, and mitigation strategies. *Seed Science and Technology*, *51*(1), 65–96. DOI: 10.15258/sst.2023.51.1.07

Ma, J., Wang, Y., Chen, S., & Xu, T. (2025). The Central Role of Auxin in Orchestrating Apical Stem Cells in Plants. *Plant, Cell & Environment*, *48*(7), 5053–5072. DOI: 10.1111/pce.15464 PMID: 40152539

Major, D. J. (2023). Light and photoperiod. In CRC Handbook of plant science in agriculture (pp. 195-216). CRC Press. Major, D. J. (2023). Light and photoperiod. In CRC Handbook of plant science in agriculture (pp. 195-216). CRC Press.

Majumder, A., Roychowdhury, D., and Ray, S. (2023). Biotechnological approaches for ex situ conservation of medicinal plants. *Medicinal Plants: Biodiversity, Biotechnology and Conservation*, 729-800.

Majumder, S., Igamberdiev, A. U., & Debnath, S. C. (2025). Somaclonal Variation and Clonal Fidelity in Commercial Micropropagation: Challenges and Perspectives. *Agronomy (Basel)*, *15*(6), 1489. DOI: 10.3390/agronomy15061489

Malabadi, R. B., Chalannavar, R. K., & Kolkar, K. P.Ravindra B. MalabadiRaju K. ChalannavarKiran P. Kolkar. (2025). Plant cell totipotency: Plant tissue culture applications-An updated review. *World Journal of Advanced Engineering Technology and Sciences*, *16*(02), 112–135. DOI: 10.30574/wjaets.2025.16.2.1262

Malviya, R., & Gayen, D. (2024). Seed Deterioration: Unraveling the Role of Phytohormones on Seed Germination Under Aging Condition. *Journal of Plant Growth Regulation*, •••, 1–17.

Mammadov, J., Aggarwal, R., Buyyarapu, R., & Kumpatla, S. (2012). SNP markers and their impact on plant breeding. *International Journal of Plant Genomics*, *2012*, 728398. DOI: 10.1155/2012/728398 PMID: 23316221

Mangena, P. (2022). Plant transformation via Agrobacterium tumefaciens: culture conditions, recalcitrance and advances in soybean.

Manickam, S., Rajagopalan, V. R., Kambale, R., Rajasekaran, R., Kanagarajan, S., & Muthurajan, R. (2023, November 8). Plant Metabolomics: Current Initiatives and Future Prospects. *Current Issues in Molecular Biology*, *45*(11), 8894–8906. DOI: 10.3390/cimb45110558 PMID: 37998735

Manohar, K. A., Shukla, G., Shahina, N. N., Sivasankarreddy, K., Ravuther, S. S., & Chakravarty, S. (2024). Conventional Versus Non-Conventional Methods of Propagation of Forest Tree Species: Applications and Limitations. In *Biotechnological Approaches for Sustaining Forest Trees and Their Products* (pp. 483–517). Springer Nature Singapore. DOI: 10.1007/978-981-97-4363-6_18

Martínez de Azagra Paredes, A., Del Río San José, J., Reque Kilchenmann, J., Diez Hernández, J. M., & Sanz Ronda, F. J. (2022). Methods for watering seedlings in arid zones. *Forests, 13*(2), 351. DOI: 10.3390/f13020351

Martins, J. P. R., Wawrzyniak, M. K., Ley-López, J. M., Kalemba, E. M., Mendes, M. M., & Chmielarz, P. (2022). 6-Benzylaminopurine and kinetin modulations during in vitro propagation *of Quercus robur* (L.): An assessment of anatomical, biochemical, and physiological profiling of shoots. [PCTOC]. *Plant Cell, Tissue and Organ Culture*, *151*(1), 149–164. DOI: 10.1007/s11240-022-02339-9

Ma, Y., & Li, Y. (2024). The research progress on genomic selection and breeding for drought resistance in wheat: From genetic analysis to smart breeding applications. *Geographical Research Bulletin*, *3*, 625–646.

McAfee, K. (2003). Corn culture and dangerous DNA: Real and imagined consequences of maize transgene flow in Oaxaca. *Journal of Latin American Geography*, *2*(1), 18–42. DOI: 10.1353/lag.2004.0008

McEowen, R. A. (2003). Legal Issues Related to the Use and Ownership of Genetically Modified Organisms. *Washburn Law Journal*, *43*, 611.

McIntyre, I. (2021). *Environmental blockades: Obstructive direct action and the history of the environmental movement*. Routledge. DOI: 10.4324/9781003037781

Megersa, H. G. (2017). Propagation methods of selected horticultural crops by specialized organs [review]. *J Hortic, 4*(198), 2376–0354. DOI: 10.4172/2376-0354.1000198

Mehmood, M., Tanveer, N. A., Joyia, F. A., Ullah, I., & Mohamed, H. I. (2025). Effect of high temperature on pollen grains and yield in economically important crops: A review. *Planta, 261*(6), 141. DOI: 10.1007/s00425-025-04714-0 PMID: 40374974

Mermer, J. E. (2018). The impact of intellectual property in fostering innovation. *Available at SSRN* 3379331. DOI: DOI: 10.2139/ssrn.3379331

Mgbeoji, I. (2001). Patents and traditional knowledge of the uses of plants: Is a communal patent regime part of the solution to the scourge of bio piracy? *Indiana Journal of Global Legal Studies*, 163–186.

Michael, G. S. (2010). *Plant reproductive biology*. Science Publishers.

Mishra, R. K. (2023). Fresh water availability and its global challenge. *British Journal of Multidisciplinary and Advanced Studies, 4*(3), 1–78. DOI: 10.37745/bjmas.2022.0208

Moeller, D. A., Geber, M. A., Eckhart, V. M., & Tiffin, P. (2012). Reduced pollinator service and elevated pollen limitation at the geographic range limit of an annual plant. *Ecology, 93*(5), 1036–1048. DOI: 10.1890/11-1462.1 PMID: 22764490

Moeslund, J. E., Zlinszky, A., Ejrnæs, R., Brunbjerg, A. K., Bøcher, P. K., Svenning, J. C., & Normand, S. (2019). Light detection and ranging explains diversity of plants, fungi, lichens, and bryophytes across multiple habitats and large geographic extent. *Ecological Applications, 29*(5), e01907. DOI: 10.1002/eap.1907 PMID: 31002436

Monteiro, C. A., Cannon, G., Levy, R., Moubarac, J. C., Jaime, P., Martins, A. P., & Parra, D. (2016). NOVA. The star shines bright. *World Nutrition, 7*(1-3), 28–38.

Moraes, R. M., Cerdeira, A. L., & Lourenço, M. V. (2021, March 21). Using Micropropagation to Develop Medicinal Plants into Crops. *Molecules (Basel, Switzerland), 26*(6), 1752. DOI: 10.3390/molecules26061752 PMID: 33800970

Morris, S. J., & Blackwood, C. B. (2024). The ecology of soil biota and their function. In *Soil microbiology, ecology and biochemistry* (pp. 275–302). Elsevier. DOI: 10.1016/B978-0-12-822941-5.00010-7

Mota-Vargas, C., & Rojas-Soto, O. R. (2012). The importance of defining the geographic distribution of species for conservation: The case of the Bearded Wood-Partridge. *Journal for Nature Conservation, 20*(1), 10–17. DOI: 10.1016/j.jnc.2011.07.002

Movahedi, A., Aghaei-Dargiri, S., Li, H., Zhuge, Q., & Sun, W. (2023, November 13). CRISPR Variants for Gene Editing in Plants: Biosafety Risks and Future Directions. *International Journal of Molecular Sciences*, *24*(22), 16241. DOI: 10.3390/ijms242216241 PMID: 38003431

Mu, H., Wang, B., & Yuan, F. (2022, November 15). Bioinformatics in Plant Breeding and Research on Disease Resistance. *Plants*, *11*(22), 3118. DOI: 10.3390/plants11223118 PMID: 36432847

Muhammad Aslam, M., Waseem, M., Jakada, B. H., Okal, E. J., Lei, Z., Saqib, H. S. A., Yuan, W., Xu, W., & Zhang, Q. (2022, January 19). Mechanisms of Abscisic Acid-Mediated Drought Stress Responses in Plants. *International Journal of Molecular Sciences*, *23*(3), 1084. DOI: 10.3390/ijms23031084 PMID: 35163008

Murlidhar, J., Sadawarti, Pandey, K. K. Singh, S. P. Singh, Y. P. (2017). A review on comparison of tissue culture v/s conventional system of seed potato production. *Indian Journal of Hill Farming.*, *30*(2), 160–167.

Musacchi, S., & Neri, D. (2019). Optimizing production of quality nursery plants for fruit tree cultivation. In *Achieving sustainable cultivation of temperate zone tree fruits and berries* (pp. 183–242). Burleigh Dodds Science Publishing.

Muthazhagu Kuppuraj, S. A., Chokkalingam, Y., Jothiganapathy, K., Vedachalam, V., Bisht, D. S., Cholin, S., & Saminadane, T. (2025). Maternal effect on the inheritance of pericarp colour and grain dimension in rice (*Oryza sativa* L.). *Journal of Genetics*, *104*(1), 5. DOI: 10.1007/s12041-024-01490-w PMID: 39959938

Nabuuma, D., Reimers, C., Hoang, K. T., Stomph, T., Swaans, K., & Raneri, J. E. (2022). Impact of seed system interventions on food and nutrition security in low and middle-income countries: A scoping review. *Global Food Security*, *33*, 100638. DOI: 10.1016/j.gfs.2022.100638

Nahar, K. (2025). Propagation Technologies: Seedling, Stem Cutting and Grafting. In *Jatropha curcas L: A Potential 2G Energy Crop to Produce Biofuel in Bangladesh: Agronomy, Biotechnology, Biodiesel and Byproducts* (pp. 33–55). Springer Nature Switzerland. DOI: 10.1007/978-3-031-77642-7_3

Nair, A., Chauhan, P., Saha, B., & Kubatzky, K. F. (2019). Conceptual evolution of cell signaling. *International Journal of Molecular Sciences*, *20*(13), 3292. DOI: 10.3390/ijms20133292 PMID: 31277491

Nathan, R., Katul, G. G., Bohrer, G., Kuparinen, A., Soons, M. B., Thompson, S. E., Trakhtenbrot, A., & Horn, H. S. (2011). Mechanistic models of seed dispersal by wind. *Theoretical Ecology*, *4*(2), 113–132. DOI: 10.1007/s12080-011-0115-3

Nautiyal, P. C., Sivasubramaniam, K., & Dadlani, M. (2023). Seed dormancy and regulation of germination. *Seed Science and Technology*, *52*, 39–66.

Nazarov, A., Kulikova, E., & Molokova, E. (2024). Economic security through technological advancements in agriculture: A pathway to sustainable agro-industrial growth. In *BIO Web of Conferences* (Vol. 121, p. 02012). EDP Sciences. DOI: 10.1051/bioconf/202412102012

Nazir, N., Yaseen, I., Been, T. J., Khalil, A., Iqbal, U., Mir, M. M., & Bhat, S. (2023). Improved Propagation Techniques in Temperate Nuts. In *Temperate Nuts* (pp. 113–148). Springer Nature Singapore. DOI: 10.1007/978-981-19-9497-5_5

Nega, A. (2014). Review on concepts in biological control of plant pathogens. *Journal of Biology, Agriculture and Healthcare*, *4*(27), 33–54.

Nemhauser, J. L. (2018). Back to basics: What is the function of an Aux/IAA in auxin response? *The New Phytologist*, *218*(4), 1295–1297. DOI: 10.1111/nph.15172 PMID: 29738089

Netanel, N. W. (2008). Introduction: The WIPO Development Agenda and its development policy context.

Neves, M., Correia, S., Cavaleiro, C., & Canhoto, J. (2021). Modulation of organogenesis and somatic embryogenesis by ethylene: An overview. *Plants*, *10*(6), 1208. DOI: 10.3390/plants10061208 PMID: 34198660

Nomani, M. Z. M., & Rahman, F. (2016). Bio Piracy of Traditional Knowledge Related Geographical Indications: A Select Study of Some Indian Cases. [MIPR]. *Manupatra Intellectual Property Reports*, *3*(3), 135–152.

Nongdam, P. (2016). Development of synthetic seed technology in plants and its applications: A review. *International Journal of Current Science*, *19*(4), 86–101.

Nowakowska, K., Nongdam, P., Samsurizal, N. A., & Pacholczak, A. (2023). An efficient micropropagation protocol for the endangered European shrub February daphne (Daphne mezereum L.) and Identification of bacteria in culture. *Agriculture*, *13*(9), 1692. DOI: 10.3390/agriculture13091692

Ochar, K., & Kim, S. H. (2023). Conservation and global distribution of onion (Allium cepa L.) germplasm for agricultural sustainability. *Plants*, *12*(18), 3294. DOI: 10.3390/plants12183294 PMID: 37765458

Okediji, R. L. (2008). WIPO-WTO relations and the future of global intellectual property norms. *Netherlands Yearbook of International Law*, *39*, 69–125. DOI: 10.1017/S016767680800069X

Okocha, O. I., and Ogbu, J. U. (2021). Horticultural crops production. *Agricultural technology for colleges, 151.*

Ollinger, S. V. (2003). *Forest ecosystems. Encyclopedia of Life Sciences.* Nature Publishing Group.

Olusegun, O. O., & Olubiyi, I. A. (2017). Implications of genetically modified crops and intellectual property rights on agriculture in developing countries. *Journal of African Law, 61*(2), 253–271. DOI: 10.1017/S0021855317000134

Onwuka, B., & Mang, B. J. A. P. A. R. (2018). Effects of soil temperature on some soil properties and plant growth. *Advances in Plants & Agriculture Research, 8*(1), 34–37. DOI: 10.15406/apar.2018.08.00288

Opoku, A., & Fortune, C. (2011) The implementation of sustainable practices through leadershipin construction organizations In: Egbu C, Lou ECW (eds) Proceedings 27th annual ARCOM Bosch, C. (2009). Agenda 21 for Culture: An Agreement by Cities and Local Governments for Cultural Development. Available at SSRN: https://ssrn.com/abstract=2056382 or Packalén, S. (2010). Culture and sustainability. Corporate social responsibility and environmental management, 17(2), 118-121.

Opstad, N., Sønsteby, A., Myrheim, U., & Heide, O. M. (2011). Seasonal timing of floral initiation in strawberry: Effects of cultivar and geographic location. *Scientia Horticulturae, 129*(1), 127–134. DOI: 10.1016/j.scienta.2011.03.022

Osip, C., Ballescas, A., Osip, L. P., Besarino, N. L., Bagayna, A. D., & Jumalon, C. B. (2000). Philippine council for Agr. Forestry and Natural Resources. *Resources and Technology, 134*, 17–18.

Osland, M. J., Chivoiu, B., Feher, L. C., Dale, L. L., Lieurance, D., Daniel, W. M., & Spencer, J. E. (2023). Plant migration due to winter climate change: Range expansion of tropical invasive plants in response to warming winters. *Biological Invasions, 25*(9), 2813–2830. DOI: 10.1007/s10530-023-03075-7

Ovcharenko, O. O., Rudas, V. A., & Kuchuk, M. V. (2023). Protoplast Fusion for Cellular Engineering of the Brassicaceae. *Cytology and Genetics, 57*(5), 432–450. DOI: 10.3103/S0095452723050043

Ozyigit, I. I., Dogan, I., Hocaoglu-Ozyigit, A., Yalcin, B., Erdogan, A., Yalcin, I. E., Cabi, E., & Kaya, Y. (2023, June 29). Production of secondary metabolites using tissue culture-based biotechnological applications. *Frontiers in Plant Science, 14*, 1132555. DOI: 10.3389/fpls.2023.1132555 PMID: 37457343

Pacini, E., & Dolferus, R. (2019). Pollen developmental arrest: Maintaining pollen fertility in a world with a changing climate. *Frontiers in Plant Science*, *10*, 679. DOI: 10.3389/fpls.2019.00679 PMID: 31178886

Padmanabhan, A., Reddy, R. S., & Sharma, S. (2017). *MODERN BIOTECHNOLOGY AND INDIA'S GOVERNANCE IMPERATIVES*. Carnegie Endowment for International Peace.

Padmavathiamma, P. K., & Li, L. Y. (2007). Phytoremediation technology: Hyper-accumulation metals in plants. *Water, Air, and Soil Pollution*, *184*(1-4), 105–126. DOI: 10.1007/s11270-007-9401-5

Pamies, D., Leist, M., Coecke, S., Bowe, G., Allen, D., Gstraunthaler, G., ... and Stacey, G. (2022). Guidance document on good cell and tissue culture practice 2.0 (GCCP 2.0).

Pandey, S., & Husen, A. (2022). Role of plant stem or shoot cutting positions and hormone treatments in adventitious root formation. In *Environmental, Physiological and Chemical Controls of Adventitious Rooting in Cuttings* (pp. 367-379). Academic Press. DOI: 10.1016/B978-0-323-90636-4.00009-X

Pandey, D. K., & Chaudhary, B. (Eds.). (2025). *Regeneration Potential in Plants*. CRC Press. DOI: 10.1201/9781032702391

Pandey, S. C., Modi, P., Pereira, V., & Fosso Wamba, S. (2024). Empowering small farmers for sustainable agriculture: A human resource approach to SDG-driven training and innovation. *International Journal of Manpower*. Advance online publication. DOI: 10.1108/IJM-11-2023-0655

Pandey, V. K., Shafi, Z., Choudhary, P., Pathak, A., & Rustagi, S. (2025). Navigating the Functional and Bioactive Landscape of Sodium Alginate-Based Edible Coatings for Modernized Fruit Preservation. *Journal of Food Safety*, *45*(3), e70020. DOI: 10.1111/jfs.70020

Pandey, V. K., Tripathi, A., Srivastava, S., Dar, A. H., Singh, R., Farooqui, A., & Pandey, S. (2023, March 24). Exploiting the bioactive properties of essential oils and their potential applications in food industry. *Food Science and Biotechnology*, *32*(7), 885–902. DOI: 10.1007/s10068-023-01287-0 PMID: 37123062

Pant, M., & Husen, A. (2022). Micropropagation in mature trees by manipulation of phase change, stress, and culture environment. In *Environmental, Physiological and Chemical Controls of Adventitious Rooting in Cuttings* (pp. 421-437). Academic Press. DOI: 10.1016/B978-0-323-90636-4.00002-7

Panthee, D. R. (2023). Application of Molecular Markers in Crop Improvement and Beyond. *Agronomy (Basel)*, *13*(8), 2041. DOI: 10.3390/agronomy13082041

Parihar, P., Singh, S., Singh, R., Singh, V. P., & Prasad, S. M. (2015). Effect of salinity stress on plants and its tolerance strategies: A review. *Environmental Science and Pollution Research International*, *22*(6), 4056–4075. DOI: 10.1007/s11356-014-3739-1 PMID: 25398215

Parikh, H. S., Dave, G., Singh, P. K., & Tiwari, A. (2025). Exploring the significance of xenic and axenic cultures for diatoms: Methodological insights, scope, and biotechnological implications. *Systems Microbiology and Biomanufacturing*, *5*(3), 1–21. DOI: 10.1007/s43393-025-00363-2

Park, H. S., Lee, W. K., Lee, S. C., Lee, H. O., Joh, H. J., Park, J. Y., Kim, S., Song, K., & Yang, T. J. (2021, January 28). Inheritance of chloroplast and mitochondrial genomes in cucumber revealed by four reciprocal F_1 hybrid combinations. *Scientific Reports*, *11*(1), 2506. DOI: 10.1038/s41598-021-81988-w PMID: 33510273

Parodi, O. (2011). Three steps towards a culture of sustainability. In Parodi, O., Ayestaran, I., & Banse, G. (Eds.), *Sustainable development-relationships to culture, knowledge and ethics* (pp. 76–92). KIT Scientific Publishing.

Parzymies, M., & Dabski, M. (2015). Methods of eliminating contaminations in tissue cultures on the example of selected ornamental plant species. *BioTechnologia. Journal of Biotechnology Computational Biology and Bionanotechnology*, *96*(1).

Pasternak, T. P., & Steinmacher, D. (2024). Plant growth regulation in cell and tissue culture in vitro. *Plants*, *13*(2), 327. DOI: 10.3390/plants13020327 PMID: 38276784

Patel, B., & Sahay, N. S. (2025). Strategies for the improvement of micropropagation of banana through in vitro culture. *Plant Tissue Culture & Biotechnology*, *35*(1), 185–198. DOI: 10.3329/ptcb.v35i1.82675

Pathak, S., Singh, J., Kumar, K., & Bisht, V. (2025). Breeding for the Future: A Comprehensive Review of Recurrent Selection. *Journal of Scientific Research and Reports*, *31*(7), 694–708. DOI: 10.9734/jsrr/2025/v31i73289

Paul, M. (2024). BREEDING TECHNIQUES FOR ASEXUALLY PROPAGATED CROPS. *Fundamentals of plant breeding*, 154.

Paulikienė, S., Benesevičius, D., Benesevičienė, K., & Ūksas, T. (2025). Review—Seed Treatment: Importance, Application, Impact, and Opportunities for Increasing Sustainability. *Agronomy (Basel)*, *15*(7), 1689. DOI: 10.3390/agronomy15071689

Pellissier, L., Fournier, B., Guisan, A., & Vittoz, P. (2010). Plant traits co-vary with altitude in grasslands and forests in the European Alps. *Plant Ecology, 211*(2), 351–365. DOI: 10.1007/s11258-010-9794-x

Peña-Iglesias, A., & Ayuso, P. (1982, September). Characterisation of Spanish garlic viruses and their elimination by in vitro shoot apex culture. In *IV Conference on recent Advances in Vegetable Virus Research 127* (pp. 183-193). .

Pereira, A. M., & Coimbra, S. (2019). Advances in plant reproduction: From gametes to seeds. *Journal of Experimental Botany*, *70*(11), 2933–2936. DOI: 10.1093/jxb/erz227 PMID: 31152172

Pereira, F. M. P., Usman, M., Mayer, N. A., Nachtigal, J. C., Maphanga, O. R. M., & Willemse, S. (2017). Advances in guava propagation. *Revista Brasileira de Fruticultura, 39*(4), e-358.

Phillips, G. C., & Garda, M. (2019). Plant tissue culture media and practices: An overview. *In Vitro Cellular & Developmental Biology. Plant, 55*(3), 242–257. DOI: 10.1007/s11627-019-09983-5

Phogat, V. K., Tomar, V. S., & Dahiya, R. I. T. A. (2015). Soil physical properties. Soil science: An introduction, 135-171.

Pill, W. G. (2020). Low water potential and presowing germination treatments to improve seed quality. In *Seed quality* (pp. 319–359). CRC Press. DOI: 10.4324/9781003075226-10

Pimentel, D., Hunter, M. S., LaGro, J. A., Efroymson, R. A., Landers, J. C., Mervis, F. T., McCarthy, C. A., & Boyd, A. E. (1989). Benefits and risks of genetic engineering in agriculture. *Bioscience, 39*(9), 606–614. DOI: 10.2307/1311090

Pincelli-Souza, R. P., Tang, Q., Miller, B. M., & Cohen, J. D. (2024). Horticultural potential of chemical biology to improve adventitious rooting. *Horticulture Advances, 2*(1), 12. DOI: 10.1007/s44281-024-00034-7

Pithiya, M., Sharma, S., Sharma, M., Sharma, M., & Kotwal, N. (2022). Advancements and Challenges in Plant Tissue Culture: a Comprehensive Overview. Journal of Plant Biota. DOI: 10.51470/JPB.2022.1.1.12

Polunin, N., & Grinevald, J. (1988). Vernadsky and biospheral ecology. *Environmental Conservation, 15*(2), 117–122.

Posch, B. C., Kariyawasam, B. C., Bramley, H., Coast, O., Richards, R. A., Reynolds, M. P., Trethowan, R., & Atkin, O. K. (2019). Exploring high temperature responses of photosynthesis and respiration to improve heat tolerance in wheat. *Journal of Experimental Botany*, *70*(19), 5051–5069. DOI: 10.1093/jxb/erz257 PMID: 31145793

Poschlod, P., Mašková, T., Chen, S. C., Phartyal, S., Rosbakh, S., Silveira, F. A., ... and Vandelook, F. (2025). A handbook for standardised measurements of regenerative plant functional traits.

Prakash, A., & Paul, P. (2021). *Recent trends in propagation of horticulture crops and application of Genomics tools in plant Breeding. Recent Trends in Propagation of Horticulture Crops*. Taran Publication.

Prasad, R., Bagde, U. S., & Varma, A. (2012). An overview of intellectual property rights in relation to agricultural biotechnology. *African Journal of Biotechnology*, *11*(73), 13476–13752. DOI: 10.5897/AJB12.262

Prasanna, P. A., Rao, L. V., Prasad, S. H., Amtul, W., Meera, S. N., Nirmala, B., & Syamaladevi, D. P. (2020). Intellectual property rights protection for plant varieties in India: Status, emerging issues, and challenges. *Agricultural Economics Research Review*, *32*(2), 259. DOI: 10.5958/0974-0279.2019.00037.5

Prasath, D. K., Kandiannan, R., Chitra, V. A., Muhammed Nissar, J., Suresh, K., & Babu, N. (2017). Quality seed production in ginger and turmeric: Present status and future prospects. *Shodh Chintan*, *9*, 163–170.

Preece, J. E. (2003). A century of progress with vegetative plant propagation. *HortScience*, *38*(5), 1015–1025. DOI: 10.21273/HORTSCI.38.5.1015

Prip, C. (2024). Farmers' Rights in the Plant Treaty: Interrelations and recent interactions with other international regimes and processes. *Development Studies Research*, *11*(1), 2357095. DOI: 10.1080/21665095.2024.2357095

Ptushenko, O. S., Ptushenko, V. V., & Solovchenko, A. E. (2020). Spectrum of light as a determinant of plant functioning: A historical perspective. *Life (Chicago, Ill.)*, *10*(3), 25. DOI: 10.3390/life10030025 PMID: 32192016

Pueyo, Y., Kéfi, S., Díaz-Sierra, R., Alados, C. L., & Rietkerk, M. (2010). The role of reproductive plant traits and biotic interactions in the dynamics of semi-arid plant communities. *Theoretical Population Biology*, *78*(4), 289–297.

Pultrone, C. (2012). An overview of contract farming: Legal issues and challenges. *Uniform Law Review-Revue De Droit Uniforme*, *17*(1-2), 263–289. DOI: 10.1093/ulr/17.1-2.263

Qandil, A. M. A., & Ibrahim, A. K. (2007). *Production of medicinal, aromatic and ornamental plants*. Open Education Center, Faculty of Agriculture, Ain Shams University.

Qian, D., Wang, M., Niu, Y., Yang, Y., & Xiang, Y. (2025). Sexual reproduction in plants under high temperature and drought stress. *Cell Reports*, *44*(3), 115390. DOI: 10.1016/j.celrep.2025.115390 PMID: 40056418

Qiu, J., Bai, Y., Fu, Y. B., & Wilmshurst, J. F. (2010). Spatial variation in temperature thresholds during seed germination of remnant Festuca hallii populations across the Canadian prairie. *Environmental and Experimental Botany*, *67*(3), 479–486. DOI: 10.1016/j.envexpbot.2009.09.002

Rachappanavar, V. (2025). Utilizing CRISPR-based genetic modification for precise control of seed dormancy: Progress, obstacles, and potential directions. *Molecular Biology Reports*, *52*(1), 204. DOI: 10.1007/s11033-025-10285-w PMID: 39907946

Radi, H., Koufan, M., Belkoura, I., Koussa, T., & Mazri, M. A. (2025). In Vitro Mycorrhization for Plant Propagation and Enhanced Resilience to Environmental Stress: A Review. *Plants*, *14*(14), 2097. DOI: 10.3390/plants14142097 PMID: 40733334

Raghavan, M., Pradhan, V., Singh, S., Jain, S., Kumari, N., Kaur, R., Kumar, G., & Kaur, R. (2025). Advances in Genetic Plant Breeding for Horticultural Crop Improvement. *Plant Cell Biotechnology and Molecular Biology*, *26*(7-8), 432–444. DOI: 10.56557/pcbmb/2025/v26i7-89576

Rahbaran, M., Razeghian, E., Maashi, M. S., Jalil, A. T., Widjaja, G., Thangavelu, L., Kuznetsova, M. Y., Nasirmoghadas, P., Heidari, F., Marofi, F., & Jarahian, M. (2021, November 30). Cloning and Embryo Splitting in Mammalians: Brief History, Methods, and Achievements. [Erratum in: Stem Cells Int. 2024 Jul 17;2024:9837676. doi: 10.1155/2024/9837676. PMID: 34887927; PMCID: PMC8651392]. *Stem Cells International*, *2347506*, 1–11. Advance online publication. DOI: 10.1155/2021/2347506 PMID: 34887927

Rahman, S. U., Khan, M. O., Ullah, R., Ahmad, F., & Raza, G. (2024). Agrobacterium-mediated transformation for the development of transgenic crops; present and future prospects. *Molecular Biotechnology*, *66*(8), 1836–1852. DOI: 10.1007/s12033-023-00826-8 PMID: 37573566

Rahmawati, D., Kartiwijaya, R., & Yuliawati, L. B. (2023). Micropropagation of banana plant Cavendish (*Musa acuminata* L.) using shoot-tip culture to support food security. *BIODIVERS–BIOTROP Science Magazine*, *2*(2), 16–21. DOI: 10.56060/bdv.2023.2.2.2092

Rajaseger, G., Chan, K. L., Yee Tan, K., Ramasamy, S., Khin, M. C., Amaladoss, A., & Kadamb Haribhai, P. (2023, September 30). Hydroponics: Current trends in sustainable crop production. *Bioinformation, 19*(9), 925–938. DOI: 10.6026/97320630019925 PMID: 37928497

Ramadan, E. A., Ramadan, N. A., & Mohammed, A. A. H. (2022). Aflatoxigenic fungi in nuts and dried fruits in Mosul and Duhok city. [Vojenske Zdravotnicke Listy]. *Vojenske Zdravotnicke Listy, 91*(3), 224–234. DOI: 10.31482/mmsl.2021.049

Ramankutty, N., Mehrabi, Z., Waha, K., Jarvis, L., Kremen, C., Herrero, M., & Rieseberg, L. H. (2018). Trends in global agricultural land use: Implications for environmental health and food security. *Annual Review of Plant Biology, 69*(1), 789–815. DOI: 10.1146/annurev-arplant-042817-040256 PMID: 29489395

Raman, R. (2017, October 2). The impact of Genetically Modified (GM) crops in modern agriculture: A review. *GM Crops and Food: Biotechnology in Agriculture and the Food Chain, 8*(4), 195–208. DOI: 10.1080/21645698.2017.1413522 PMID: 29235937

Ranaware, A. S., Kunchge, N. S., Lele, S. S., & Ochatt, S. J. (2023). Protoplast technology and somatic hybridisation in the family Apiaceae. *Plants, 12*(5), 1060. DOI: 10.3390/plants12051060

Rapinski, M., Raymond, R., Davy, D., Herrmann, T., Bedell, J. P., Ka, A., Odonne, G., Chanteloup, L., Lopez, P. J., Foulquier, É., da Silva, E. F., El Deghel, N., Boëtsch, G., Coxam, V., Joliet, F., Guihard-Costa, A.-M., Tibère, L., Nazare, J.-A., & Duboz, P. (2023). Local food systems under global influence: The case of food, health and environment in five socio-ecosystems. *Sustainability (Basel), 15*(3), 2376. DOI: 10.3390/su15032376

Rathour, A., & Kumar, R. (2023). *Plant Propagation Techniques in Horticulture. Recent Trend in Agriculture*. Integrated Publications New Delhi.

Rayment, G. E., & Lyons, D. J. (2011). *Soil chemical methods: Australasia* (Vol. 3). CSIRO publishing.

Razzaq, K., & Du, J. (2024). Phytohormonal regulation of plant development in response to fluctuating light conditions. *Journal of Plant Growth Regulation*, •••, 1–34.

Reed, K. M., & Bargmann, B. O. (2021). Protoplast regeneration and its use in new plant breeding technologies. *Frontiers in Genome Editing, 3*, 734951. DOI: 10.3389/fgeed.2021.734951 PMID: 34713266

Reed, R. C., Bradford, K. J., & Khanday, I. (2022). Seed germination and vigor: Ensuring crop sustainability in a changing climate. *Heredity*, *128*(6), 450–459. DOI: 10.1038/s41437-022-00497-2 PMID: 35013549

Reid, J. (2009). Biopiracy: The struggle for traditional knowledge rights. *Am. Indian L. Rev.*, *34*, 77.

Rinalducci, S., Egidi, M. G., Mahfoozi, S., Godehkahriz, S. J., & Zolla, L. (2011). The influence of temperature on plant development in a vernalization-requiring winter wheat: A 2-DE based proteomic investigation. *Journal of Proteomics*, *74*(5), 643–659. DOI: 10.1016/j.jprot.2011.02.005 PMID: 21320650

Rivard, C. L., & Louws, F. J. (2008). Grafting to manage soilborne diseases in heirloom tomato production. *HortScience*, *43*(7), 2104–2111. DOI: 10.21273/HORTSCI.43.7.2104

Robinson, G. M., & Carson, D. A. (2015). The globalisation of agriculture: Introducing the Handbook. In *Handbook on the globalisation of agriculture* (pp. 1–28). Edward Elgar Publishing., DOI: 10.4337/9780857939838.00006

Rodda, D. (2020). Biopiracy Analysis of Existing International and Indian Legal Framework. Issue 2 *Int'l JL Mgmt. &. Human.*, *3*, 227.

Roddy, A. B., Théroux-Rancourt, G., Abbo, T., Benedetti, J. W., Brodersen, C. R., Castro, M., Castro, S., Gilbride, A. B., Jensen, B., Jiang, G.-F., Perkins, J. A., Perkins, S. D., Loureiro, J., Syed, Z., Thompson, R. A., Kuebbing, S. E., & Simonin, K. A. (2020). The scaling of genome size and cell size limits maximum rates of photosynthesis with implications for ecological strategies. *International Journal of Plant Sciences*, *181*(1), 75–87. DOI: 10.1086/706186

Rodrigues, B. R. A., Nietsche, S., Mercadante-Simões, M. O., Pereira, M. C. T., & Ribeiro, L. M. (2018). Climatic seasonality influences the development of pollen grains and fruiting in Annona squamosa. *Environmental and Experimental Botany*, *150*, 240–248. DOI: 10.1016/j.envexpbot.2018.03.025

Rojo, J., Rapp, A., Lara, B., Fernández-González, F., & Pérez-Badia, R. (2015). Effect of land uses and wind direction on the contribution of local sources to airborne pollen. *The Science of the Total Environment*, *538*, 672–682. DOI: 10.1016/j.scitotenv.2015.08.074 PMID: 26327635

Roychowdhury, R., Das, S. P., Gupta, A., Parihar, P., Chandrasekhar, K., Sarker, U., Kumar, A., Ramrao, D. P., & Sudhakar, C. (2023, June 16). Multi-Omics Pipeline and Omics-Integration Approach to Decipher Plant's Abiotic Stress Tolerance Responses. *Genes*, *14*(6), 1281. DOI: 10.3390/genes14061281 PMID: 37372461

Rteba, M. M., & Hussain, W. S. (2024). Sustainable Agriculture: Way to Achieve Food Security. In Achieving Food Security Through Sustainable Agriculture (pp. 182-202). IGI Global.

Rubin, H., & Rubin, A. L. (2018). Phenotypic selection as the biological mode of epigenetic conversion and reversion in cell transformation. *Proceedings of the National Academy of Sciences of the United States of America*, *115*(4), E725–E732. DOI: 10.1073/pnas.1717299115 PMID: 29311337

Ruxton, G. D., & Schaefer, H. M. (2012). The conservation physiology of seed dispersal. *Philosophical Transactions of the Royal Society of London. Series B, Biological Sciences*, *367*(1596), 1708–1718. DOI: 10.1098/rstb.2012.0001 PMID: 22566677

Ruzzante, S., Labarta, R., & Bilton, A. (2021). Adoption of agricultural technology in the developing world: A meta-analysis of empirical literature. *World Development*, *146*, 105599. DOI: 10.1016/j.worlddev.2021.105599

Sabir, I. A., Hu, X., Khan, I., & Qin, Y. (2025). Regulatory Mechanisms of Bud Dormancy: Environmental, Hormonal, and Genetic Perspectives. *International Journal of Molecular Sciences*, *26*(6), 2517. DOI: 10.3390/ijms26062517 PMID: 40141161

Sadh, P. K., Duhan, S., & Duhan, J. S. (2018). Agroindustrial wastes and their utilization using solid state fermentation. *Bioresources and Bioprocessing*, *5*(1), 1–15. DOI: 10.1186/s40643-017-0187-z

Sagoff, M. (2001). Biotechnology and agriculture: The common wisdom and its critics. *Indiana Journal of Global Legal Studies*, *9*, 13.

Sahin, E. C., Aydin, Y., & Uncuoglu, A. A. (2024). Molecular Marker Applications in the Selection of Elite Genotypes for Plant Stress Tolerance and Genetic Fidelity. *OBM Genetics*, *8*(3), 1–25. DOI: 10.21926/obm.genet.2403260

Saini, N. (2024). The potential of plant growth regulator (auxin) in enhancing fruit crop propagation via Stem Cuttings. International Journal of Advanced Biochemistry Research, 8(5), 1001-1007. DOI: 10.33545/26174693.2024.v8.i5l.1230

Salgotra, R. K., & Chauhan, B. S. (2023). Genetic diversity, conservation, and utilization of plant genetic resources. *Genes*, *14*(1), 174. DOI: 10.3390/genes14010174 PMID: 36672915

Salman, A. H. (2000).The Effect of Interaction Between Saline Water Irrigation and Organic Manures on Some Soil Properties and Onion Yield (*Allium cepa L.*).MSc. Thesis. Agriculture College–Baghdad University.Pp.107.

Salman, N. D. and Abd Al-Whaab N. R.(2016) Effect of bacterial biofertilizer and compost on the growth of Tomato plant The Iraqi Journal of Agricultural Sciences – 74(2):0261-0264.

Salse, J., Barnard, R. L., Veneault-Fourrey, C., & Rouached, H. (2024). Strategies for breeding crops for future environments. *Trends in Plant Science*, *29*(3), 303–318. DOI: 10.1016/j.tplants.2023.08.007 PMID: 37833181

Sami, A., Xue, Z., Tazein, S., Arshad, A., He Zhu, Z., Ping Chen, Y., Hong, Y., Tian Zhu, X., & Jin Zhou, K. (2021, December). CRISPR-Cas9-based genetic engineering for crop improvement under drought stress. *Bioengineered*, *12*(1), 5814–5829. DOI: 10.1080/21655979.2021.1969831 PMID: 34506262

Sampaio, B. L., Edrada-Ebel, R., & Da Costa, F. B. (2016). Effect of the environment on the secondary metabolic profile of Tithonia diversifolia: A model for environmental metabolomics of plants. *Scientific Reports*, *6*(1), 29265. DOI: 10.1038/srep29265 PMID: 27383265

Sansavini, S. (2014). *A short history of scholarship in horticulture and pomology. Horticulture: Plants for People and Places* (Vol. 3). Social Horticulture.

Sardar, A. (2023, September 29). Genetic amelioration of fruit and vegetable crops to increase biotic and abiotic stress resistance through CRISPR Genome Editing. [Erratum in: Front Plant Sci. 2024 Apr 25;15:1418620. doi: 10.3389/fpls.2024.1418620. PMID: 37841604; PMCID: PMC10570431]. *Frontiers in Plant Science*, *14*, 1260102. DOI: 10.3389/fpls.2023.1260102 PMID: 37841604

Saunders, M. E. (2018). Insect pollinators collect pollen from wind-pollinated plants: Implications for pollination ecology and sustainable agriculture. *Insect Conservation and Diversity*, *11*(1), 13–31. DOI: 10.1111/icad.12243

Sawhney, N. (2003). *Cooperative innovation in the commons: rethinking distributed collaboration and intellectual property for sustainable design innovation* (Doctoral dissertation, Massachusetts Institute of Technology).

Schweingruber, F. H., Börner, A., & Schulze, E. D. (2012). Anatomical adaptations to environmental conditions. In *Atlas of Stem Anatomy in Herbs, Shrubs and Trees* (Vol. 2, pp. 355–370). Springer Berlin Heidelberg.

Sekhar, P. R., Savithri, Y., Kavitha, N., & Kumari, D. A. (2024). Role of biodiversity and its importance in sustainable development: A comprehensive review. *Uttar Pradesh Journal of Zoology*, *45*(8), 184–188.

Sekhukhune, M. K., & Maila, Y. M. (2025). Growth Regulator Indole-3-Butyric Acid on Rooting Potential of *Actinidia deliciosa* Rootstock and Actinidia arguta Female Scion Species Stem Cuttings. *Horticulturae, 11*(1), 53. DOI: 10.3390/horticulturae11010053

Sell, S. K. (2009). Corporations, seeds, and intellectual property rights governance.

Seneviratne, S. I., Corti, T., Davin, E. L., Hirschi, M., Jaeger, E. B., Lehner, I., Orlowsky, B., & Teuling, A. J. (2010). Investigating soil moisture–climate interactions in a changing climate: A review. *Earth-Science Reviews*, *99*(3-4), 125–161. DOI: 10.1016/j.earscirev.2010.02.004

Senula, A., Keller, E. R. J., & Leseman, D. E. (2000). Elimination of viruses through meristem culture and thermotherapy for the establishment of an in vitro collection of garlic (*Allium sativum*). *Acta Horticulturae*, (530), 121–128. DOI: 10.17660/ActaHortic.2000.530.12

Shah, S. H., Islam, S., Mohammad, F., & Siddiqui, M. H. (2023). Gibberellic acid: A versatile regulator of plant growth, development and stress responses. *Journal of Plant Growth Regulation*, *42*(12), 7352–7373. DOI: 10.1007/s00344-023-11035-7

Shandilya, U., & Khanduja, V. (2020, October). Intelligent farming system with weather forecast support and crop prediction. In *2020 5th International Conference on Computing, Communication and Security (ICCCS)* (pp. 1-6). IEEE. DOI: DOI: 10.1109/ICCCS49678.2020.9277437

Sharma, I. P., Kanta, C., Dwivedi, T., & Rani, R. (2020). Indigenous agricultural practices: A supreme key to maintaining biodiversity. *Microbiological advancements for higher altitude agro-ecosystems & sustainability,* 91-112.

Sharma, P., Roy, M., Roy, B., & Deka, S. D. (2023). Post harvest management strategies and storage approaches for quality seed production. Emerging issues in agricultural sciences, 2, 110–129.

Sharma, P., Gupta, A. K., Meena, H. M., Dubey, A., Singh, M., & Mallick, S. (2023). Inoculation of plant-growth-promoting rhizobacteria and earthworms in the rhizosphere reinstates photosynthetic attributes and secondary metabolites in *Brassicajuncea* L. under chromium toxicity. *Plant and Soil*, *483*(1), 573–587. DOI: 10.1007/s11104-022-05765-y

Shaw, N., Barak, R. S., Campbell, R. E., Kirmer, A., Pedrini, S., Dixon, K., & Frischie, S. (2020). Seed use in the field: Delivering seeds for restoration success. *Restoration Ecology*, *28*(S3), S276–S285. DOI: 10.1111/rec.13210

Shetty, M. J., Chandan, K., Krishna, H. C., & Aparna, G. S. (2018). Genetically modified crops: An overview. *Journal of Pharmacognosy and Phytochemistry, 7*(1), 2405–2410.

Shilo, E., Ziv, B., Shamir, E., & Rimmer, A. (2015). Evaporation from Lake Kinneret, Israel, during hot summer days. *Journal of Hydrology (Amsterdam), 528*, 264–275. DOI: 10.1016/j.jhydrol.2015.06.042

Shilpa, P., Subbalakshmi, G., & Anuradha, M. (2025). In Vitro Production of Colors and Pigments. In *In Vitro Production of Plant Secondary Metabolites: Theory and Practice* (pp. 233–251). Springer Nature Singapore. DOI: 10.1007/978-981-97-8808-8_18

Shishkova, S., Rost, T. L., & Dubrovsky, J. G. (2008). Determinate root growth and meristem maintenance in angiosperms. *Annals of Botany, 101*(3), 319–340. DOI: 10.1093/aob/mcm251 PMID: 17954472

Shiva, V. (1996). Agricultural biodiversity, intellectual property rights and farmers' rights. *Economic and Political Weekly*, 1621–1631.

Shivran, M., Sharma, N., Dubey, A. K., Singh, S. K., Sharma, N., Sharma, R. M., Singh, N., & Singh, R. (2022). Scion–rootstock relationship: Molecular mechanism and quality fruit production. *Agriculture, 12*(12), 2036. DOI: 10.3390/agriculture12122036

Shiwani, K., Sharma, D., & Kumar, A. (2022). Improvement of plant survival and expediting acclimatization process. In *Commercial scale tissue culture for horticulture and Plantation crops* (pp. 277–291). Springer Nature Singapore. DOI: 10.1007/978-981-19-0055-6_12

Shreni Agrawal, E. R. (2022). A review: Agrobacterium-mediated gene transformation to increase plant productivity. *J Phytopharm, 11*(2), 111–117. DOI: 10.31254/phyto.2022.11211

Şimşek, Ö., Dalda Şekerci, A., Isak, M. A., Bulut, F., İzgü, T., Tütüncü, M., & Dönmez, D. (2024). Optimizing Micropropagation and Rooting Protocols for Diverse Lavender Genotypes: A Synergistic Approach Integrating Machine Learning Techniques. *Horticulturae, 10*(1), 52. DOI: 10.3390/horticulturae10010052

Singh, D. P., Singh, A. K., & Singh, A. (2021). *Plant breeding and cultivar development*. Academic Press.

Singh, K., & Kolar, P. (Eds.). (2024). *Digital Agricultural Ecosystem: Revolutionary Advancements in Agriculture*. John Wiley & Sons., DOI: 10.1002/9781394242962

Sirangi, S. K. (2025). *Tissue Culture and Conservation Strategies for Indian Flora.* Deep Science Publishing. DOI: 10.70593/978-93-7185-376-7

Smidt, H. J., & Jokonya, O. (2022). Factors affecting digital technology adoption by small-scale farmers in agriculture value chains (AVCs) in South Africa. *Information Technology for Development, 28*(3), 558–584. DOI: 10.1080/02681102.2021.1975256

Smith, T. (2009). Going to Seed: Using Monsanto as a Case Study to Examine the Patent and Antitrust Implications of the Sale and Use of Genetically Modified Seed. *Alabama Law Review, 61*, 629.

Sofiev, M., Belmonte, J., Gehrig, R., Izquierdo, R., Smith, M., Dahl, Å., & Siljamo, P. (2012). Airborne pollen transport. Allergenic pollen: A review of the production, release, distribution and health impacts, 127-159.

Songnuan, W. (2013). Wind-pollination and the roles of pollen allergenic proteins. *Asian Pacific Journal of Allergy and Immunology, 31*(4), 261. PMID: 24383968

Souza, R., Ambrosini, A., & Passaglia, L. M. (2015). Plant growth-promoting bacteria as inoculants in agricultural soils. *Genetics and Molecular Biology, 38*(4), 401–419. DOI: 10.1590/S1415-475738420150053 PMID: 26537605

Sripathy, K. V., & Groot, S. P. (2023). Seed development and maturation. In *Seed science and technology: Biology, production, quality* (pp. 17–38). Springer Nature Singapore. DOI: 10.1007/978-981-19-5888-5_2

Srivastava, P. S., Bharti, N., Pande, D., & Srivastava, S. (2002). In Mukerji, K. G., Manoharachary, C., & Chamola, B. P. (Eds.), *"Role of mycorrhiza in in vitro micro-propagation of plants" in Techniques in mycorrhizal studies* (pp. 443–468). Springer.

Stein, H. (2005). Intellectual property and genetically modified seeds: The United States, trade, and the developing world. *Northwestern Journal of Technology and Intellectual Property, 3*(2), 151.

Stoian-Dod, R. L., Dan, C., Morar, I. M., Sestras, A. F., Truta, A. M., Roman, G., & Sestras, R. E. (2023). Seed germination within genus Rosa: The complexity of the process and influencing factors. *Horticulturae, 9*(8), 914. DOI: 10.3390/horticulturae9080914

Stuepp, C. A., Wendling, I., Trueman, S. J., Koehler, H. S., & Zuffellato-Ribas, K. C. (2017). The use of auxin quantification for understanding clonal tree propagation. *Forests, 8*(1), 27. DOI: 10.3390/f8010027

Subeesh, A., & Mehta, C. R. (2021). Automation and digitization of agriculture using artificial intelligence and the internet of things. *Artificial Intelligence in Agriculture*, *5*, 278–291. DOI: 10.1016/j.aiia.2021.11.004

Sultan, S. J., Qaddawi, Z. T., & Mohammed, A. A. H. (2024). Molecular diagnosis of bacteria isolated from *Trifolium repens* root nodules. *Journal of Applied and Natural Science*, *16*(3), 1153–1157. DOI: 10.31018/jans.v16i3.5404

Suman, S. (2017). Plant tissue culture: A promising tool of quality material production with special reference to micropropagation of banana. *Biochemical and Cellular Archives*, *17*, 1–26.

Sun, L., Lai, M., Ghouri, F., Nawaz, M. A., Ali, F., Baloch, F. S., Nadeem, M. A., Aasim, M., & Shahid, M. Q. (2024). Modern plant breeding techniques in crop improvement and genetic diversity: From molecular markers and gene editing to artificial intelligence—a critical review. *Plants*, *13*(19), 2676. DOI: 10.3390/plants13192676 PMID: 39409546

Sussex, I. M. (2008, May). The scientific roots of modern plant biotechnology. *The Plant Cell*, *20*(5), 1189–1198. DOI: 10.1105/tpc.108.058735 PMID: 18515500

Su, Y. H., Tang, L. P., Zhao, X. Y., & Zhang, X. S. (2021, January). Plant cell totipotency: Insights into cellular reprogramming. *Journal of Integrative Plant Biology*, *63*(1), 228–243. DOI: 10.1111/jipb.12972 PMID: 32437079

Svolacchia, N., & Sabatini, S. (2023). Cytokinins. *Current Biology*, *33*(1), R10–R13. DOI: 10.1016/j.cub.2022.11.022 PMID: 36626853

Tafolla-Arellano, J. C., Báez-Sañudo, R., & Tiznado-Hernández, M. E. (2018). The cuticle as a key factor in the quality of horticultural crops. *Scientia Horticulturae*, *232*, 145–152. DOI: 10.1016/j.scienta.2018.01.005

Taha, A. M., & Khalifa, H. E. H. (2024). Olive in Egypt: Cultural Practices. *Olives and Olive Related Products-Innovations in Production and Processing: Innovations in Production and Processing, 35.*

Taji, A., Kumar, P., & Lakshmanan, P. (2024). *In vitro plant breeding*. CRC Press. DOI: 10.1201/9781003578529

Tamimi, S. M., & Othman, H. (2023). Silver nanoparticles for enhancing the efficiency of micropropagation of banana (Musa acuminata L.). *Tropical Life Sciences Research*, *34*(2), 161–175. DOI: 10.21315/tlsr2023.34.2.8 PMID: 38144380

Tang, L., Syed, A. U. A., Otho, A. R., Junejo, A. R., Tunio, M. H., Hao, L., Asghar Ali, M. N. H., Brohi, S. A., Otho, S. A., & Channa, J. A. (2024). Intelligent Rapid Asexual Propagation Technology—A Novel Aeroponics Propagation Approach. *Agronomy (Basel), 14*(10), 2289. DOI: 10.3390/agronomy14102289

Tang, M., Zhao, W., Xing, M., Zhao, J., Jiang, Z., You, J., Ni, B., Ni, Y., Liu, C., Li, J., & Chen, X. (2021). Resource allocation strategies among vegetative growth, sexual reproduction, asexual reproduction and defense during growing season of Aconitum kusnezoffii Reichb. *The Plant Journal, 105*(4), 957–977. DOI: 10.1111/tpj.15080 PMID: 33180340

Tantalaki, N., Souravlas, S., & Roumeliotis, M. (2019). Data-driven decision making in precision agriculture: The rise of big data in agricultural systems. *Journal of Agricultural & Food Information, 20*(4), 344–380. DOI: 10.1080/10496505.2019.1638264

Tasila Konja, D., Mabe, F. N., & Oteng-Frimpong, R. (2019). Profitability and profit efficiency of certified groundnut seed and conventional groundnut production in Northern Ghana: A comparative analysis. *Cogent Economics & Finance, 7*(1), 1631525. DOI: 10.1080/23322039.2019.1631525

Teiba, I. I., El-Bilawy, E. H., Elsheery, N. I., & Rastogi, A. (2023). Microbial allies in agriculture: harnessing plant growth-promoting microorganisms as guardians against biotic and abiotic stresses. *Horticulturae, 10*(1), 12. DOI: 10.3390/horticulturae10010012

Tepperman, J. M., Hwang, Y. S., & Quail, P. H. (2006). phyA dominates in transduction of red-light signals to rapidly responding genes at the initiation of Arabidopsis seedling de-etiolation. *The Plant Journal, 48*(5), 728–742. DOI: 10.1111/j.1365-313X.2006.02914.x PMID: 17076805

Tesfay, S. Z., Modi, A. T., & Mohammed, F. (2016). The effect of temperature in moringa seed phytochemical compounds and carbohydrate mobilization. *South African Journal of Botany, 102*, 190–196. DOI: 10.1016/j.sajb.2015.07.003

Thakur, S. S., Hashmi, S., Mishra, S., Ekka, S., Kushwaha, A., & Kujur, R. (2024). *A Review on Plant Tissue Culture*. Asian Journal of Biology., DOI: 10.9734/ajob/2024/v20i2387

Thamburaj, S., & Singh, N. (2000). *Textbook of Vegetables, Tubercrops and Spices*. Indian Council of Agricultural Research.

Thanoon, A., & Khalil Younes, A. (2023). Fusarium oxysporum f. sp. tuberosi causing wilt disease on potato plants in Iraq. (2023). *NTU Journal of Agriculture and Veterinary Science, 3*(4). Advance online publication. DOI: 10.56286/ntujavs.v3i4.706

Thilakarathne, A. S., Liu, F., & Zou, Z. (2025). Plant Signaling Hormones and Transcription Factors: Key Regulators of Plant Responses to Growth, Development, and Stress. *Plants, 14*(7), 1070. DOI: 10.3390/plants14071070 PMID: 40219138

Thokchom, R., & Singh, S. S. (2024). Plant propagation techniques. In *Horticulture research* (1st ed., pp. 125–137). ND Global Publication House.

Thrupp, L. A., & Giling, R. (1996). *New partnerships for sustainable agriculture* (pp. viii+-136). Washington, DC: World Resources Institute.

Tilman, D., Clark, M., Williams, D. R., Kimmel, K., Polasky, S., & Packer, C. (2017). Future threats to biodiversity and pathways to their prevention. *Nature, 546*(7656), 73–81.

Timerman, D., & Barrett, S. C. (2021). The biomechanics of pollen release: New perspectives on the evolution of wind pollination in angiosperms. *Biological Reviews of the Cambridge Philosophical Society, 96*(5), 2146–2163. DOI: 10.1111/brv.12745 PMID: 34076950

Tipu, M. M., & Sherif, S. M. (2024). Ethylene and its crosstalk with hormonal pathways in fruit ripening: Mechanisms, modulation, and commercial exploitation. *Frontiers in Plant Science, 15*, 1475496. DOI: 10.3389/fpls.2024.1475496 PMID: 39574438

Touch, V., Tan, D. K., Cook, B. R., Liu, L. D., Cross, R., Tran, T. A., Utomo, A., Yous, S., Grunbuhel, C., & Cowie, A. (2024). Smallholder farmers' challenges and opportunities: Implications for agricultural production, environment and food security. *Journal of Environmental Management, 370*, 122536. DOI: 10.1016/j.jenvman.2024.122536 PMID: 39299125

Trigiano, R. N., & Gray, D. J. (2016). *Plant tissue culture, development, and biotechnology*. CRC Press. DOI: 10.1201/9781439896143

Tripathi, A., & Shukla, S. (2024). Methods of genetic transformation: Major emphasis to crop plants. *Journal of Microbiology, Biotechnology and Food Sciences, 13*(4), e10276–e10276.

Tripathi, L., Ntui, V. O., & Tripathi, J. N. (2022). Control of Bacterial Diseases of Banana Using CRISPR/Cas-Based Gene Editing. *International Journal of Molecular Sciences, 23*(7), 3619. DOI: 10.3390/ijms23073619 PMID: 35408979

Tripathi, L., Ntui, V. O., & Tripathi, J. N. (2024, August 16). Application of CRISPR/Cas-based gene-editing for developing better banana. *Frontiers in Bioengineering and Biotechnology*, *12*, 1395772. DOI: 10.3389/fbioe.2024.1395772 PMID: 39219618

Twaij, B. M., Jazar, Z. H., & Hasan, M. N. (2020). Trends in the use of tissue culture, applications and future aspects. *International Journal of Plant Biology*, *11*(1), 8385. DOI: 10.4081/pb.2020.8385

Tyagi, S., Kumar, R., Kumar, V., Won, S. Y., & Shukla, P. (2021, January 2). Engineering disease resistant plants through CRISPR-Cas9 technology. *GM Crops and Food: Biotechnology in Agriculture and the Food Chain*, *12*(1), 125–144. DOI: 10.1080/21645698.2020.1831729 PMID: 33079628

UNESCO. (2012a). *Culture: a driver and an enabler of sustainable development. UN systemtask team on the post-2015 development agenda, The United Nations Scientific, Educational and Cultural Organization*. UNESCO.

United Nations. (1992). Convention on Biological Diversity, June 5, 1992. https://www.cbd.int/doc/legal/cbd-en.pdf

Uysal, O., Sevimli, T., Sevimli, M., Gunes, S., & Sariboyaci, A. E. (2018). Cell and tissue culture: the base of biotechnology. In *Omics technologies and bio-engineering* (pp. 391–429). Academic Press., DOI: 10.1016/B978-0-12-804659-3.00017-8

Van Eck, J. (2020, September 18). Applying gene editing to tailor precise genetic modifications in plants. *The Journal of Biological Chemistry*, *295*(38), 13267–13276. DOI: 10.1074/jbc.REV120.010850 PMID: 32723863

van Esse, H. P., Reuber, T. L., & van der Does, D. (2020, January). Genetic modification to improve disease resistance in crops. *The New Phytologist*, *225*(1), 70–86. DOI: 10.1111/nph.15967 PMID: 31135961

Vanneste, S., Pei, Y., & Friml, J. (2025). Mechanisms of auxin action in plant growth and development. *Nature Reviews. Molecular Cell Biology*, *26*(9), 1–19. DOI: 10.1038/s41580-025-00851-2 PMID: 40389696

Velasquez, S. M., Barbez, E., Kleine-Vehn, J., & Estevez, J. M. (2016). Auxin and cellular elongation. *Plant Physiology*, *170*(3), 1206–1215. DOI: 10.1104/pp.15.01863 PMID: 26787325

Venema, J. H., Dijk, B. E., Bax, J. M., van Hasselt, P. R., & Elzenga, J. T. M. (2008). Grafting tomato (Solanum lycopersicum) onto the rootstock of a high-altitude accession of Solanum habrochaites improves suboptimal-temperature tolerance. *Environmental and Experimental Botany*, *63*(1-3), 359–367. DOI: 10.1016/j.envexpbot.2007.12.015

Ventura, J. A., Lima, I. D. M., Martins, M. V. V., Culik, M. P., & Costa, H. (2019). Impact and management of diseases in the propagation of fruit plants. *Revista Brasileira de Fruticultura, 41*(4), e-647. DOI: 10.1590/0100-29452019647

Verkey, E. (2007). Shielding farmers' rights. *Journal of Intellectual Property Law & Practice*, *2*(12), 825–831. DOI: 10.1093/jiplp/jpm189

Verma, K. (2025). *Plant Biotechnology and Genetic Advances*. Educohack Press.

Verma, M., Mishra, J., & Arora, N. K. (2019). In Sobti, R. C., Arora, N. K., & Kothari, R. (Eds.), *"Plant growth-promoting rhizobacteria: diversity and applications" in Environmental biotechnology: For sustainable future* (pp. 129–173). Springer.

Verma, R. C., Waseem, M. A., Sharma, N., Bharathi, K., Singh, S., Anto Rashwin, A., & Singh, B. V. (2023). The role of insects in ecosystems, an in-depth review of entomological research. *International Journal of Environment and Climate Change*, *13*(10), 4340–4348. DOI: 10.9734/ijecc/2023/v13i103110

Vibha, B., Oza, D. N., and Hathi, H. S. (2024). Specialized Propagation Structure for Ornamental Plants.

Volovnik, S. V. (2025). Phytochory–the dispersal of animals by terrestrial and aquatic plants. *Journal of Natural History*, *59*(21-24), 1469–1539. DOI: 10.1080/00222933.2025.2475536

Wang, G., Zeng, J., Du, C., Tang, Q., Hua, Y., Chen, M., Yang, G., Tu, M., He, G., Li, Y., He, J., & Chang, J. (2024). Divergent roles of the auxin response factors in lemongrass (Cymbopogon flexuosus (Nees ex Steud.) W. Watson) during plant growth. *International Journal of Molecular Sciences, 25*(15), 8154. DOI: 10.3390/ijms25158154

Wang, M., Li, R., & Zhao, Q. (2023). Multi-Omics Techniques in Genetic Studies and Breeding of Forest Plants. *Forests*, *14*(6), 1196. DOI: 10.3390/f14061196

Wang, M.-R., Bi, W., Shukla, M. R., Ren, L., Hamborg, Z., Blystad, D.-R., Saxena, P. K., & Wang, Q.-C. (2021). Epigenetic and Genetic Integrity, Metabolic Stability, and Field Performance of Cryopreserved Plants. *Plants*, *10*(9), 1889. DOI: 10.3390/plants10091889 PMID: 34579422

Wang, P., Si, H., Li, C., Xu, Z., Guo, H., Jin, S., & Cheng, H. (2025). Plant genetic transformation: Achievements, current status and future prospects. *Plant Biotechnology Journal*, *23*(6), 2034–2058. DOI: 10.1111/pbi.70028 PMID: 40052992

Wang, X. (2022). Managing Land Carrying Capacity: Key to Achieving Sustainable Production Systems for Food Security. *Land (Basel), 11*(4), 484. DOI: 10.3390/land11040484

Wang, X., Morin, X., Zhang, J., Chen, G., Mao, L., Chen, Y., Song, Z., Du, Y., & Ma, K. (2023). Geographical patterns and determinants in plant reproductive phenology duration. *Frontiers in Plant Science, 14*, 1199316. DOI: 10.3389/fpls.2023.1199316 PMID: 37396633

Wang, Y., He, Z., Wang, J., Wu, C., Yu, G., & Tang, Y. (2018). Experiment on transplanting performance of automatic vegetable pot seedling transplanter for dry land. *Nongye Gongcheng Xuebao (Beijing), 34*(3), 19–25. DOI: 10.11975/j.issn.1002-6819.2018.03.003

Wani, A. W., Yaseen, I., Fayaz, S., Khursheed, A., Gupta, A., Hassan, G. I., Rajan, R., Qayoom, S., Gani, I., & Jayarajan, S. (2022). Advances in Propagation and Nursery Management: Methods and Techniques. In *Apples* (pp. 45-57). CRC Press.

Watt, M. S., & Bloomberg, M. (2012). Key features of the seed germination response to high temperatures. *The New Phytologist, 196*(2), 332–336. DOI: 10.1111/j.1469-8137.2012.04280.x PMID: 22882304

Weerasekara, I., Sinniah, U. R., Namasivayam, P., Nazli, M. H., Abdurahman, S. A., & Ghazali, M. N. (2021). The influence of seed production environment on seed development and quality of soybean (Glycine max (L.) Merrill). *Agronomy (Basel), 11*(7), 1430. DOI: 10.3390/agronomy11071430

Weiss, J., & Gruda, N. S. (2025). Novel breeding techniques and strategies for enhancing greenhouse vegetable product quality. *Agronomy (Basel), 15*(1), 207. DOI: 10.3390/agronomy15010207

Wendt Dos Santos, A. L., Souza Reis, R., Schuabb Heringer, A., Segal Floh, E. I., Santa-Catarina, C., & Silveira, V. (2018). Proteomics as a Tool to Study Molecular Changes During Plant Morphogenesis In Vitro. *Methods in Molecular Biology (Clifton, N.J.), 1815*, 339–349. DOI: 10.1007/978-1-4939-8594-4_24 PMID: 29981134

Wen, Z., Lu, X., Wen, J., Wang, Z., & Chai, M. (2024). Physical seed dormancy in legumes: Molecular advances and perspectives. *Plants, 13*(11), 1473. DOI: 10.3390/plants13111473 PMID: 38891282

Wikandari, R., Manikharda, , Baldermann, S., Ningrum, A., & Taherzadeh, M. J. (2021, December). Manikharda, Baldermann S, Ningrum A, Taherzadeh MJ. Application of cell culture technology and genetic engineering for production of future foods and crop improvement to strengthen food security. *Bioengineered*, *12*(2), 11305–11330. DOI: 10.1080/21655979.2021.2003665 PMID: 34779353

Wilson, S. (2013). Biodiversity loss, natural disasters, and small island developing states [Briefing paper]. Foundation for International Environmental Law and Development. https://www.field.org.uk/wp-content/uploads/2013/07/SIDS-Vulnerability-Brief-2013.pdf

Winter, L. Cultivating Farmers' Rights: Reconciling Food Security, Indigenous Agriculture, and TRIPS' (2010). *Vanderbilt Journal of Transnational Law, 43*, 223.

Wu, W., Chen, L., Liang, R., Huang, S., Li, X., Huang, B., Luo, H., Zhang, M., Wang, X., & Zhu, H. (2025). The role of light in regulating plant growth, development and sugar metabolism: A review. *Frontiers in Plant Science*, *15*, 1507628. DOI: 10.3389/fpls.2024.1507628 PMID: 39840366

Wu, Z. Y., Milne, R. I., Liu, J., Nathan, R., Corlett, R. T., & Li, D. Z. (2023). The establishment of plants following long-distance dispersal. *Trends in Ecology & Evolution*, *38*(3), 289–300. DOI: 10.1016/j.tree.2022.11.003 PMID: 36456382

Xu, X., & Hernandez, R. (2020). The effect of light intensity on vegetative propagation efficacy, growth, and morphology of "Albion" strawberry plants in a precision indoor propagation system. *Applied Sciences (Basel, Switzerland)*, *10*(3), 1044. DOI: 10.3390/app10031044

Xu, X., Zhang, C., Xu, X., Cai, R., Guan, Q., Chen, X., Chen, Y., Zhang, Z., Xu-Han, X., Lin, Y., & Lai, Z. (2023). Riboflavin mediates m6A modification targeted by miR408, promoting early somatic embryogenesis in longan. *Plant Physiology*, *192*(3), 1799–1820. DOI: 10.1093/plphys/kiad139 PMID: 36930572

Yadav, D., & Singh, S. P. (2018). Vegetative methods of plant propagation: I-cutting layering and budding. *Journal of Pharmacognosy and Phytochemistry*, *7*(2), 3267–3273.

Yahya, R. T., Qaddawi, Z. T., Rasheed, A. A., & Sultan, S. J. (2024). Effect of diode laser rays in stimulating anthocyanin pigment levels in radish (*Raphanus sativus* L.) plant tissues. Journal of Applied and Natural Science., 16(3): 1158 – 1163. ISSN: 0974-9411 (Print), 2231-5209 (Online).

Yamori, W., Hikosaka, K., & Way, D. A. (2014). Temperature response of photosynthesis in C3, C4, and CAM plants: Temperature acclimation and temperature adaptation. *Photosynthesis Research, 119*(1), 101–117. DOI: 10.1007/s11120-013-9874-6 PMID: 23801171

Yan, D., Duermeyer, L., Leoveanu, C., & Nambara, E. (2014). The functions of the endosperm during seed germination. *Plant & Cell Physiology, 55*(9), 1521–1533. DOI: 10.1093/pcp/pcu089 PMID: 24964910

Yang, L., Liu, S., & Lin, R. (2020). The role of light in regulating seed dormancy and germination. *Journal of Integrative Plant Biology, 62*(9), 1310–1326. DOI: 10.1111/jipb.13001 PMID: 32729981

Yang, P., Zhang, S., Hu, D., Li, X., Guo, Y., Guo, H., Zhang, L., & Ding, X. (2024). Research Progress on the Mechanism and Application of the Type I CRISPR-Cas System. *International Journal of Molecular Sciences, 25*(23), 12544. DOI: 10.3390/ijms252312544 PMID: 39684256

Yan, S., Bhawal, R., Yin, Z., Thannhauser, T. W., & Zhang, S. (2022, July 23). Recent advances in proteomics and metabolomics in plants. *Molecular Horticulture, 2*(1), 17. DOI: 10.1186/s43897-022-00038-9 PMID: 37789425

Yasemin, S., & Beruto, M. (2024). A review on flower bulb micropropagation: Challenges and opportunities. *Horticulturae, 10*(3), 284. DOI: 10.3390/horticulturae10030284

Youngquist, J., Krzysik, A., English, B., & Spelter, H. P, Chow (1993): Forest Agricultural fibers for use in building components. Proc7268 Madison, WI: product Society: 123 – 134.

Yuan, P., Usman, M., Liu, W., Adhikari, A., Zhang, C., Njiti, V., & Xia, Y. (2024, December). Advancements in Plant Gene Editing Technology: From Construct Design to Enhanced Transformation Efficiency. *Biotechnology Journal, 19*(12), e202400457. DOI: 10.1002/biot.202400457 PMID: 39692063

Yu, H., Jia, H., Shen, N., Gang, D., Yuan, W., Yang, Y., Hu, C., & Qu, J. (2024). Can "risk-sharing" mechanisms help clonal aquatic plants mitigate the stress of nanoplastics? *Environmental Science & Technology, 58*(6), 2984–2997. DOI: 10.1021/acs.est.3c09436 PMID: 38306608

Zaka, M., Morsi, Y. El-Din Mohamed, B. and Amin, I. F. (2018) The Economic Impacts of Recycling Agricultural Waste on Agricultural Production in Egypt, Journal of the Association of Arab Universities for Agricultural Sciences, AinShams University, 26(2B) (Special Issue): 1289-1302.

Zandalinas, S. I., Mittler, R., Balfagón, D., Arbona, V., & Gómez-Cadenas, A. (2018). Plant adaptations to the combination of drought and high temperatures. *Physiologia Plantarum*, *162*(1), 2–12. DOI: 10.1111/ppl.12540 PMID: 28042678

Zhang, C., Wang, S., Wang, Y., Wang, H., Qin, M., Dai, X., Yan, B., Guo, X., Zhou, L., Lin, H., & Guo, L. (2023).[Application of tissue culture technology of medicinal plants in sustainable development of Chinese medicinal resources] 48 5, 1186-1193. https://doi.org/DOI: 10.19540/j.cnki.cjcmm.20221017.104

Zhang, M., Ming, Y., Wang, H. B., & Jin, H. L. (2024). Strategies for adaptation to high light in plants. *aBIOTECH*, *5*(3), 381–393. DOI: 10.1007/s42994-024-00164-6 PMID: 39279858

Zhao, C., Zhang, H., Song, C., Zhu, J. K., & Shabala, S. (2020). Mechanisms of plant responses and adaptation to soil salinity. *Innovation (Cambridge (Mass.))*, *1*(1), 100017. DOI: 10.1016/j.xinn.2020.100017 PMID: 34557705

Zhao, J., Yang, W., Cai, H., Cao, G., & Li, Z. (2025). Current Progress and Future Trends of Genomics-Based Techniques for Food Adulteration Identification. *Foods*, *14*(7), 1116. DOI: 10.3390/foods14071116 PMID: 40238250

Zhi, H., Zhou, S., Pan, W., Shang, Y., Zeng, Z., & Zhang, H. (2022). The Promising Nanovectors for Gene Delivery in Plant Genome Engineering. *International Journal of Molecular Sciences*, *23*(15), 8501. DOI: 10.3390/ijms23158501 PMID: 35955636

Zhou, M., & Yang, J. (2023). Delaying or promoting? Manipulation of leaf senescence to improve crop yield and quality. *Planta*, *258*(3), 48. DOI: 10.1007/s00425-023-04204-1 PMID: 37477756

Zirbel, C. R., Bassett, T., Grman, E., & Brudvig, L. A. (2017). Plant functional traits and environmental conditions shape community assembly and ecosystem functioning during restoration. *Journal of Applied Ecology*, *54*(4), 1070–1079. DOI: 10.1111/1365-2664.12885

Źróbek-Sokolnik, A. (2011). Temperature stress and responses of plants. In *Environmental adaptations and stress tolerance of plants in the era of climate change* (pp. 113–134). Springer New York.

Zuzarte, M., Salgueiro, L., & Canhoto, J. (2024). Plant Tissue Culture: Industrial Relevance and Future Directions. *Advances in Biochemical Engineering/Biotechnology*, *188*, 1–15. Advance online publication. DOI: 10.1007/10_2024_254 PMID: 38796640

About the Contributors

Waleed Abobatta is an associate professor at the Horticulture Research Institute. He's the Egyptian PI of the project "Fig Waste Into Green Ultra-Resilient Eco-coatings (FIGURE)" (2025-2027) funded by PRIMA; he is a member of the scientific team of "Rationalization of water using super absorbance hydrogels prepared by gamma irradiation technology as conditioners for agricultural soil for water and fertilizer management in different crops" from 2022 to 2024, Cairo, Egypt; and he is a member of Scientific Team for the National Campaign for Navel Orange (H.R.I.) 2015-2018, Egypt. Abobatta is a member of the International Society for Horticultural Science (ISHS), the Asian Council of Science Editors, the Association of Agricultural Research Institutions in the Near East and North Africa (AARINENA), and a board member of the Arab Federation for the Protection of Wild and Marine Life—Arab League. He is a member of organizing committee of the Conferences of HRI and a member of the Scientific Committee for Greenhouse Plantation (A R C) from 2014 to 2020. He is a supervisor of M.Sc. and Ph.D. theses in various branches of agricultural science. He has over 113 publications in the form of research papers, books, and book chapters, including 17 Scopus-indexed papers and an H-index of 5.

Fahad Al-dulaimi An accomplished academic and researcher in the life sciences, specializing in mycology and botany. He has published more than 40 papers in international scientific journals and has contributed significantly to advancing knowledge in his field. With extensive experience in teaching at both undergraduate and postgraduate levels, he has supervised over 25 master's and doctoral theses. His research interests focus on fungal cultivation, plant tissue culture, and sustainable agricultural practices, while he is also engaged in developing modern curricula and training programs that foster innovation and sustainability in higher education.

Janan AL-tarjuman Assistant Professor and a faculty member in the Department of Plant Production Techniques at the Agricultural Technical College, part of the Northern Technical University in Mosul, Iraq. She earned her Bachelor's degree in Biology in 2004, followed by a Master's degree in Mycology in 2011, and completed her Ph.D. in the same field in 2017—all from the University of Mosul. Dr. Jinan has made significant scientific contributions, having published 15 research papers and participated in more than 30 conferences and scientific seminars. Her research interests focus on mycology and plant production techniques. She has taken part in scientific studies concerning fungal species and their genetic classification, with some of her findings listed in the global biotechnology database (NCBI). One of her notable achievements includes co-discovering and classifying a new fungal strain named Macrophomina phaseolina, registered under the accession number MHJF23. In addition to her research activities, Dr. Jinan has contributed to academic writing, co-authoring a chapter titled "Food Security Challenges" in a scientific book on sustainable agriculture and food security, published by the international academic publisher IGI Global. She has also been involved in supervising postgraduate students and served as a committee member for several Master's thesis defenses, covering topics such as biofuel production from plants using yeast fermentation, and the biosynthesis of gold nanoparticles using mustard plants and the fungus Fusarium oxysporum.

Waad Faizy a faculty member at the Department of Plant Production Technologies, Technical Agricultural College – Mosul, Northern Technical University. He holds a B.Sc. (1997) and M.Sc. (2015) in Horticulture and Landscape Engineering from the University of Mosul. He has published several scientific papers, participated in academic workshops, and currently serves as Head of the Plant Production Technologies Department.

Oindrilla Ghosh is a Research Scholar hailing from the School of Law at Christ (Deemed to be University) Bangalore, India, focuses her research on health of tribal communities. Her work includes total 22 publications on Neuroscience on Organizational Behavior at IGI Global and also Environmental Disasters that are indexed in SCOPUS and Web of Sciences conference proceedings. She completed her early education at Loreto Day School, Bowbazar and went on to achieve distinction in her undergraduate (B.A/LL.B) and postgraduate (LL.M) studies at Calcutta University. Having served as an Assistant Professor of Law at Kingston Law College,Barasat, Kolkata, Ghosh has actively participated in numerous National Conferences, International Conferences, UN Global Summits, webinars, and seminars organized by prestigious academic institutions. Her expertise lies in health perspectives, environmental studies, human behavior, and related areas. Additionally, she has conducted various workshops, lectures in educational institutions, and is engaged with NGO's to help the needy.

Binod Kumar is a graduate student known for his academic excellence, with his total 17 scholarly works indexed in Scopus database on Neuroscience in Organizational Behavior, Neuromarketing and AI Applications on Wildlife Conservation. Hailing from Ranchi, Jharkhand in India, Binod is currently pursuing a Master of Laws (LL.M) at the prestigious Radha Govind University. His passion for technology and innovation along with climate changes is evident through his high academic performance and numerous projects. Binod has a keen interest in artificial intelligence, machine learning, neuropsychology, climate changes and disaster management, fields in which he has undertaken several research projects. He is particularly fascinated by the potential of AI to solve real-world problems and is dedicated to learning and applying new techniques in these areas. He volunteers at local NGOs, teaching underprivileged children basic computer skills. With a strong foundation in law and a commitment to making a positive impact, Binod Kumar is poised to become a future leader in the legal industry.

Muheddin Mahmud Rteba, Libyan Authority for Scientific Research, Associate Professor in Crops Production, Production of cereals and legume seed crops, Weeds and Weed control, Crop technology, Production of oil and industrial crops, Plant response to environmental stress, Special studies, Research Methods . have more than 15 research, 17years of experience in the field of university teaching. A workshop entitled (Sustainable agriculture...a way to eliminate desertification) at the University of Mosul - College of Science - Department of Life Sciences2022, A workshop entitled (Modern scientific methods to increase agricultural crop production and achieve agricultural sustainability) Libyan Authority for Scientific Research 2023,Providing courses for postgraduate students and faculty members at the College of Agriculture, Omar Al-Mukhtar Uni, Libya, on the End note program. A training course on how to analyze data using the statistical software package SPSS, which was held under the supervision of the Training and Development Committee at Omar Al-Mukhtar Uni, Member of the committee preparing the list of graduate studies, College of Agriculture, Omu Uni

Index

S

T

V

W

www.ingramcontent.com/pod-product-compliance
Ingram Content Group UK Ltd.
Pitfield, Milton Keynes, MK11 3LW, UK
UKHW050844101026
15624UKWH00018B/513